The Haynes
Emissions Control Manual

by Mike Stubblefield
and John H Haynes
Member of the Guild of Motoring Writers

The Haynes Automotive Repair Manual for emissions control systems

(10W3 - 1667)

ABCDE
FGHIJ

Haynes Publishing Group
Sparkford Nr Yeovil
Somerset BA22 7JJ England

Haynes North America, Inc
861 Lawrence Drive
Newbury Park
California 91320 USA

Acknowledgements

We are grateful to the following companies for providing test equipment shown in this manual:

Rinda Technologies
P.O. Box 860
Prospect Heights, IL 60070

The John Fluke Manufacturing Company
P.O. Box 9090
Everett, WA 98206

The Actron Manufacturing Company
9999 Walford Avenue
Cleveland, OH 44102 – 4696

Equus Products Inc.
17291 Mt. Hermann
Fountain Valley, CA 92708

In addition, thanks are due to the Chrysler Corporation, Ford Motor Company, General Motors Corporation, Mazda Motor Corporation, Mitsubishi Motors Corporation, Nissan Motor Company, Toyota Motor Corporation and the Isuzu Motor Company for providing technical information and certain illustrations. Technical writers who contributed to this project include Larry Warren and Robert Maddox.

A book in the **Haynes Automotive Repair Manual Series**

Printed in the U.S.A.

ISBN 1 85010 667 3

Library of Congress Catalog Card Number 92-70518

92 –192

Contents

Chapter 1 Introduction

Chapter 2 Troubleshooting

Chapter 3 System descriptions and servicing

Chapter 4 Vacuum diagrams and VECI labels

Chapter 5 Acronym list and glossary

Index

1 Introduction

1 Introduction to emissions control and engine management systems

Automobiles and trucks are the number one cause of air pollution in this country; they contribute over *half* of the airborne pollutants – hydrocarbons, carbon monoxide, oxides of nitrogen and others – collectively referred to as "smog." The constituents of smog irritate the eyes, the nose, the throat and the lungs; some of them – such as carbon monoxide and lead – are toxic enough to cause serious illness; some are even carcinogenic. Ironically, modern vehicles (those built since about 1980) are almost 100 percent "cleaner" than '60s and '70s vehicles. So why do cars and trucks continue to produce such a disproportionate percentage of air pollution? Well, certainly, there are more vehicles on the road now than there were 10 years ago. But there's another reason. Recent studies indicate that *over one-third of the vehicles on the road right now are poorly maintained;* they're out of tune and their emissions control systems are broken, worn out, disconnected or improperly maintained. If every newer vehicle was well-tuned and its emission control systems were hooked up and well-maintained, its tailpipe emissions would be significantly less. And older vehicles? True, they will never run as efficiently as newer models, but when they're well tuned – and their first-generation smog equipment is in good working order – they pollute a lot less.

Haynes Automotive Repair Manuals already show you how to keep your vehicle well-tuned; in this manual, we're going to show you how easy it is to understand and maintain the emission control systems on your vehicle, and we're going to show you how to troubleshoot and fix them when they malfunction.

It's time to start thinking of emission controls as an integral part of the modern vehicle. We will never clean up our skies until we take better care of our vehicles' engines and their emission control systems. Of course, not everyone is ready to accept the proposition that the vehicle owner must accept personal responsibility for the proper maintenance of his or her vehicle's emission control systems simply for the good of the environment. Fair enough. Let's put the matter in terms anyone can understand – dollars and cents. Many of you live in states such as California where regular emissions certification testing is part of owning a car or truck. These tests cost money, and repairing emission systems or replacing emission devices that malfunction costs even more money. Most states put a cap on how much money you can be forced to spend on restoring your emission control systems to their original condition, but that figure is climbing. In some states, it's already $500 or more. In many cases, these costs can be avoided, or significantly reduced, by simply maintaining your emission control systems as they age, instead of ignoring them.

Finally, there's another economic incentive for proper maintenance: A well-tuned vehicle with all its emission control systems in place and in good shape not only pollutes less, it often gets better mileage (and produces more horsepower!). Your vehicle's engine was *designed* to operate with its emission control systems intact.

2 Understanding the material in this manual

Are you one of those mechanics who's perfectly capable of understanding and fixing the most complex mechanical systems on a vehicle but intimidated when confronted by electrical or vacuum-operated devices with few or no moving parts? Is it because you like to *see* what it is you're dealing with?

In this manual we're going to show you a lot of information about this mysterious "invisible" world of emission control systems, information sensors, computers and actuators. It's essential that you develop a logical approach to this stuff so it doesn't overwhelm you.

First, get involved! Pop the hood and have a look at the maze of vacuum lines and the weird gadgets located all over the engine compartment. This will stimulate your thinking: "What does this do? Where does that go?"

Now have a look at the VECI label (more on the VECI later). See if you can match the actual routing of the vacuum hoses and the location of the components in the engine compartment to the schematic on the VECI.

Once you've stimulated your curiosity, open this manual and start reading. But make frequent trips back to the engine compartment. There are five things you need to learn about each emission control component or system, the computer and each information sensor or output actuator used on your vehicle:

1) *What's it called?* (AIR, EGR, EVAP, PCV, etc.)
2) *How does it work?* (electrically-actuated; vacuum-operated; both; etc.)
3) *Where is it located?* (engine; throttle body; intake manifold; etc.)
4) *How do I check or test it?* (ohmmeter; vacuum gauge; voltmeter; etc.)
5) *How do I replace it?* (usually self-evident)

As you repeat this process over and over again on various vehicles, you'll discover that they often have the same names (and acronyms), work the same way, are located in the same general area, are checked/tested the same way and have similar replacement procedures! Once you understand how one EGR or PCV valve or EVAP canister works, you understand how all of them work (though their actual vacuum plumbing may vary somewhat).

When you're trying to understand how an engine management system works, visualize the computer as the "brain," the information sensors as its "nervous system" and the output actuators as its limbs. This analogy will help you ask the right questions about the role of the myriad components used in a typical management system. The computer "asks questions" about the engine through its information sensors: "What's the temperature of the intake air?" "What's the temperature of the coolant?" "What's the engine speed?" "What's the angle of the throttle valve?" etc. This stream of data never stops as long as the engine is running. Inside the computer, the data is compared to the computer's "map." The map is the computer's programmed memory of what the engine should be doing under each specific combination of circumstances – engine warming up at idle, engine warming up under load, engine accelerating, engine decelerating, engine idling with the air conditioner turned on, etc. If what the engine is actually doing (as indicated by the sensors) doesn't match what it should be doing (according to the map), the computer directs one, some or all its output actuators to alter the operating conditions of the engine until the sensors indicate that it's operating the way it should.

If you can visualize this model of the management system, you'll be able to think logically about the role of each component in the system. And thinking logically about each component enables you to ask the right questions about a part when it malfunctions and you have to troubleshoot it. In most cases, simply knowing the name of a sensor and its location tells you what it does or should be doing.

3 The Federally-mandated emissions warranty – questions and answers

Before you dive under the hood to troubleshoot or fix a problem related to emissions, there are some things you should know about the Federally-mandated extended warranty **(see illustration)** designed to protect you from the cost of repairs to any emission-related failures beyond your control.

There are actually TWO emission control warranties – the "Design and Defect Warranty" and the "Performance Warranty." We will discuss them separately.

The Design and Defect Warranty

Basically, the Design and Defect Warranty covers the repair of all emission control related parts which fail during the first five years or the first 50,000 miles of service. **Note:** *This 50,000-mile figure may increase. In fact, at the time this manual was being written, a 100,000-mile warranty was being considered.* According to Federal law, the manufacturer must repair or replace the defective part free of charge if:

1) Your car is less than five years old and has less than 50,000 miles;
2) An original equipment part or system fails because of a defect in materials or workmanship; and
3) The failure would cause your vehicle to exceed Federal emissions standards.

If these three conditions are present, the manufacturer must honor the warranty. All manufacturers have established procedures to provide owners with this coverage. The Design and Defect Warranty applies to used vehicles too. It doesn't matter whether you bought the vehicle new or used; if the vehicle hasn't exceeded the warranty time or mileage limitations, the warranty applies.

Note that the length of the warranty is five years and 50,000 miles for *cars.* The Design and Defect Warranty applies to all vehicles manufactured in the last five years, including cars, pickups, recreational vehicles, heavy-duty trucks and motorcycles. The length of the warranty varies somewhat with the type of vehicle. If you own some type of vehicle other than a car, read the description of the emissions warranty in your owner's manual or warranty booklet to determine the length of the warranty on your vehicle.

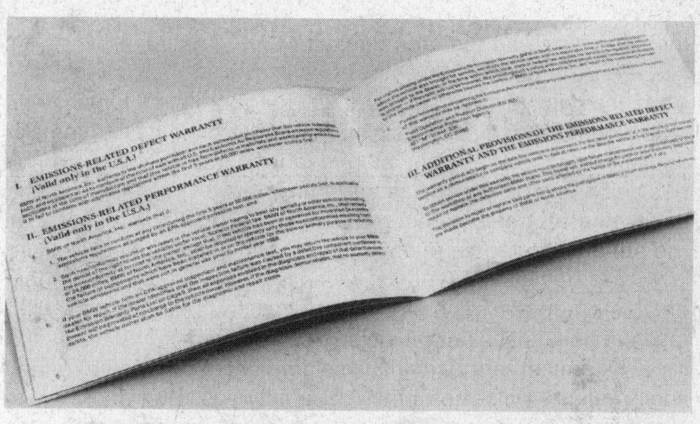

3.1 You'll find the details of your vehicle's Federally-mandated extended warranty coverage in your owner's manual or in a separate booklet like this one, in the glove box

What parts or repairs are covered by the warranty?

Coverage includes all parts whose primary purpose is to control emissions and all parts that have an effect on emissions. Let's divide these two types of parts in two categories – emissions-control parts and emissions-related parts – then divide the parts within each category into systems. Our list would look something like this:

Primary emissions control parts

Air induction system

1. Thermostatically controller air cleaner
2. Air box

Air injection system

1. Diverter, bypass or gulp valve
2. Reed valve
3. Air pump
4. Anti-backfire or deceleration valve

Early Fuel Evaporative (EFE) system

1. EFE valve
2. Heat riser valve
3. Thermal vacuum switch

Primary emissions control parts (continued)

Evaporative emission control system

1 Purge valve
2 Purge solenoid
3 Fuel filler cap
4 Vapor storage canister and filter

Exhaust gas conversion systems

1 Oxygen sensor
2 Catalytic converter
3 Thermal reactor
4 Dual-walled exhaust pipe

Exhaust Gas Recirculation (EGR) system

1 EGR valve
2 EGR solenoid
3 EGR backpressure transducer
4 Thermal vacuum switch
5 EGR spacer plate
6 Sensor and switches used to control EGR flow

Fuel metering systems

1 Electronic control module or computer command module
2 Deceleration controls
3 Fuel injectors
4 Fuel injection rail
5 Fuel pressure regulator
6 Fuel pressure dampener
7 Throttle body
8 Mixture control solenoid or diaphragm
9 Air flow meter
10 Air flow module or mixture control unit
11 Electronic choke
12 Altitude compensator sensor
13 Mixture settings on sealed carburetors
14 Other feedback control sensors, switches and valves

Ignition systems

1 Electronic spark advance
2 High energy electronic ignition
3 Timing advance/retard systems

Miscellaneous parts

Hoses, gaskets, brackets, clamps and other accessories used in these systems

Positive Crankcase Ventilation (PCV) system

1 PCV valve
2 PCV filter

Emissions-related parts

The following parts have a primary purpose other than emissions control, but they still have a significant effect on your vehicle's emissions. If they break or malfunction, your vehicle's emissions may exceed Federal standards, so they're also covered by the Design and Defect Warranty. They include:

Air induction system

1 Turbocharger
2 Intake manifold

Carburetor systems

1 Carburetor
2 Choke

Exhaust system

Exhaust manifold

Fuel injection system

Fuel distributor

Ignition system

1 Distributor
2 Ignition wires and coil
3 Spark plugs

Miscellaneous parts

Hoses, gaskets, brackets, clamps and other accessories used in the above systems

If – after reading the list above and the manufacturer's description of your warranty coverage in your owner's manual or warranty booklet – you're confused about whether certain parts are covered, contact your dealer service department or the manufacturer's zone or regional representative.

Can any part of a warranty repair be charged to you?

No! You can't be charged for any labor, parts or miscellaneous items necessary to complete the job when a manufacturer repairs or replaces any part under the emissions warranty. For example, if a manufacturer agrees to replace a catalytic converter under the emissions warranty, you shouldn't be charged for the catalyst itself or for any pipes, brackets, adjustments or labor needed to complete the replacement.

How long does the warranty apply?

Parts which don't have a replacement interval stated in the maintenance instructions are warranted for what the EPA calls the "useful life" of the vehicle, which, for cars, is, as stated before, five years or 50,000 miles. For other types of vehicles, read your warranty description in the owner's manual or the warranty booklet to determine the length of the warranty coverage.

Other parts, for example those with a stated replacement interval such as "15,000 miles or 12 months," are warranted only up to the first replacement.

Any parts that are the subject of a maintenance instruction that requires them to be "checked and replaced if necessary," or the subject of any similar requirement, are warranted for the entire period of warranty coverage.

How do you know if you're entitled to coverage?

If you or a reliable mechanic can show that a part in one of the listed systems is defective, it's probably covered under the emissions warranty. When you believe you've identified a defective part that might be covered, you should make a warranty claim to the person identified by the manufacturer in your owner's manual or warranty booklet.

What should you do if your first attempt to obtain warranty coverage is denied?

1) Ask for the complete reason – in writing – for the denial of emissions warranty coverage;
2) Ask for the name(s) of the person(s) who determined the denial of coverage;
3) Ask for the name(s) of the person(s) you should contact to appeal the denial of coverage under the emissions warranty.

Once you've obtained this information, look in your owner's manual or warranty booklet for the name of the person designated by the manufacturer for warranty assistance and contact this person.

How does maintenance affect your warranty?

Performance of scheduled maintenance is YOUR responsibility. You're expected to either perform scheduled maintenance yourself, or have a qualified repair facility perform it for you. If a part failure can be directly attributed to poor maintenance of your vehicle or vehicle abuse (proper operation of the vehicle is usually spelled out in your owner's manual or maintenance booklet), the manufacturer might not be liable for replacing that part or repairing any damage caused by its failure. To assure maximum benefit from your emissions control systems in reducing air pollution, as well as assuring continued warranty coverage, you should have all scheduled maintenance performed, or do it yourself.

Do you have to show any maintenance receipts before you can make a warranty claim?

No! Proof of maintenance isn't required to obtain coverage under the emissions warranty. If a listed part is defective in materials or workmanship, the manufacturer must provide warranty coverage. Of course, not all parts fail because of defects in materials or workmanship.

Though you're not automatically required to show maintenance receipts when you make a warranty claim, there is one circumstance in which you will be asked for proof that scheduled maintenance has been performed. If it looks as if a part failed because of a lack of scheduled maintenance, you can be required to prove that the maintenance was performed.

How is your warranty affected if you use leaded gasoline in your vehicle?

When leaded gas is used in vehicles designed to run on unleaded, the emissions controls – particularly the catalytic converter – can be damaged. And lead deposits inside the engine can lead to the failure of certain engine parts. The emissions warranty does not cover ANY part failures that result from the use of leaded fuel in a vehicle that requires unleaded fuel.

Can anyone besides dealers perform scheduled maintenance recommended by the manufacturer?

Absolutely! Scheduled maintenance can be performed by anyone who is qualified to do so, *including you* (as long as the maintenance is performed in accordance with the manufacturer's instructions). If you're going to take the vehicle to a repair facility, refer to your owner's manual or maintenance booklet and make a list of all scheduled maintenance items before you go. When you get there, don't simply ask for a "tune-up" or a "15,000 mile servicing." Instead, specify exactly what you want done. Then make sure the work specified is entered on the work order or receipt that you receive. This way, you'll have a clear record that all scheduled maintenance has been done.

If you buy a used vehicle, how do you know whether it's been maintained properly?

Realistically, you don't. But it never hurts to ask the seller to give you the receipts which prove the vehicle has been properly maintained according to the schedule. These receipts are proof that the work was done properly and on time, if the question of maintenance ever arises.

And once you buy a used vehicle, you should continue to maintain it in accordance with the maintenance schedule in the owner's manual or warranty booklet (If the seller doesn't have these items anymore, buy new ones at the dealer).

What should you do if the manufacturer won't honor what you feel is a valid warranty claim?

As we said earlier, if an authorized warranty representative denies your claim, you should contact the person designated by the manufacturer for further warranty assistance. Additionally, you're free to pursue any independent legal actions you deem necessary to obtain coverage. Finally, the EPA is authorized to investigate the failure of manufacturers to comply with the terms of this warranty. If you've followed the manufacturer's procedure for making a claim and you're still not satisfied with the manufacturer's determination, contact the EPA by writing:

Warranty Complaint
Field Operations and Support Division (EN-397F)
U.S. Environmental Protection Agency
Washington, D.C. 20460

The Performance Warranty

The Performance Warranty covers those repairs required because the vehicle has failed an emission test. If you reside in an area with an Inspection/Maintenance program that meets Federal guidelines, you may be eligible for this additional Performance Warranty. For more information on the Performance Warranty, ask your local Inspection/Maintenance program official or call or write the nearest EPA office and ask for a copy of the pamphlet *"If Your Car Just Failed An Emission Test...You May Be Entitled To Free Repairs,"* which describes the Performance Warranty in detail.

You may be eligible for coverage under this warranty if:

1) Your 1981 or later car or light truck fails an approved emissions test; and
2) Your state or local government requires that you repair the vehicle; and
3) The test failure didn't result from misuse of the vehicle or a failure to follow the manufacturer's written maintenance instructions; and
4) You present the vehicle to a warranty-authorized manufacturer representative, along with evidence of the emission test failure, during the relevant warranty period; then . . .
   a) for the first two years or 24,000 miles, whichever comes first, the manufacturer must pay for all repairs necessary to pass the emissions test and . . .
   b) for the first five years or 50,000 miles, the manufacturer must pay for all repairs to primary emission control parts which are necessary to pass the emissions test.

What vehicles are covered by the Performance Warranty?

The Federally mandated Performance Warranty covers all 1981 and later cars and light duty trucks produced in the last five years. And it doesn't matter whether you bought your vehicle new or used, from a dealer or from a private party. As long as it hasn't exceeded the warranty time or mileage limitations, and has been properly maintained, the Performance Warranty applies.

What types of repairs are covered by the Performance Warranty?

Two types of repairs are covered by the Performance Warranty, depending on the age of your vehicle:

1) Any repair or adjustment which is necessary to make your vehicle pass an approved locally-required emission test is

covered if your vehicle is less than two years old and has less than 24,000 miles.

2) Any repair or adjustment of a "primary emissions control" part (see "The Design and Defect Warranty") which is necessary to make your vehicle pass an approved locally-required test is covered if your vehicle is less than five years old and has less than 50,000 miles. Although coverage is limited after two years/24,000 miles to primary emission control parts, repairs must still be complete and effective. If the complete and effective repair or a primary part requires that non-primary parts be repaired or adjusted, these repairs are also covered.

What if the dealer claims your vehicle can pass the emissions test without repair?

The law doesn't require you to fail the emissions test to trigger the warranty. If any test shows that you have an emissions problem, get it fixed while your vehicle is still within the warranty period. Otherwise, you could end up failing a future test because of the same problem – and paying for the repairs yourself. If you doubt your original test results or the dealer's results, get another opinion to support your claim.

What kinds of reasons can the manufacturer use to deny a claim?

As long as your vehicle is within the age or mileage limits explained above, the manufacturer can deny coverage under the Performance Warranty only if you've failed to properly maintain and use your vehicle. Proper use and maintenance of the vehicle are *your* responsibilities. The manufacturer can deny your claim if there's evidence that your vehicle failed an emissions test as a result of:

a) Vehicle abuse, such as off-road driving, or overloading; or
b) Tampering with emission control parts, including removal or intentional damage; or
c) Improper maintenance, including failure to follow maintenance schedules and instructions, or use of replacement parts which aren't equivalent to the originally installed part; or
d) Misfueling: The use of leaded fuel in a vehicle requiring "unleaded fuel only" or use of other improper fuels.

If any of the above have taken place, and seem likely to have caused the particular problem which you seek to have repaired, then the manufacturer can deny coverage.

If your claim is denied for a valid reason, you may have to pay the costs of the diagnosis. Therefore, you should always ask for an estimate of the cost of the diagnosis before work starts.

Can anyone besides a dealer perform scheduled maintenance?

Yes! Scheduled maintenance can be done by anyone with the knowledge and ability to perform the repair. For your protection, we recommend that you refer to your owner's manual to specify the necessary items to your mechanic. And get an itemized receipt or work order for your records.

You can also maintain the vehicle yourself, as long as the maintenance is done in accordance with the manufacturer's instructions included with the vehicle. Make sure you keep receipts for parts and a maintenance log to verify your work.

Why maintenance is important to emissions control systems

Emission control has led to many changes in engine design. As a result, most vehicles don't require tune-ups and other maintenance as often. But some of the maintenance that is required enables your vehicle's emission controls to do their job properly.

Failure to do this emissions-related maintenance can cause problems. For example, failure to change your spark plugs during a 30,000-mile tune-up can lead to misfiring and eventual damage to your catalytic converter.

Vehicles that are well-maintained and tamper-free don't just pollute less – they get better gas mileage. Which saves you money. Regular maintenance also gives you better performance and catches engine problems early, *before* they get serious – and costly.

How do you make a warranty claim?

Bring your vehicle to a dealer or any facility authorized by the manufacturer to perform warranty repairs to the vehicle or its emissions control system. Notify them that you wish to obtain a repair under the Performance Warranty. You should have with you a copy of your emissions test report as proof of your vehicle's failure to pass the emissions test. And bring your vehicle's warranty statement for reference. The warranty statement should be in your owner's manual or in a separate booklet provided by the manufacturer with the vehicle.

How do you know if your claim has been accepted as valid?

After presenting your vehicle for a Performance Warranty claim, give the manufacturer 30 days to either repair the vehicle or notify you that the claim has been denied. If your inspection/maintenance program dictates a shorter deadline, the manufacturer must meet that shorter deadline. Because of the significance of these deadlines, you should get written verification when you present your vehicle for a Performance Warranty claim.

The manufacturer can accept your claim and repair the vehicle, or deny the claim outright, or deny it after examining the vehicle. In either case, the reason for denial must be provided *in writing* with the notification.

What happens if the manufacturer misses the deadline for a written claim denial?

You can agree to extend the deadline, or it may be automatically extended if the delay is beyond the control of the manufacturer. Otherwise, a missed deadline means the manufacturer forfeits the right to deny the claim. You are then entitled to have the repair performed at the facility of your choice, at the manufacturer's expense.

If your claim is accepted, do you have to pay for either the diagnosis or the repair?

You can't be charged for any costs for diagnosis of a valid warranty claim. Additionally, when a manufacturer repairs, replaces or adjusts any part under the Performance Warranty, you may not be charged for any parts, labor or miscellaneous items necessary to to complete the repair. But if your vehicle needs other repairs that aren't covered by your emissions warranty, you can have that work performed by any facility you choose.

What happens to your warranty if you use leaded gasoline?

When leaded gas is used in vehicles requiring unleaded, some emission controls (especially the catalyst) are quickly damaged. Lead deposits also form inside the engine, decreasing spark plug life and increasing maintenance costs.

If your use of leaded fuel leads to an emissions failure, your warranty won't cover the repair costs. So using leaded fuel will not only ruin some of your emission controls, it will cost you money.

Can your regular repair facility perform warranty repairs?

If you want to have the manufacturer pay for a repair under the Performance Warranty, you MUST bring the vehicle to a facility authorized by the vehicle manufacturer to repair either the vehicle or its emission control systems. If your regular facility isn't authorized by the manufacturer, tell your mechanic to get your "go-ahead" before performing any repair that might be covered by the Performance Warranty.

Do you have to provide proof of maintenance when you make a warranty claim?

You're not automatically required to show maintenance receipts when you make a warranty claim. But if the manufacturer feels your failure to perform scheduled maintenance has caused your emissions failure, you can be required to present your receipts or log as proof that the work was in fact done.

If you buy a used vehicle, how do you know whether it's been properly maintained?

When you buy a used vehicle, try to get the maintenance receipts or log book from the previous owner. Also ask for the owner's manual, warranty or maintenance booklet, and any other information that came with the vehicle when it was new. If the seller doesn't have these documents, you can buy them from the manufacturer.

To guarantee future warranty protection for your vehicle, conform to the maintenance schedule provided by the manufacturer.

Does the warranty cover parts that must be replaced as a part of regularly scheduled maintenance?

Parts with a scheduled replacement interval that's less than the length of the warranty, such as "replace at 15,000 miles or 12 months," are warranted only up to the first replacement point. Parts with a maintenance instruction that requires them to be "checked and replaced if necessary," or some similar classification, receive full coverage under the warranty. However, should you fail to check a part at the specified interval, and should that part cause another part to fail, the second part will NOT be covered, because your failure to maintain the first part caused the failure.

The manufacturer may or may not require that such replacement parts be a specific brand. But if a test failure is caused by the use of a part of inferior quality to the original equipment part, the manufacturer may deny your warranty claim.

What if the manufacturer won't honor a claim you believe to be valid?

First, use the information contained above to make your case to the dealer. Then follow the appeals procedure outlined in your vehicle's warranty statement or owner's manual. Every manufacturer employs warranty representatives who handle such appeals. The manufacturer must either allow your claim or give you a *written* denial, including the specific reasons for denying your claim, within 30 days, or you are entitled to free repairs.

Also, the Environmental Protection Agency is authorized to investigate the failure of manufacturers to comply with the terms of this warranty. If you've followed the manufacturer's procedures and you're still unimpressed with the the reason for denial of your claim, contact the EPA at:

Warranty Complaint
Field Operations and Support Division (EN-397-F)
U.S. Environmental Protection Agency
Washington, D.C. 20460

Finally, you're also entitled to pursue any independent legal actions which you consider appropriate to obtain coverage under the Performance Warranty.

4 Emissions-related routine maintenance

The cheapest and easiest way to keep your emissions system operating properly is simply to check it over on a regular basis. This is the best to way spot trouble because many of these systems have few (if any) symptoms as they wear out or degrade in operation. Because the emissions, fuel and ignition systems are interrelated, a minor problem in one can have a ripple effect on others **(see illustration)**. These minor malfunctions among several systems can eventually lead to a breakdown which could have been avoided by a simple check and maintenance program. When the vehicle is new, the emissions system should only be serviced by a factory authorized dealer service department to protect the factory warranty. Refer to Chapter 3 for information on the design and operation of these systems.

Note: *The following maintenance schedule is generalized. Your particular vehicle may require additional maintenance and/or different intervals. Refer to the* Haynes Automotive Repair Manual *for your vehicle for more specific information. Also, the following schedule only concerns equipment that affects emissions. Your vehicle will require additional maintenance for its other components.*

Every 15,000 miles or 12 months, whichever comes first

Check the operation of the heated air intake system (Chapter 3, Section 5)
Check and, if necessary, replace the air filter
Check and, if necessary, replace the PCV filter (if equipped)
Replace the ignition points and adjust the dwell (early models with point-type ignition systems only)
Check and adjust, if necessary, the ignition timing*
Check and adjust, if necessary, the engine idle speed*

Every 30,000 miles or 24 months, whichever comes first

Inspect the catalytic converter system (Chapter 3, Section 7)
Inspect the evaporative emissions control system (Chapter 3, Section 2)
Inspect the Positive Crankcase Ventilation (PCV) system (Chapter 3, Section 3)
Check the Exhaust Gas Recirculation (EGR) system (Chapter 3, Section 1)
Inspect the air injection system (Chapter 3, Section 4)
Check all emissions system hoses for cracking, disconnections and deterioration
Replace the spark plugs
Check the operation of the carburetor choke (if equipped)
Inspect the spark plug wires, and, on vehicles with distributor-type ignition systems, inspect the distributor cap and rotor
Check the carburetor or fuel injection throttle body mounting nut or bolt torque

**Many modern vehicles do not require ignition timing and idle speed checks. The Haynes manual for your particular vehicle will tell you whether or not these checks are required.*

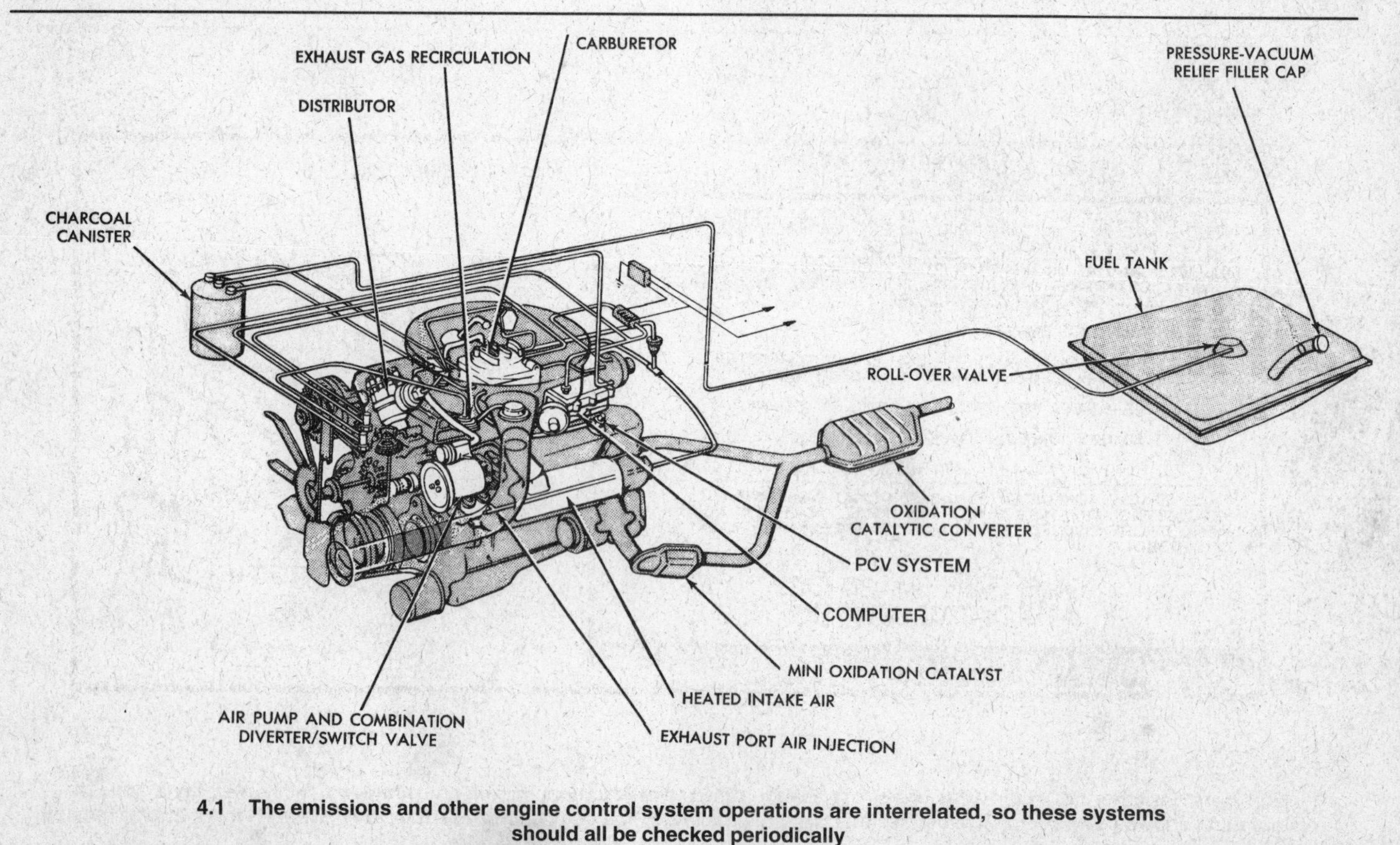

4.1 The emissions and other engine control system operations are interrelated, so these systems should all be checked periodically

5 The Vehicle Emissions Control Information (VECI) label

What is a VECI label?

The Vehicle Emissions Control Information (VECI) label **(see illustrations)** identifies the engine, the fuel system and the emission control systems used on your specific vehicle. It also provides essential tune-up specifications, such as the spark plug gap, slow-idle speed, fast-idle speed, enriched-idle speed, initial ignition timing setting (if adjustable), breaker point dwell and carburetor adjustment instructions.

Some VECI labels simply provide the specs for these adjustments; others even include brief step-by-step adjustment procedures. Most labels also provide a simplified vacuum diagram of the emissions control devices used on your vehicle and the vacuum lines connecting them to each other and to engine vacuum. You won't find everything you need to know about your emissions systems on the VECI label, but it's a very good place to start.

The information on the VECI label is specific to your vehicle. Any changes or modifications authorized by the manufacturer will be marked on the label by a technician making the modification. He may also indicate the change with a special modification decal and place it near the VECI label.

What does a VECI label look like?

The VECI label is usually a small, white, adhesive-backed, plastic-coated label about 4 X 6 inches in size, located some-

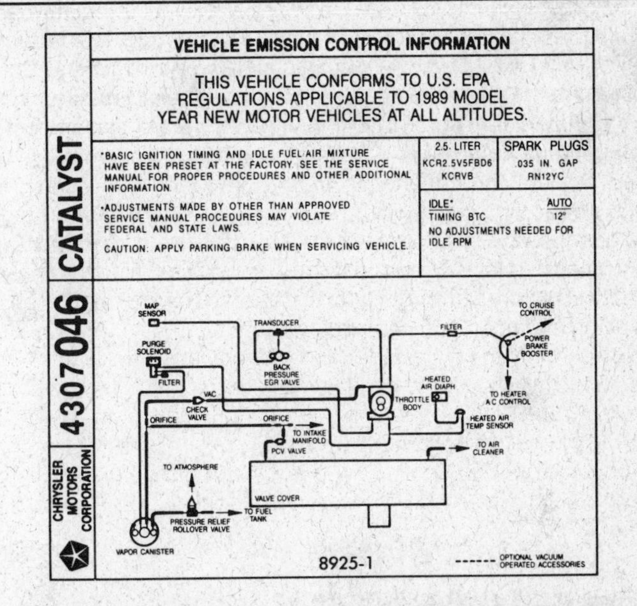

5.1 Here's a typical Vehicle Emission Control Information (VECI) label (this one's for a Chrysler) – note the warning against ignition timing adjustments and the schematic-style vacuum diagram that shows you the major emission control components and their relationship to each other, but not their location

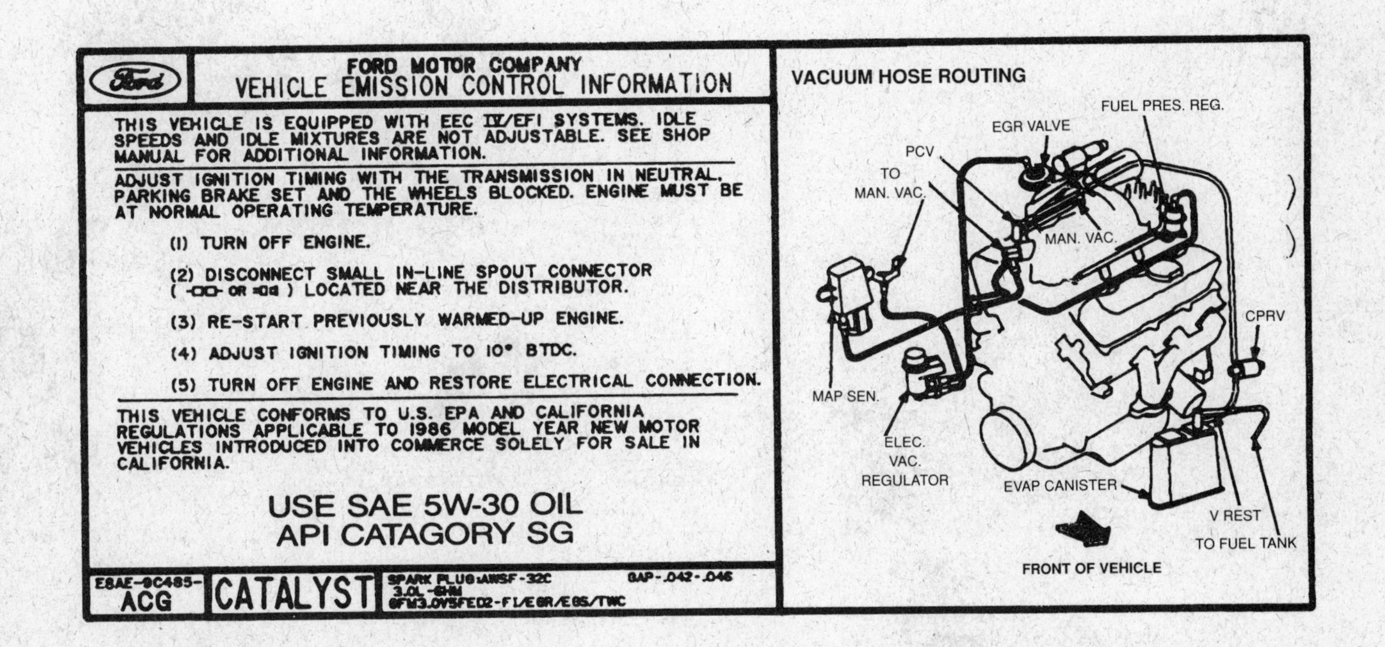

5.2 Here's another VECI label (this one's for a Ford) – note the brief ignition timing procedure and the line-art style vacuum diagram that shows you not only what emissions components are on the engine, but what they look like and where they're located

where in the engine compartment. It's usually affixed to the underside of the hood, the radiator support, the front upper crossmember, the firewall or one of the inner wheel wells.

What if you can't find the VECI label?

If you're not the original owner of your vehicle, and you can't find the VECI label anywhere, chances are it's been removed, or the body part to which it was affixed has been replaced. Don't worry – you can buy a new one at a dealer service department.

Be sure to give the parts department the VIN number, year, model, engine, etc. of your vehicle; be as specific as possible. For instance, if it's a high-altitude model, a 49-state model or a California model, be sure to tell them, because each of these models may have a different fuel system and a unique combination of emission control devices.

Don't just forget about the VECI label if you don't have one. Intelligent diagnosis of the emissions control systems on your vehicle begins here. Without the VECI label, you can't be sure everything is still installed and connected.

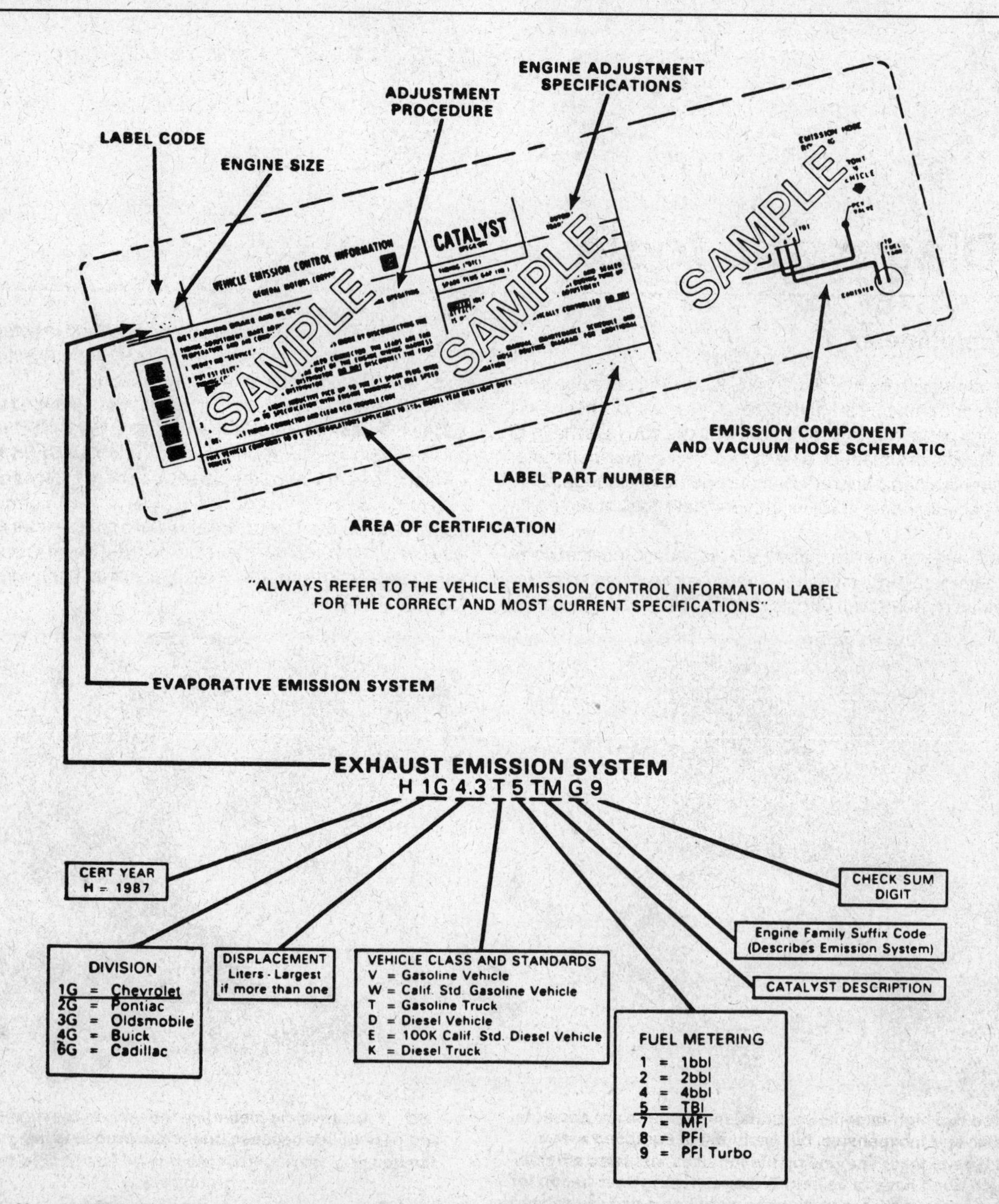

5.3 No matter what vehicle they're used on, General Motors VECI labels are all formatted the same way, even though the information provided at each location on the label is specific to the make and model

6 Diagnostic tools

Digital multimeter

Lots of interesting high-tech gadgets for testing emission control devices and systems are available. Simple visual checks will identify many problems, but there are two tools you *must* have to do all the tests in this manual. One of them is a digital multimeter and the other is a hand-operated vacuum pump and gauge. We'll get to the vacuum pump in a minute. First, let's look at the multimeter.

The multimeter is a small, hand-held diagnostic tool that combines an ohmmeter and voltmeter – and sometimes an ammeter (which you won't need for the tests in this book) – into one handy unit. A multimeter can measure the voltage and resistance in a circuit. Many emission devices and systems are electrically powered, so the multimeter is an essential tool.

There are two types of multimeters: Conventional units (a box with two leads) and probe types (small, hand-held units with a built-in probe and one flexible lead) **(see illustrations)**. Probes – which are about the same size as a portable soldering pen – are easier to use in tight spaces because of their compact dimensions. And you don't need three hands to hold a meter and two test leads all at the same time (you can hold the meter in one hand and the single lead in the other). But probes usually have less features than conventional units.

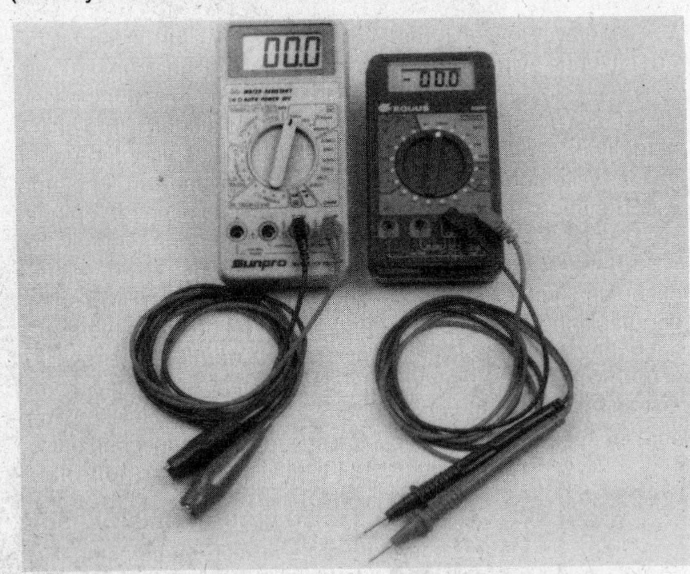

6.1 These two high-impedance digital multimeters are accurate, versatile and inexpensive, but each unit is equipped with a different type of lead: The one on the left uses insulated alligator clips which don't have to be held in place, freeing your hands for using the meter itself; the unit on the right has a pair of probes, which are handy for testing wires and terminals inside connectors (Our advice? Buy both types of leads, or make your own)

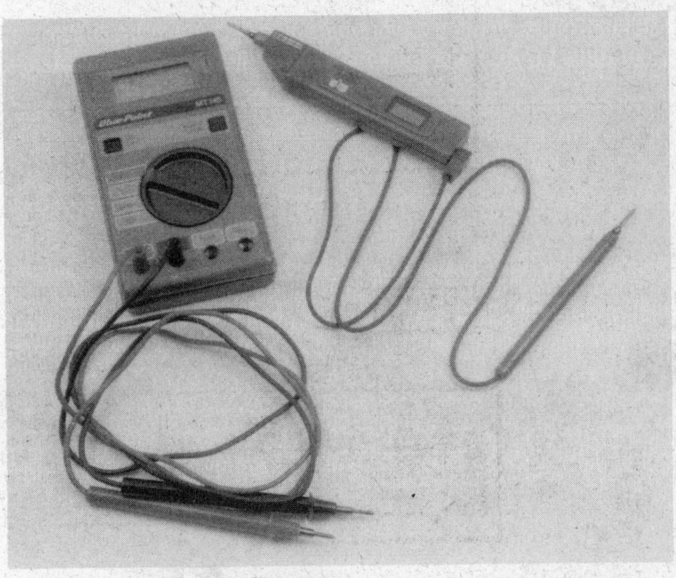

6.2 A probe-style meter like the unit on the right is small and easy to use because one of the probes is integrated into the housing, leaving your other hand free to hold the single ground lead

6.3 To make a voltage measurement, turn the mode switch or knob on your multimeter to the Volts DC position and hook up the meter in PARALLEL to the circuit being tested; if you hook it up in series, like an ammeter or ohmmeter, you won't get a reading and you could damage something (note how the positive probe is being used to make contact with a wire through the backside of the connector without actually unplugging the connector)

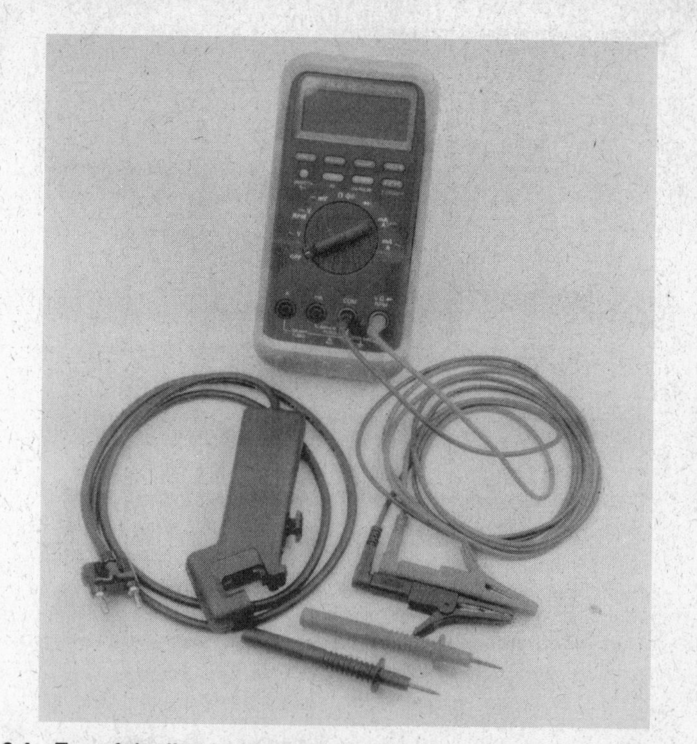

6.4 Top-of-the-line multimeters like this Fluke Model 88 can do a lot of things besides just measure volts, amps and ohms – using a wide array of adapters and cables, most of which are included in the basic kit, they can check the status of all the important information sensors, measure the duty cycle of feedback carburetors and idle air control motors, and even measure the pulse width of the fuel injectors

Why a *digital* multimeter? Partly because digital meters are easier to read, particularly when you're trying to read tenths of a volt or ohm. But mainly you need a digital meter – instead of an "analog" (needle type) – because digital multimeters are more accurate than analog meters.

Using a multimeter to read voltage is simply a matter of selecting the voltage range and hooking up the meter IN PARALLEL **(see illustration)** to the circuit being checked. Older analog (needle-type) meters have always allowed a certain amount of voltage to "detour" through this parallel circuit, which affects the accuracy of the measurement being taken.

This leaking voltage isn't that important when you're measuring 12-volt circuits – and you just want to know if a circuit has 12 or 13 or 14 volts present. If some of the voltage trickles through the meter itself, your judgment call about the health of the circuit is unaffected. But many emission control and engine management circuits operate at five volts or less; and some of them operate in the millivolt (thousandths of a volt) range. So voltage readings must be quite accurate – in many cases to the tenth, hundredth or even thousandth of a volt. Even if an older analog meter could measure voltage values this low (and even if you could read them!), the readings would be inaccurate because of the voltage detouring out of the circuit into the meter.

Digital meters have 10-Meg ohms (10 million ohms) resistance built into their circuitry to prevent voltage leaks through the meter. And this is the main reason we specify a digital voltmeter. When you shop around for a good meter, you may find a newer analog type meter with a high-resistance circuit design similar to that of a digital meter, but it will still be difficult to read when performing low-voltage tests, so don't buy it – get a digital model!

Some of the more sophisticated multimeters **(see illustration)** can perform many of the same functions as scanners, such as checking camshaft and crankshaft position sensors, feedback carburetors, fuel injection on-time, IAC motors, MAF sensors, MAP sensors, oxygen sensors, temperature sensors and throttle position sensors.

Ohmmeters

So why don't we just specify a digital *voltmeter*? Because you'll also need to use an ohmmeter a lot too: Some solenoids and other devices have specific resistance values under specified conditions, so you'll need an ohmmeter to test them. And sometimes the engine won't start, so there's no voltage available to test. When these situations arise, you'll need a good digital ohmmeter to measure resistance (expressed in ohms). But don't buy a separate ohmmeter; get a digital multimeter with an ohmmeter built in.

An ohmmeter has its own voltage source (a low-voltage DC power supply, usually a dry-cell battery). It measures the resistance of a circuit or component and is always connected to an open circuit or a part removed from a circuit. **Caution:** *Don't connect an ohmmeter to a "live" (hot) circuit; current from an outside source will damage an ohmmeter.*

Because an ohmmeter doesn't use system voltage, it's not affected by system polarity. You can hook up the test leads to either side of the part you want to test **(see illustration)**. When you use an ohmmeter, start your test on the lowest range, then switch to a higher range that gives you a more precise reading. Voltage and current are limited by the power supply and internal resistance, so you won't damage the meter by setting it on a low or high scale.

Temperature and the condition of the battery affect an ohmmeter's accuracy. Digital ohmmeters are self-adjusting, but if you're using an analog meter, you must adjust it every time you use it:

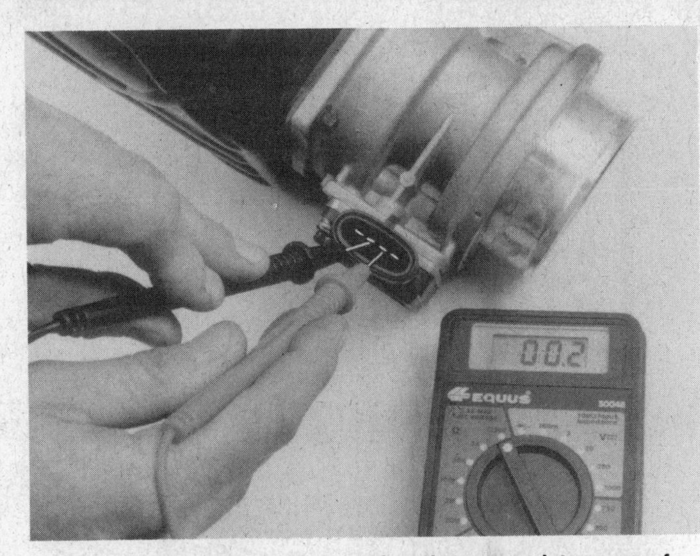

6.5 To measure resistance, select the appropriate range of resistance and touch the meter probes to the terminals you're testing; the polarity (which terminal you touch with which lead) makes no difference on an ohmmeter because it's self-powered and the circuit is turned off

6.6 Get a thermometer with a range from zero to about 220 degrees – there are automotive-specific thermometers available, but a cooking thermometer will work

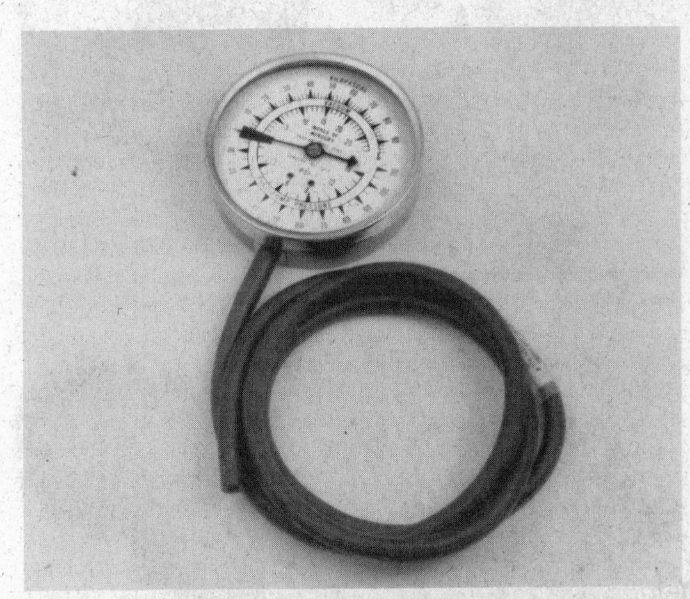

6.7 A vacuum gauge can tell you whether the engine is producing good intake vacuum, help you determine whether the catalytic converter is blocked and help you diagnose a wide variety of engine-related problems

Simply touch the two test leads together and turn the zero adjustment knob until the needle indicates zero ohms, or continuity, through the meter on the lowest scale.

Thermometer

If you're going to be testing coolant temperature sensors, get a good automotive thermometer **(see illustration)** capable of reading from zero to about 220-degrees F. If you can't find an automotive-specific unit, a good cooking thermometer will work.

Vacuum gauge

Measuring intake manifold vacuum is a good way to diagnose all kinds of things about the condition of an engine. Manifold vacuum is tested with a vacuum gauge **(see illustration)**, which measures the difference in pressure between the intake manifold and the outside atmosphere. If the manifold pressure is lower than the atmospheric pressure, a vacuum exists. Vacuum is measured in inches of mercury (in-Hg) and in kiloPascals (kPa) or in millimeters of mercury (mm-Hg). The Atmospheric pressure at sea level is 30 in-Hg. For every 1000 foot increase in altitude, the atmospheric pressure drops one inch.

In this book, we'll show you how to diagnose a restricted exhaust system with a vacuum gauge. To hook up the gauge, connect the flexible connector hose to the intake manifold, air intake plenum, or any vacuum port below the carburetor or throttle body. On some models, you can simply remove a plug from the manifold or carburetor/throttle body; on others, you'll have to disconnect a vacuum hose or line from the manifold, carb or throttle body and hook up the gauge inline with a tee fitting (included with most vacuum gauge kits).

A good vacuum reading is about 15 to 20 in-Hg (50 to 65 kPa) at idle (engine at normal operating temperature). Low or fluctuating readings can indicate many different problems. For instance, a low and steady reading may be caused by retarded ignition or valve timing. A sharp vacuum drop at intervals may be caused by a burned intake valve. Refer to the instruction manual that comes with your gauge for a complete troubleshooting chart showing the possible causes of various readings.

Vacuum leak detector

Determining whether you've got a vacuum leak is one thing; finding it is another. In this Chapter, we'll show you how to find a vacuum leak using simple, inexpensive tools (see "Finding vacuum leaks").

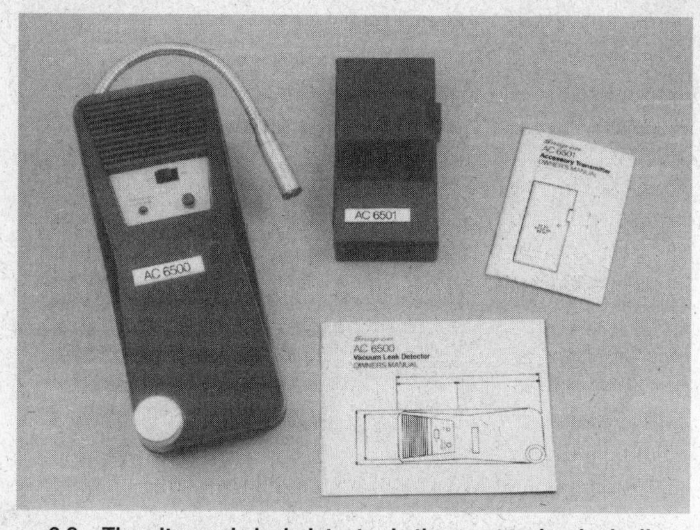

6.8 The ultrasonic leak detector is the most technologically sophisticated tool for finding vacuum leaks, but it's also expensive and not required for the tests in this manual

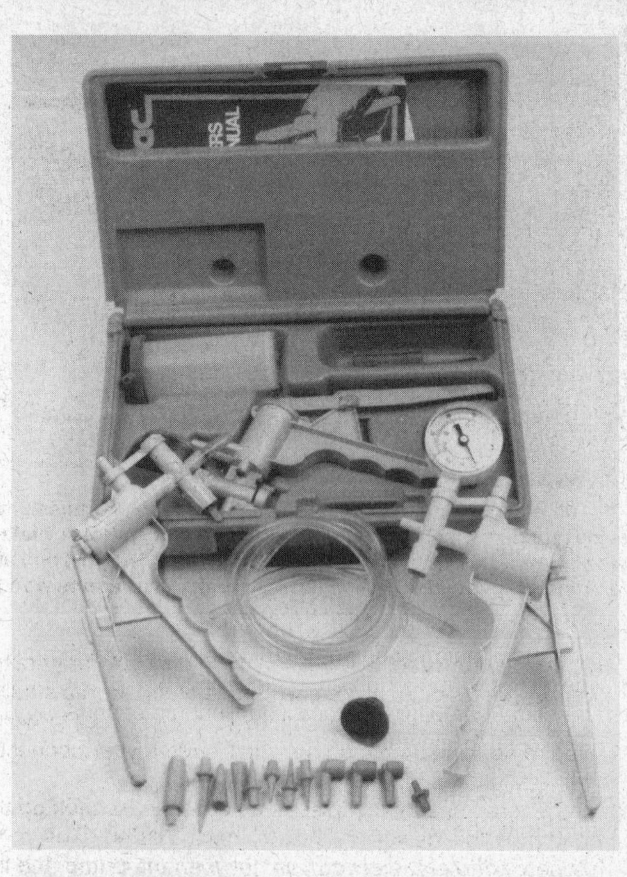

6.9 A hand-operated vacuum pump and gauge tool is indispensable for troubleshooting emission systems – it can help you track down vacuum leaks and test all vacuum-operated devices; Mityvac pumps (shown) are available as inexpensive plastic models, like the two in the foreground (one of which can be purchased without a gauge), and sturdier metal units like the one in the box; they come with a variety of fittings and adapters, and can be used for a host of other applications besides emissions tests

The most technologically advanced method of detecting leaks is the ultrasonic leak detector **(see illustration)**. Air rushing through a vacuum leak creates a high-frequency sound. An ultrasonic leak detector can "hear" these high frequencies. When its probe is passed over a leak, the detector responds to the high-frequency sound by emitting a warning beep. Some detectors also have a series of LEDs that light up as the frequencies are received. The closer the detector is moved to the leak, the more LEDs light up, or the faster the beeping occurs. This allows you to zero in on the leak. An ultrasonic leak detector can sense leaks as small as 1/500th of an inch and accurately locate the leak to within 1/16-inch. These detectors are accurate, but they're also expensive and hard to find.

Vacuum pump/gauge

Two tools are indispensable for troubleshooting emission control systems. One is a digital multimeter; the other is a hand-operated vacuum pump equipped with a vacuum gauge **(see illustration)**.

Many underhood emission control system components are either operated by intake manifold vacuum, or they use it to control other system components. Devices such as check valves, dashpots, purge control valves, solenoids, vacuum control valves, vacuum delay valves, vacuum restrictors, etc. – all these devices control vacuum in some way, or are controlled by it. They amplify, block, delay, leak, reroute or transmit vacuum. Some of them must control a specified amount of vacuum for a certain period of time, or at a certain rate. A vacuum pump applies vacuum to such devices to test them for proper operation.

Suitable vacuum pump/gauges are sold by most specialty tool manufacturers. Inexpensive plastic-bodied pump/gauges – available at most auto parts stores – are perfectly adequate for diagnosing vacuum systems. Make sure the scale on the pump gauge is calibrated in "in-Hg" (inches of mercury). And buy a rebuildable pump (find out whether replacement piston seals are available). When the seals wear, the pump won't hold its vacuum and vacuum measurements will be inaccurate. At this point, you'll have to rebuild the pump.

Using a vacuum pump is simple enough. Most pump kits include an instruction manual that describes how to use the pump in a variety of situations. They also include a variety of adapters (tee-fittings, conical fittings which allow you to connect two lines of different diameters, etc.) and some vacuum hose, to help you hook up the pump to vacuum, hoses, lines, fittings, pipes, ports, valves, etc. Manufacturers also sell replacements for these adapters and fittings in case they wear out, or you lose them. Sometimes, you may need to come up with a really specialized fitting for a more complicated hook-up. A good place to find weird fittings is the parts department of your local dealer. A well-stocked parts department has dozens of special purpose vacuum line fittings designed for various makes and models. Draw a picture of what you want for the partsman and chances are, he'll have the fitting you need.

Here are a few simple guidelines to keep in mind when using a vacuum pump:

1) When hooking up the pump **(see illustration)**, make sure the connection is airtight, or the test result will be meaningless.
2) Most factory-installed vacuum lines are rubber tubing (some are nylon). Make sure you're using the right-diameter connector hose when hooking up the pump to the device you wish to test. When you attach a connector hose with a larger inside diameter (I.D.) than the outside diameter (O.D.) of the fitting, pipe, port, etc. to which you're attaching

6.10 When connecting a vacuum pump/gauge to an emissions device (such as the Corvette EGR solenoid in this photo), make sure you've got airtight connections at the pump (arrow) and at the fitting or pipe of the device (arrow), or the test results won't mean much

it, the vacuum reading will be inaccurate, or you may not get a vacuum reading. If you use a hose or line with a smaller I.D. than the O.D. of the fitting, pipe, port, etc. to which you're hooking up the pump, you'll stretch your connector hose and it will be useless in future tests.

3) In general, use as few pieces as possible to hook up the pump to the device or system being tested. The more hoses, adapters, etc. you use between the pump and the device or system being tested, the more likely the possibility of a loose connection, and a leak.
4) Don't apply more vacuum than necessary to perform a test, or you could damage something. If the pump won't build up the amount of vacuum specified for the test, or won't hold it for the specified period of time because the piston seal is leaking, discontinue the test and rebuild the pump.
5) When you're done with the test, always break the vacuum in the pump before you detach the line or hose from the system. Breaking a connection while vacuum is still applied could cause a device to suck dirt or moisture into itself when exposed to the atmosphere.
6) Always clean the fitting, pipe or port to which you've hooked up the pump and reattach the factory hose or line. Inspect the end of the factory hose or line. It it's flared, frayed or torn, cut off the tip before you reattach it. Make sure the connection is clean and tight.
7) Clean your pump, adapter fittings and test hose, and put them away when you're done. Don't leave the pump laying around where it could be dropped and damaged.

Scanners, software and trouble-code tools

Scanners (computer analyzers)

Hand-held digital scanners **(see illustration)** are the most powerful and versatile tools for analyzing engine management systems used on later models vehicles. Unfortunately, they're also the most expensive. In this manual, we're going to show you how to troubleshoot sensors and actuators without resorting to analyzers.

Software

Software **(see illustration)** is available that enables your

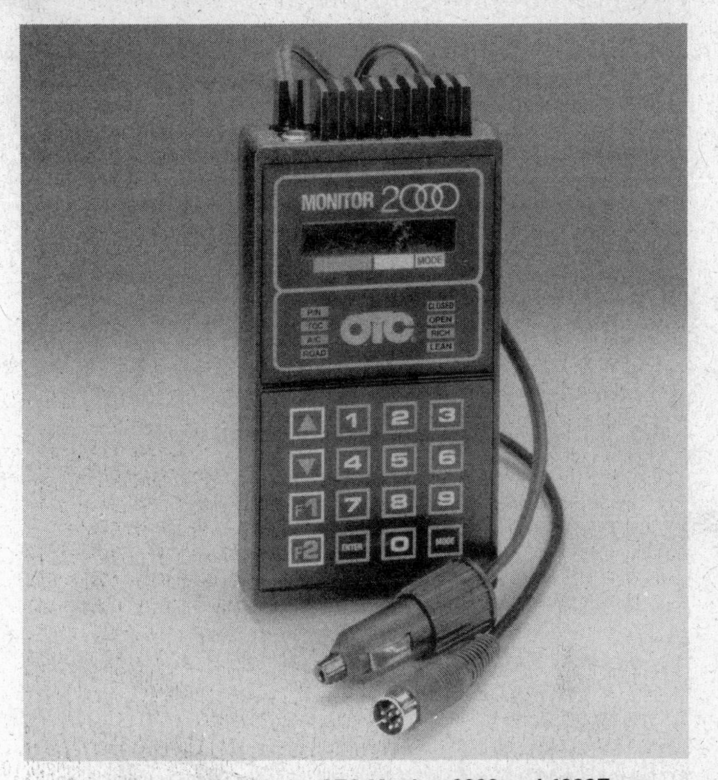

6.11 Scan tools like the OTC Monitor 2000 and 4000E are powerful diagnostic aids – using software cartridges programmed with comprehensive diagnostic information for your vehicle, they can tell you just about anything you want to know about your engine management system, but they're expensive

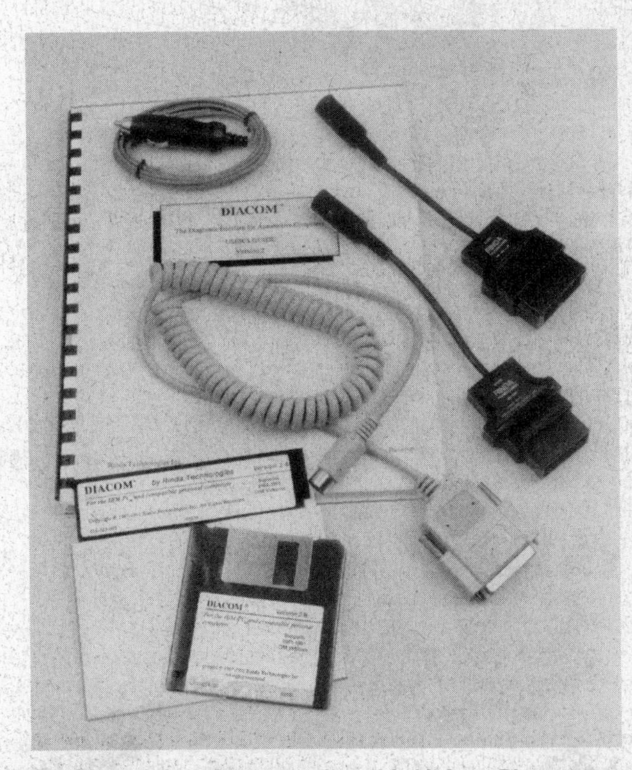

6.12 Diagnostic software, such as this kit from Diacom, turns your IBM PC, XT, AT or compatible into a scan tool, saving you the extra cost of buying a scanner but providing you with all the same information

desktop or laptop computer to interface with the engine management computer on many 1981 and later General Motors and Chrysler vehicles.

Such software can output trouble codes, identify problems without even lifting the hood, solve intermittent performance problems and even help you determine the best repair solutions with on-line technical help. We tested Rinda Technology's Diacom software. It runs on any IBM PC, XT, AT or compatible. The kit includes the software, an instruction manual and the interface cables you need to plug in your computer.

Trouble-code tools

A new type of special tool – we'll call it the trouble code tool – has recently become available to the do-it-yourselfer. These tools simplify the procedure for extracting trouble codes from your vehicle's engine management computer. Of course, you can extract trouble codes without special tools. And we'll show you how to get those codes with nothing fancier than a jumper wire or (on Fords) an analog multimeter or voltmeter. But trouble code tools do make the job a little easier and they also protect the diagnostic connector terminals and the computer itself from damage.

7 Emissions systems and components

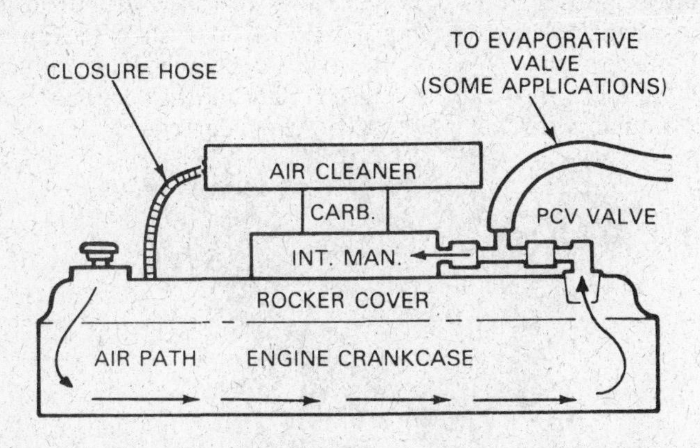

7.1 A typical Positive Crankcase Ventilation (PCV) system

Note: *For information on troubleshooting and repairing these systems, see Chapter 3.*

Positive Crankcase Ventilation (PCV) system

No piston ring can provide a perfect seal between piston and cylinder; some unburned air/fuel mixture and combustion byproducts always manages to get past the rings on the compression and power strokes. These gases – mainly hydrocarbons, or HC – are known as crankcase vapors, or blow-by gases.

Once these harmful pollutants get into the crankcase, they can mix with the engine oil, reducing its viscosity and lubricating properties. Moisture from the combustion process also condenses and makes its way into the crankcase along with unburned fuel, soot, and dust to form sludge. This condensation can also combine with unburned hydrocarbons and fuel additives, and sulfur from the original crude oil, to form carbonic acid, sulfuric acid and hydrochloric acid. These acids etch, corrode and rust the internal bearing surfaces of the engine, resulting in shorter service life.

Blow-by gases also increase crankcase pressure, which eventually builds up to a point at which engine seals or gaskets can no longer contain it, resulting in oil leakage past the seals.

The Positive Crankcase Ventilation (PCV) system **(see illustration)**, which was the first factory-installed emissions control system (introduced in 1961), prevents HC from escaping from the engine's crankcase into the atmosphere and allows it to "breathe" by permitting a charge of fresh air to enter the crankcase and mix with blow-by gases. Using intake-manifold vacuum to route air into and through the crankcase, the PCV system removes this mixture of vapors from the crankcase by venting it into the intake manifold, then into the combustion chambers where it's burned with the air/fuel mixture.

There are four general types of PCV systems:

1) *Type 1* (open system)
2) *Type 2* (restricted system)
3) *Type 3* (tube-to-air cleaner system)
4) *Type 4* (closed system)

The first three of these PCV systems are known as "open" types because the crankcase has some form of opening into the atmosphere through an unrestricted or partially restricted oil filler or breather cap. However, open type systems haven't been used since 1968 (since 1964 in California vehicles).

The fourth type, the closed system **(see illustration 3.1 in Chapter 3)**, is used today on all domestic and imported vehicles. Like the first three systems, it ventilates the engine to prevent the buildup of harmful materials like sludge, but it allows no escape of blow-by gases into the atmosphere, even during heavy acceleration.

How do you find the PCV system?

First, look for the PCV valve. Once you find the valve, the hoses are easy to identify. The valve is normally located in one of the following places:

1) A rubber grommet in the valve or rocker arm cover **(see illustration)**.
2) At the junction of the hoses **(see illustration)**.

7.2 A common location for PCV valves is right in the rubber grommet in the rocker arm cover (this one's on a 2.8L V6 in a Ford Aerostar)

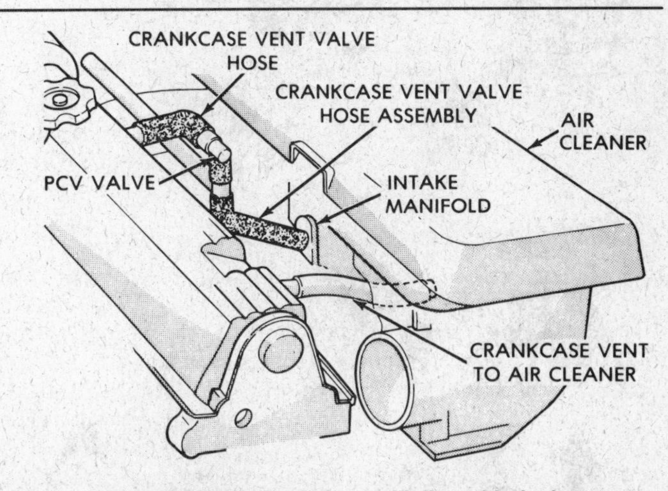

7.3 Some PCV valves are located inline with the hoses connecting the crankcase to the intake

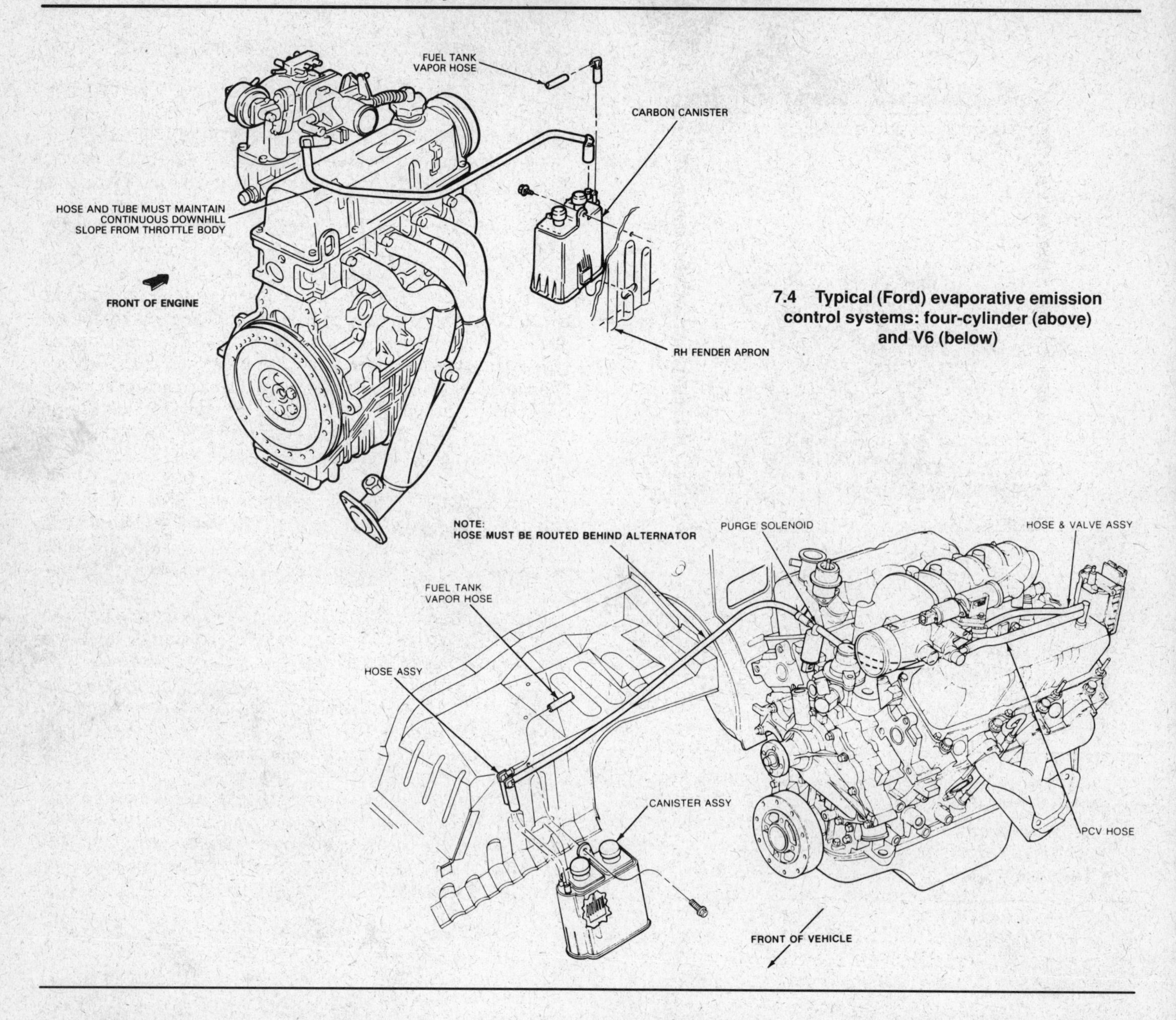

7.4 Typical (Ford) evaporative emission control systems: four-cylinder (above) and V6 (below)

3) Right in the intake manifold itself.

On some fuel-injected models, the PCV system doesn't even use a PCV valve; to find the PCV system on these models, locate the hose which connects the plenum (the chamber between the throttle body and the intake manifold) to the crankcase (usually through a camshaft, rocker arm or valve cover).

Evaporative emissions control system

The Evaporative emissions control (EVAP, EEC or ECS) system **(see illustration)** prevents the escape of gasoline vapors from the fuel tank, carburetor vents, intake manifold, etc. into the atmosphere. At one time, these vapors constituted almost 20 percent of the hydrocarbons emitted from a typical vehicle. Because they're basically raw hydrocarbons, their release into the atmosphere promotes the formation of smog. Trapping them in an evaporative emissions system and directing them into the engine combustion chambers reduces pollution and provides a slight increase in fuel economy. All 1970 and later California vehicles are equipped with an evaporative emissions system; all 1971 and later Federal vehicles have one too.

All 1970 and 1971 Chrysler vehicles and some Ford vehicles used the crankcase as the vapor storage area. In these early systems, fuel vapors from the fuel tank and carburetor accumulated within the engine's crankcase when the vehicle wasn't in operation. When the engine was started, the PCV system moved the vapors from the crankcase into the intake manifold.

Since 1972, all vehicles sold in this country have used a "carbon (charcoal) canister" **(see illustration)** as the storage receptacle for fuel vapors. This cylindrical black plastic container is usually easy to locate. Most manufacturers put it in the left or right front corner of the engine compartment.

One important thing you should know about evaporative emissions systems is their method of purging. Older vehicles with evaporative canisters simply use engine vacuum to "purge" (empty) the canister when the engine is started (and intake vacuum is high).

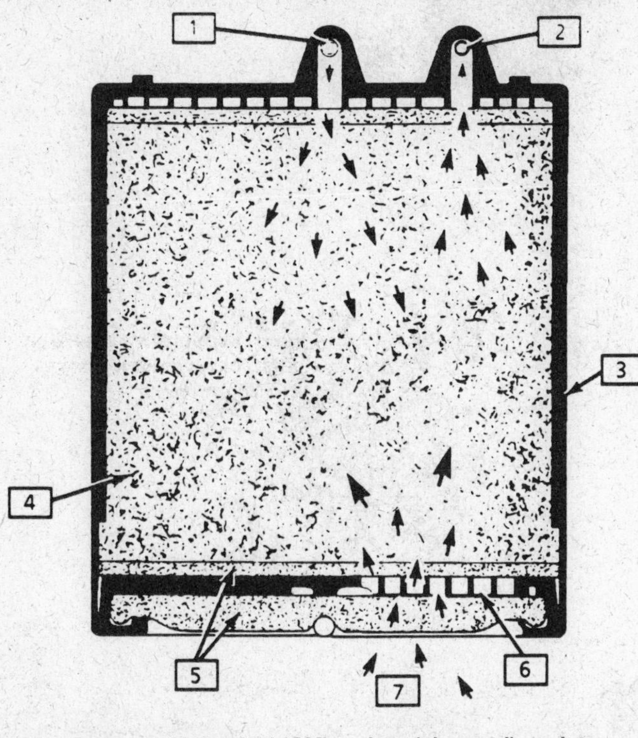

7.5 Cutaway of a typical (GM) carbon (charcoal) canister

1 Vapor inlet port
2 Canister purge vacuum
3 Canister body
4 Carbon
5 Filter
6 Grid
7 Air flow during purge

There are three basic types of vacuum-operated purging methods:

1) *Constant purge*
2) *Variable purge*
3) *Two-stage purge*

In constant-purge systems, the rate at which purging air flows through the canister is more or less fixed, regardless of how much air the engine consumes. This is usually accomplished by simply teeing into the PCV line to the carburetor, thus using intake vacuum to draw air through the charcoal granules in the canister. Even though vacuum varies with changes in engine load, an orifice in the purge line provides the system with a relatively constant air flow rate through the canister when the engine is running.

In a variable purge system, the purge line is connected to the air cleaner. So the movement of air through the canister is a result of the intake air for the carburetor passing over a tube projecting into the air cleaner snorkel. The air flow over this tube creates a vacuum that moves the vapors out of the canister and into the airstream entering the snorkel. The purge is variable because the air flow entering the air cleaner regulates its action, i.e. the amount of purge air entering the canister from the atmosphere is in proportion to the amount of air moving into the engine through the air cleaner. As a result, the more air entering the engine, the greater the vacuum on the purge line and the amount of purge air entering the canister.

If you've got a variable purge system, you'll see a purge line entering the air cleaner near the snorkel (where the velocity of the air entering the snorkel creates the necessary low-pressure area, or vacuum, which causes atmospheric pressure to force air into the canister and system).

Or, the purge line may be located on the "clean" side of the air cleaner element, where the difference in pressure, or "pressure drop," across the air filter itself is enough to permit atmospheric pressure to force air through the canister. As in the snorkel type, the amount of purge air depends on the pressure drop or vacuum at the end of the purge line inside the air cleaner. In other words, more air flows when there's more vacuum at the end of the purge line than when only a slight vacuum exists.

Some manufacturers use a two-stage purging process. A special purge valve is installed on or near the canister **(see illustration)**. This valve is operated by a ported vacuum signal that opens a second passage from the canister to the intake manifold.

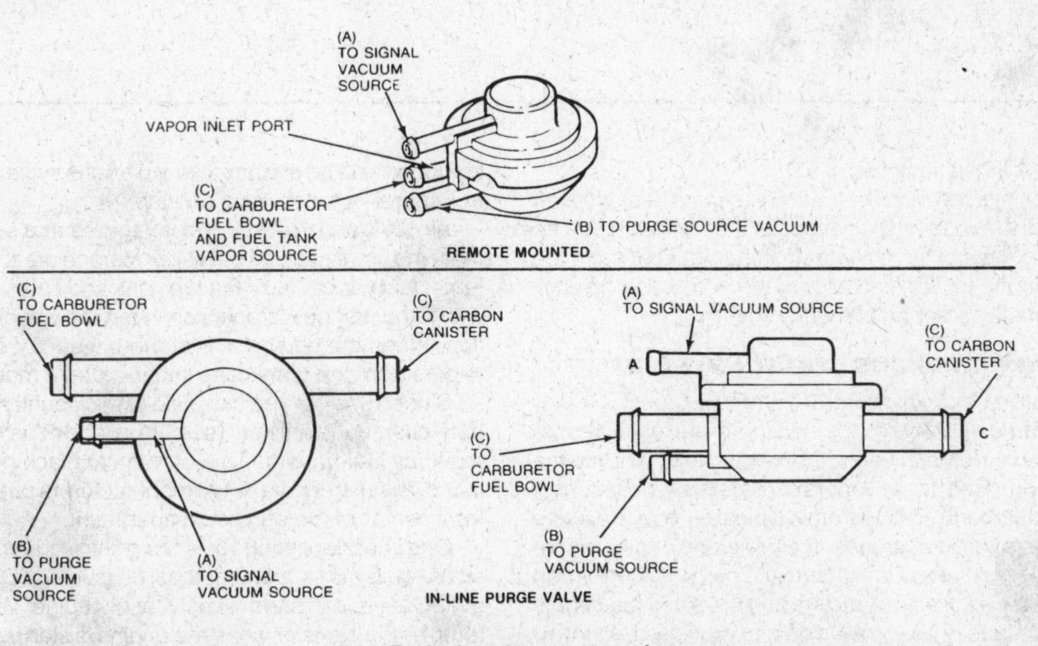

7.6 Typical canister purge valves

7.7 On carbureted vehicles, a likely location for the canister purge solenoid valve is right in the purge line between the EVAP canister and the carburetor, as shown on this Ford Aerostar van

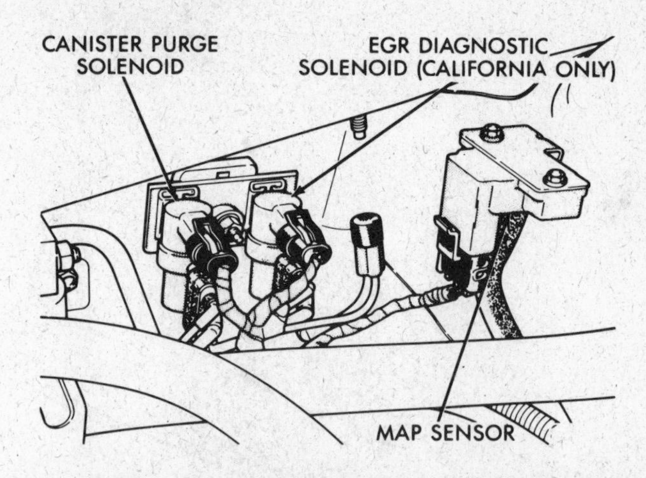

7.8 A common location for the canister purge solenoid valve on fuel-injected vehicles is the firewall or an inner fender panel, where it's often installed as part of an array of other solenoids

A ported vacuum signal is taken from a passage above the throttle valve; there's no vacuum in this passage when the throttle closes at idle, but the signal increases in proportion to the amount of throttle valve opening beyond this point. When the ported vacuum signal reaches a predetermined level, it activates the purge valve. The valve opens up a second passage from the canister to the intake manifold, allowing extra purge air to enter the canister. This air assists in reactivating the granules and carrying the vapors into the engine.

Another ported-vacuum design dispenses with the purge valve and, instead, uses an extra ported vacuum connection on the carburetor to purge the canister. The port is above the upper portion of the throttle valve, so there's no purge air through the purge line at idle. Flow begins as soon as the throttle opens above the idle position. This design improves hot-idle quality by eliminating canister purging at idle.

Some purge valves are controlled by a temperature-control valve which senses engine coolant temperature. When the temperature is lower than a predetermined value, the temperature-control valve closes the purge valve. This prevents canister vapors from entering the intake manifold or air cleaner, reducing HC and CO emissions during engine warm-up. When the temperature reaches a certain level, the valve allows the purge valve to open, and normal operation of the system begins.

Newer vehicles manufactured since the advent of computerized engine management systems use a computer-controlled canister purge solenoid **(see illustrations)** to control the canister purge valve. The computer monitors engine temperature and load to determine when to open and close the purge solenoid.

Air injection systems

The air injection system introduces oxygen (fresh air) into the hot exhaust gases when the engine is running. This promotes further oxidation (burning) of the unburned hydrocarbons and carbon monoxide in the exhaust, which reduces HC and CO emissions, respectively. The oxygen in the injected air combines with carbon monoxide to form carbon dioxide, a harmless gas, and it unites with hydrocarbons to produce water, in the form of vapor.

7.9 Sometimes, the canister purge solenoid valve is right where you'd expect to find it – right above the charcoal canister (Ford Taurus/Mercury Sable)

In some vehicles, the air injection system directs air into the exhaust manifold; in others, it injects air through the cylinder head, at the exhaust ports, allowing the oxidation process to begin a little further upstream.

There are two basic types of air injection systems – those with air pumps and those without. Systems which use air pumps are easy to identify: They use a mechanical vane-type pump **(see illustration)** driven by an accessory belt.

Here's how a typical pump-type system **(see illustration)** works: The air pump receives filtered air from the air cleaner assembly and pumps it through the air switching (or control) valve and the check valve, then transmits this filtered air into an air manifold assembly mounted on the cylinder head. As hot exhaust gases leave the combustion chamber, they meet with a blast of air from the air injection nozzles located in the exhaust ports. This added air helps to burn the unburned hydrocarbons and carbon monoxide that survived the combustion process.

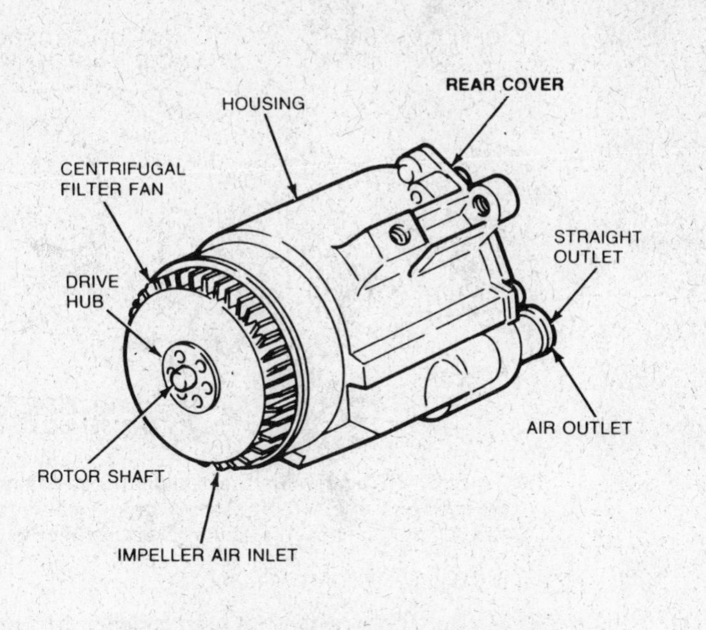

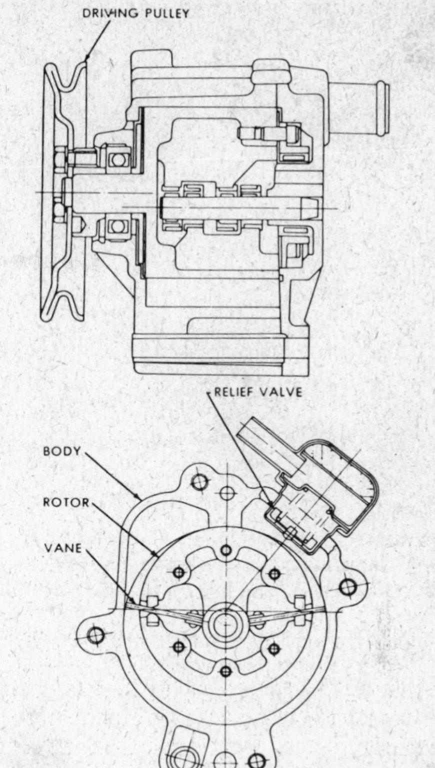

7.10 A typical Ford air injection pump (left); a cutaway view of typical Japanese (this one's from an Isuzu Pickup) air pump (right)

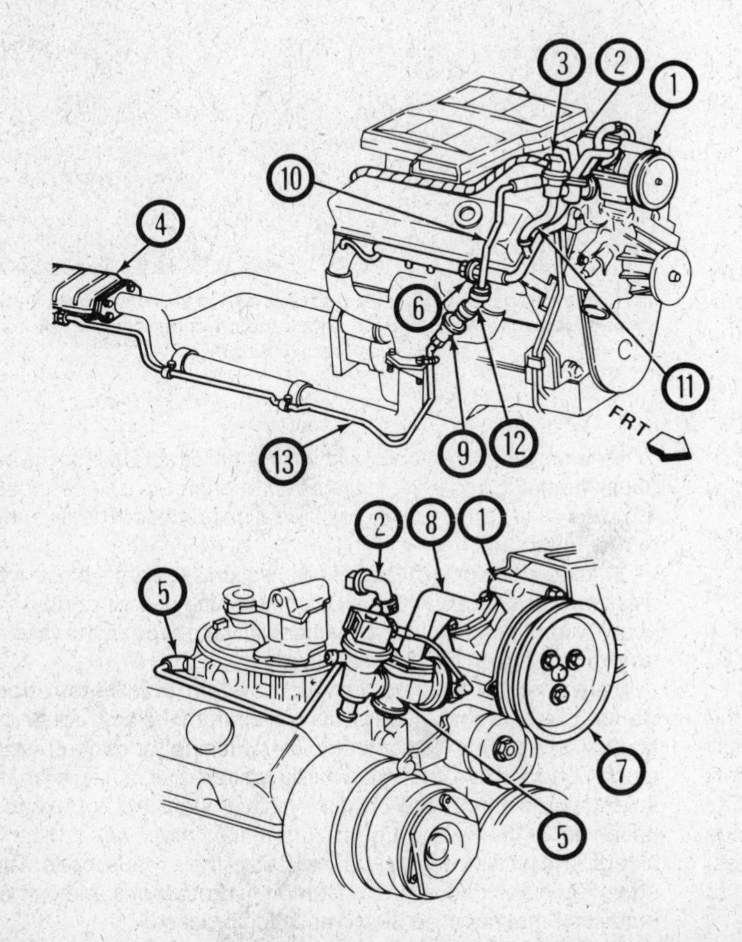

7.11 A typical pump-type air injection system (Chevrolet Corvette):

1. *Air pump*
2. *Control valve diverter hose*
3. *Control valve*
4. *Catalytic converter*
5. *Control valve vacuum harness*
6. *Exhaust air injection check valve*
7. *Pump pulley*
8. *Control valve adapter (pipe)*
9. *Catalytic converter air injection check valve*
10. *Catalytic converter air injection check valve pipe*
11. *Control valve hose*
12. *Catalytic converter air injection check valve hose*
13. *Catalytic converter air injection check valve pipe*

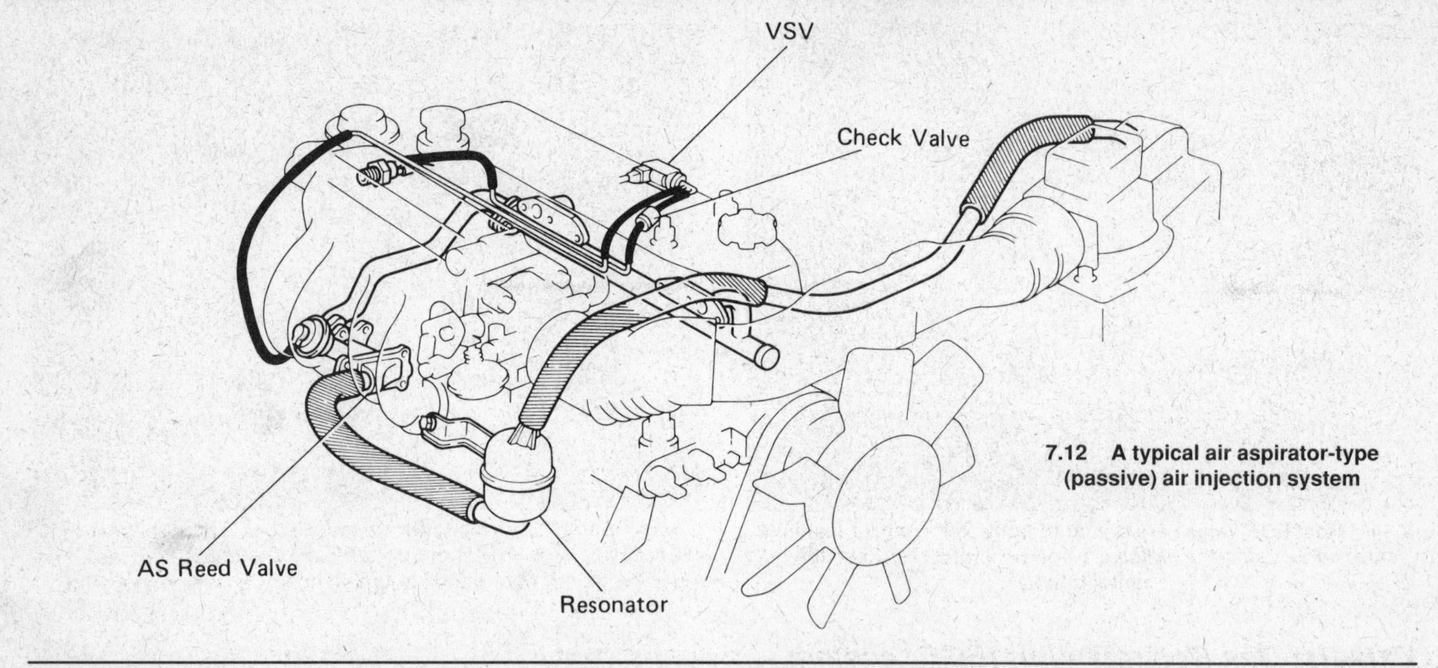

7.12 A typical air aspirator-type (passive) air injection system

Pump-type air injection systems use various types of valves to route air to the exhaust manifold, to the exhaust pipe between the oxidation and reduction catalysts (on newer models) or to atmosphere. These valves have a bewildering variety of names: mixture control valve, check valve, switching valve, diverter valve, relief valve, switch/relief valve, air bypass valve **(see illustration 4.8 in Chapter 3)**, etc. No matter what they're called, they all perform the same few tasks. Keep this in mind when reading the following description of our typical air injection system, and when trying to figure out how the air injection system works on your vehicle.

When it receives a high vacuum signal from the intake manifold, the mixture control valve introduces ambient air through the air filter into the intake manifold to dilute the momentarily rich fuel mixture that occurs on initial throttle closing, eliminating backfiring.

During heavy engine load conditions (near wide open throttle), the air switching valve diverts air from the air pump to the atmosphere to prevent overheating of the catalytic converter. This function occurs at a predetermined level of intake vacuum.

The check valve **(see illustration 4.7 in Chapter 3)** is a one-way valve which prevents exhaust gas from entering and damaging the air pump if the pump ceases operation because of a drivebelt failure.

Once you locate the air pump – which is usually supported by a bracket bolted to the block, just like the air conditioning compressor, alternator, and power steering pump – you'll be able to find the other devices (the diverter valve, relief valve, check valves, injection manifolds and tubes, etc.) by tracing the hoses that originate at the pump.

Since 1975, some Chrysler, Ford and GM – and many import – vehicles have used a much simpler air injection system. This system is referred to as an aspirator-air, pulse-air or suction-air system **(see illustration)**. Regardless of the name, all versions of this type of air system are passive – instead of an air pump, they use exhaust pressure pulsations to draw air into the exhaust system. Every time an exhaust valve closes, there's a period when the pressure inside the manifold drops below that of the atmosphere. During these low-pressure (relative vacuum) pulses, air from the clean side of the air cleaner is drawn into the exhaust manifold(s).

7.13 The valves on a Pulse Air System (Thermactor II) look like this – on a 2.3L Ford Tempo, they're down by the starter motor (on your vehicle, they might be somewhere else, but you can find them by following the hoses or lines between the air pump and the exhaust system)

The typical passive air injection system consists of a length of hose from the air cleaner to the aspirator valve, injection valve, pulse air valve, etc. **(see illustration)**, and a piece of steel tubing between the valve and the exhaust manifold. The valve itself (regardless of the fancy name) is simply a one-way check valve which uses metal reeds or a spring-loaded diaphragm to admit air into the exhaust manifold, but prevents hot exhaust gases from escaping.

To determine whether you've got a passive system, refer to the VECI label; to locate the system and its components, look for the valve first and trace the hoses back to the air cleaner.

7.14 Most EGR valves are bolted directly to the intake manifold and connected to the exhaust manifold with a short section of metal tubing

7.15 Some EGR solenoids are installed on a bracket near the EGR valve, such as this one on a Nissan Maxima (left arrow). The arrow on the right points to the air injection system solenoid

Exhaust Gas Recirculation (EGR) system

High combustion temperatures produce nitrogen oxide (NOx), a constituent of ozone. There are two ways to reduce peak combustion temperatures: Spark control systems and Exhaust Gas Recirculation (EGR) systems.

Spark control systems hold down combustion temperatures by limiting ignition timing advance during acceleration from idle to cruise (the condition under which NOx emissions are highest). Before the introduction of the catalytic converter, spark control systems were widely used; today, they're still in use, but they're no longer considered the optimal means of reducing NOx emissions.

Since the introduction of the catalytic converter, EGR systems have emerged as a better way to control NOx. EGR systems reduce peak combustion temperatures by diluting the incoming air/fuel mixture with a small amount of "inert" (won't undergo a chemical reaction) exhaust gas. A 6-to-14 percent concentration of exhaust gas, routed from the exhaust system to the intake manifold, mixes with the air/fuel mixture entering each cylinder, and reduces the mixture's ability to produce heat during combustion.

Why? Because exhaust gas contains little or no oxygen, so it dilutes the air/fuel charge with a noncombustible gas. And since this inert exhaust gas is displacing some of the the oxygen in the highly combustible air/fuel mixture, it reduces the quality of the total charge reaching each of the cylinders.

Also, the injected exhaust gases are hot, so they expand the air/fuel mixture in the intake manifold. This reduces the concentration of combustible materials swept into and compressed by the piston in each of the engine's cylinders. So the mixture of air, fuel and exhaust gas entering the combustion chambers isn't as powerful when ignited. And it therefore creates less heat than an undiluted air/fuel mixture would otherwise produce.

Not all engine operating conditions produce excessive NOx emissions. At idle, for example, the engine creates little NOx, so exhaust gas recirculation is unnecessary. The engine also operates more efficiently – and the vehicle is more driveable – if the EGR system is turned off during wide-open throttle operation. In fact, the only time the EGR system should operate is at vehicle speeds between 30 and 70 mph, when NOx emissions are high.

And at low engine temperatures, exhaust gas recirculation may not be necessary. When engine temperature is low, so is the formation of NOx. Turning off the EGR system at such times improves engine warm-up and vehicle driveability.

The amount of exhaust gas admitted to the intake air is controlled by the EGR valve, which is the primary component of the system. Finding the EGR valve is easy: Most EGR valves are bolted to the intake manifold and connected to the exhaust manifold with a short section of metal tubing or passages in the intake manifold **(see illustration)**. Most EGR valves have a vacuum hose connected to them that provides the signal to open the valve. The hose is connected to the intake manifold. Often, there's a temperature valve and/or solenoid valve in the vacuum line between the EGR valve and manifold.

The EGR valves used on modern vehicles with engine management systems are turned on and off by an EGR control solenoid valve controlled by the computer. The solenoid is usually located near the EGR valve, either on its own bracket or on the firewall **(see illustrations)**. Some computer-controlled EGR

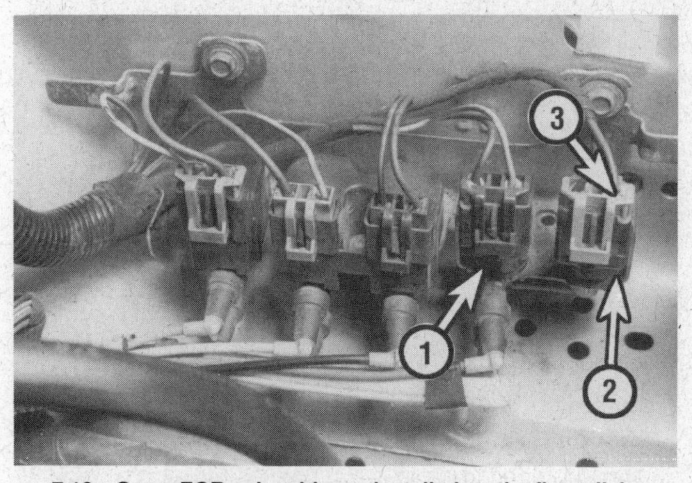

7.16 Some EGR solenoids are installed on the firewall, in an array of other solenoids, such as these units on a Ford Thunderbird – the vacuum valve (1) supplies vacuum to the electronic EGR valve when energized; when de-energized, the vent valve (2) vents the EGR valve to the atmosphere through a small vent (3)

7.17 Some EGR valves are also equipped with a position sensor like this unit on a Ford Thunderbird – the position sensor is almost always mounted right on top of the EGR valve

valves also use a position sensor, usually atop the EGR valve itself, to fine-tune the flow of gases into the intake by incremental adjustments to the actual valve inside **(see illustration)**.

Catalytic converter

The catalytic converter is probably the most effective of the principal emissions control components we'll look at. The catalytic converter, which has been installed on all vehicles since 1975, reduces the levels of HC, CO and NOx in the exhaust by providing an additional area for the oxidation or reduction of these pollutants to occur and a catalyst to promote these changes.

The catalytic converter **(see illustration)**, which looks like another muffler, is located in the engine's exhaust system somewhere between the exhaust manifold and the muffler. If the vehicle has two exhaust pipes (one for each cylinder bank), you will find a catalytic converter in each exhaust pipe. And some vehicles actually have two catalytic converters in a single exhaust pipe **(see illustration)**. These earlier dual-converter setups are simply a pair of converters, different in size but identical in function; both units, which are referred to as "mini" and "main" converters, are oxidation catalysts.

Later dual converter setups – which have become the norm on modern vehicles – actually perform different functions: One converter is a reduction catalyst and the other is an oxidation catalyst. Most recent designs incorporate both the reduction and the oxidation catalyst into one unit, known as a three-way catalyst or a hybrid converter.

How do you find the catalytic converter(s)? Raise the vehicle, place it securely on jackstands and look for a large, stainless-steel canister in the exhaust pipe. It will have heat shields and insulating pads behind it to protect the underside of the vehicle from the heat shed by the converter when it's operating.

Heated air intake systems

The air-fuel mixture must be a fine mist or vapor to burn completely in the combustion chamber. To create this vapor, the fuel must be well-dispersed, in the form of thousands of tiny, uniform droplets in the intake air. If it isn't, driveability suffers – the engine

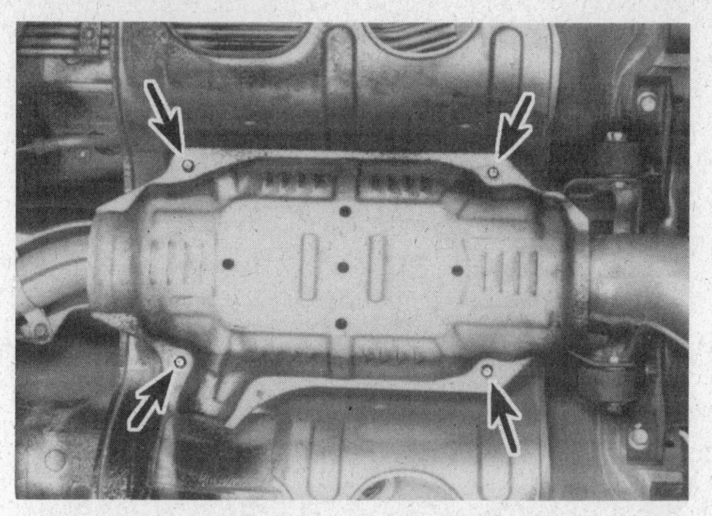

7.18 A typical catalytic converter (Nissan 300ZX) – the "cat" looks sort of like a muffler, except that it's made out of stainless steel and is surrounded by heat shields: the one above is to protect the underbody of the vehicle from the cat's high operating temperature and the one below is to prevent the cat from starting a fire or burning the home mechanic!

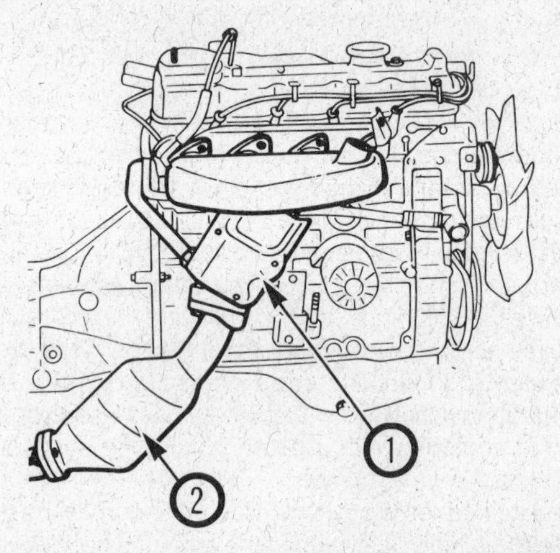

7.19 Some vehicles, such as Chryslers and Mitsubishis, use a dual converter setup: The "mini" converter (1) is up front, right at the exhaust manifold and the "main" converter (2), is further downstream in the exhaust system

runs roughly; more importantly, mileage suffers and large amounts of hydrocarbons and carbon monoxide are emitted out the tailpipe. In fact, automobiles pollute more during their initial warm-up phase than at any other time during their operation.

On modern vehicles with port fuel injection, this isn't a problem; even when the engine is cold, the injectors spray a fine mist right into the intake port, just above the intake valve, where it mixes instantly with the rush of incoming air each time the valve opens and the piston goes down.

But on older vehicles with carburetors – and to some extent, even on those vehicles which use "single-point" injectors (single or dual injectors at the throttle body itself) – the fuel entering the intake must be dispersed far upstream from the actual intake ports, and must maintain a uniform dispersion all the way to the

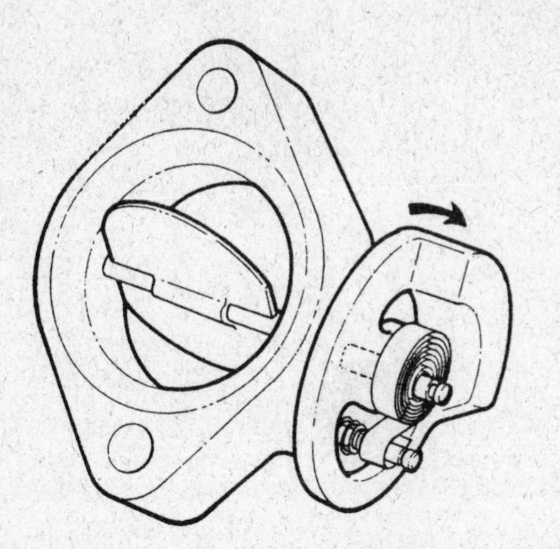

7.20 A typical thermostatically controlled heat-control valve: When the engine is cold, the thermostat (bimetal coil) closes the valve, blocking exhaust gas flow from the manifold and forcing exhaust gases to detour through the heat riser on their way to the exhaust pipe; when the engine is warm, the thermostat opens the valve, sealing off the heat riser passage and sending gases straight through the exhaust pipe

combustion chamber. Once these engines are at their normal operating temperature, the ambient temperature of the carburetor/throttle body, the runners in the intake manifold and the intake port area assures good vaporization. But when a carbureted/single-point injected engine is started cold, the air-fuel mixture doesn't vaporize well, and it tends to form irregularly sized droplets, and even fall out of suspension, sticking to the walls of the intake manifold runners. So the intake area on these vehicles needs some external heat source while the engine is still warming up; that's why they're equipped with a heated air intake system of some sort.

Heated air intakes are nothing new. The earliest designs consisted of a metal cowl fastened to the exhaust manifold(s), a flexible section of metal ducting connecting it to the cold air snorkel of the air cleaner and a heat control (damper) "door" or flap inside the snorkel that mixes air heated by the manifold(s) with cold ambient (outside) air in response to changes in the vacuum signal in a vacuum line connecting the vacuum motor to intake vacuum. This vacuum signal is controlled by a temperature sensing valve located inside the air cleaner housing. This design, generally referred to as a thermostatically controlled air cleaner, works so well that it's still with us today. There are many variations around, but they all work basically the same way **(see illustration 5.1 in Chapter 3)**.

The heated air intake system is pretty simple in operation: If the ambient temperature is below around 85-degrees F. when you start the engine, the temperature sensing vacuum valve inside the air cleaner remains closed, allowing full intake vacuum to get to the vacuum motor, which closes the door to outside air and opens it to the hot air tube from the exhaust manifold cowl.

As the engine warms up, so does the temperature of the air inside the air cleaner, since it's being heated by the exhaust manifold(s). The temperature sensing valve in the air cleaner housing begins to open, bleeding off the vacuum signal to the motor, and allowing the spring-loaded door to start closing off the heated air tube and opening the cold ambient air snorkel.

By the time the engine is fully warmed up, the vacuum sensing valve is fully open, so no vacuum gets to the door motor. The door completely shuts off the heated air tube and allows nothing but outside air into the snorkel.

On some models, a cold weather modulator traps vacuum to the motor if manifold vacuum drops off as a result of the throttle opening while the air cleaner is cold. Some systems use a retard delay valve instead of a cold weather modulator. The retard delay valve traps the vacuum for a few seconds when the throttle opens.

To further improve driveability and lower emissions, many vehicles are also equipped with an additional device: a bimetallic coil or vacuum diaphragm-operated heat-control valve **(see the accompanying illustration and illustration 5.3 in Chapter 3)**. This device provides more precise control of intake manifold heating, improving vaporization of the air-fuel mixture.

Heat-riser valves go by several names. General Motors calls them Early Fuel Evaporation (EFE) valves, Ford calls them Heat Control Valves (HCV), Chrysler calls them power heat control valves, etc. But they all look and work the same.

Here's how a typical valve works: A rotating valve is housed inside a cast iron body which fits between the exhaust manifold and the exhaust pipe. Linked to the shaft that extends through the cast iron body and attaches to the valve is a diaphragm inside a vacuum motor or a bimetallic coil.

The diaphragm type turns the shaft and valve by reacting to a vacuum signal from the intake manifold via a ported vacuum switch. On computer-controlled engines, the vacuum system also includes an electric solenoid-operated vacuum valve. When the diaphragm receives a full or high manifold vacuum signal, it closes the valve, sending hot exhaust gases up through a heat riser, through special passages in the intake manifold, and back into the exhaust system. When the signal to the motor is low or is cut off completely, a spring behind the diaphragm pushes the heat-riser valve open, and the gases go straight out the exhaust pipe.

The bimetallic-coil type works the same way, but it simply responds to the temperature of the housing itself, which is about the same temperature as the exhaust manifold.

Another device which improves vaporization by heating the mixture during cold starts is an electrically heated grid installed between the carburetor or throttle body and the intake manifold **(see illustration 5.4 in Chapter 3)**. This device is operated by a relay controlled by a temperature switch that shuts the relay off after a couple of minutes.

The engine management system

By the late 1970's, many vehicles' ignition systems were controlled by a computer. In quick succession, most other engine systems were also placed under computer control. Since 1980, most vehicles sold in the US have been equipped with computerized engine management systems to help reduce emissions.

No matter how fancy the name, no matter how sophisticated the system, all engine management systems consist of the same three basic types of components: information sensors, a computer and actuators or controls.

1) Every engine management system has a wide variety of *information sensors* (as many as a dozen or more) which monitor various operating conditions of the engine (such as coolant temperature, intake air temperature, throttle position angle, engine speed, etc.) **(see illustrations)**.

7.21 One likely place to find the air temperature sensor is the air cleaner housing, such as this Manifold Air Temperature (MAT) sensor on a Pontiac Fiero – air temperature sensors are also often located in the intake manifold or intake runners

7.22 How do you know you've got the right sensor on a vehicle such as this Ford Thunderbird, on which the air temperature sensor is installed right next to another sensor, and they're identical in appearance? Try to determine whether it's installed in a coolant passage in the intake manifold (a coolant temperature sensor) or an air intake runner (an air temperature sensor)

7.23 Coolant temperature sensors, such as this Engine Coolant Temperature (ECT) sensor on a Plymouth Sundance, are usually installed in the thermostat housing

7.24 This coolant temperature sensor on a Ford Probe is installed in the intake manifold, but note that it's protruding into a coolant passage in the manifold which leads to the thermostat housing in the foreground

7.25 Like air temperature sensors, coolant temperature sensors, such as this one (arrow) between two other sensors on a Pontiac Grand Am, can be difficult to identify – if you're unable to identify the sensor you're looking for using the "air-intake-passage-vs.-coolant-passage-installation" approach, trace the electrical leads back to the main harness or to the device to which they're attached; if that doesn't work, try counting the number and color of the electrical leads, then refer to a wiring diagram

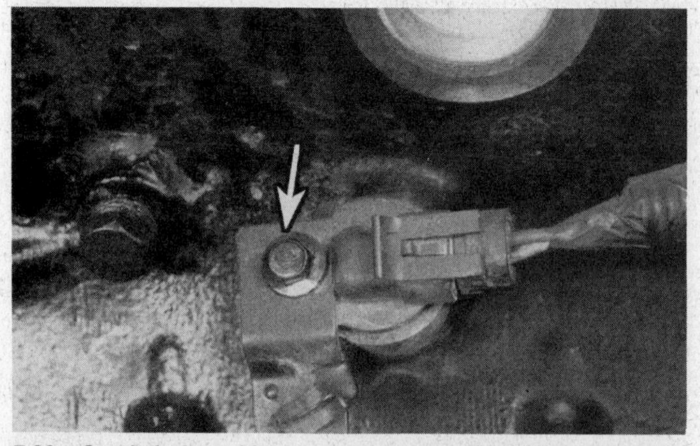

7.26 Crankshaft position sensors are installed either in the side of the block, such as this unit on a Chevrolet Corsica/Beretta, next to the crankshaft at the front of the engine. . .

7.27 . . . or inside the distributor assembly, such as this unit on a Nissan pickup truck

7.28 Knock sensors, such as this Electronic Spark Control (ESC) sensor on a Chevrolet Corvette, are usually installed in the block, where they can detect harmful detonation in the combustion chamber

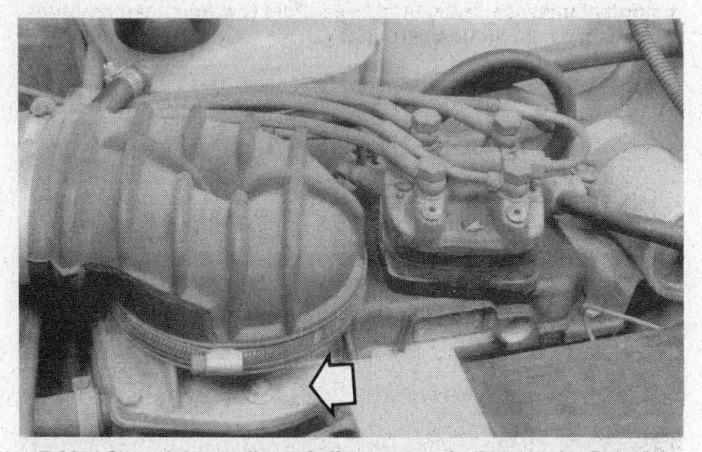

7.29 One of the earliest air flow meter designs is the Bosch plate-type unit (arrow) found on all German vehicles with continuous injection (CIS, CIS-E), such as this unit on a VW Fox with Bosch CIS-E – finding the air flow meter on one of these vehicles is easy: It's always right next to the fuel distributor, the device with all the braided-stainless-steel fuel lines attached (in fact, the air flow meter and the fuel distributor are always housed in one integral unit known as the mixture control unit)

7.30 Vane-type air flow meters, such as this unit (arrow) on a Ford Explorer, use a spring-loaded swinging door connected to a variable resistor to send a variable voltage signal to the computer – this type of air flow meter is always located downstream from the air cleaner housing and upstream from the throttle body

7.31 The latest air flow meters, such as this Mass Air Flow (MAF) unit on a Chevrolet Corsica, use a heated resistance wire to measure air flow by sending a voltage signal to the computer that varies in proportion to the mass of air passing over the wire – this type of air flow meter is also located between the air cleaner housing and the throttle body

7.32 Some air flow meters, such as this unit on a Nissan pick-up truck, are mounted right on the side of the throttle body – these units also use a hot wire to measure air flow mass, but they re-route some of the air entering the throttle body into a side passage, where the wire is located

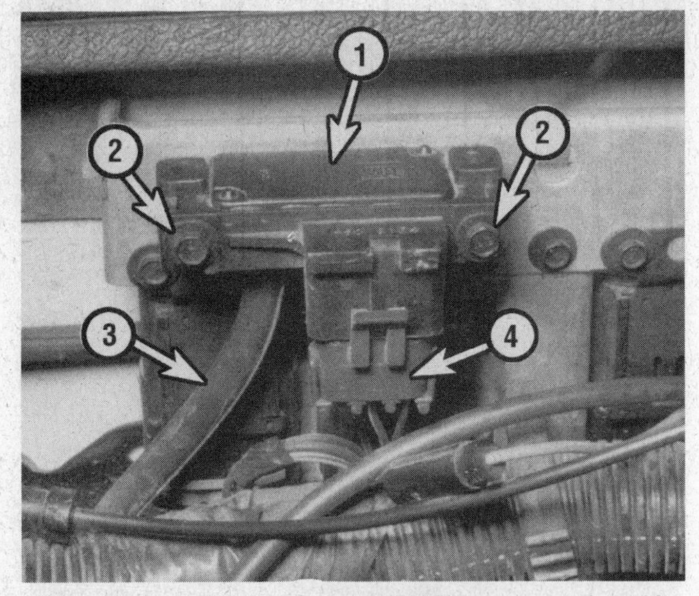

7.33 Manifold Absolute Pressure (MAP) sensors, such as this Chevrolet Corsica unit, are usually a black plastic box located on the firewall, and usually have the same parts:

1. *MAP sensor assembly*
2. *Mounting screws (some units are simply clipped onto a bracket)*
3. *MAP sensor vacuum line (goes to intake manifold vacuum)*
4. *MAP sensor electrical connector (usually goes to main engine harness)*

2) The sensors transmit this data, as a variable voltage signal, to a *computer* **(see illustrations)** which analyzes this information by comparing it to the "map" inside its memory. The map is simply a program which very specifically details how the engine should be operating under every conceivable operating condition (cold starts, warm-ups, acceleration, deceleration, etc.).

3) If the computer notes a discrepancy between what's happening and what the map says SHOULD be happening under a given set of circumstances, the computer transmits commands, again in the form of voltage data, to a smaller

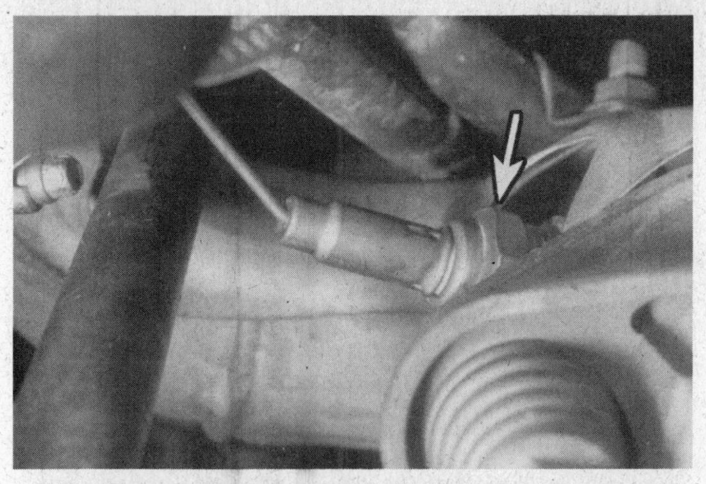

7.34 Oxygen sensors are easy to find (but not always easy to get to): They're always in the exhaust system, somewhere between the exhaust manifold(s) and the catalytic converter(s) – this unit (arrow) as seen from underneath the vehicle, is in the left exhaust manifold of a Pontiac Grand Am (there's usually another one in the other manifold on V6s and V8s)

7.35 The Throttle Position Sensor (TPS) is always mounted on the carburetor or throttle body, usually right on the end of the throttle valve shaft, such as this TBI-mounted unit on a Chevrolet full-size pickup truck

7.36 Computers can be anywhere there's room, but there are three common locations: Many are installed beneath the right side of the dash – usually right under the glove box, as on this Pontiac Grand Am

7.37 Another likely location is behind the kick panel (the small triangular area just in front of the front door and beneath the extreme right end of the dash) as on this Chevrolet Corsica

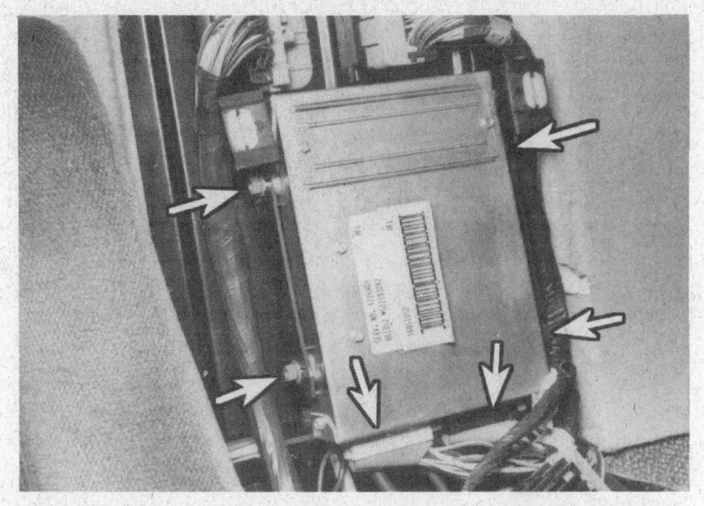

7.38 A third computer location is between the seats, as on this Pontiac Fiero, or even underneath one of the front seats (arrows point to electrical connector and mounting bolt locations)

7.39 Ignition system computers, such as this GM Electronic Spark Control (ESC) module on a Chevy pickup, are usually on the firewall – but even if they're not, you can always identify them by their large multi-pin connector which connects them to a knock sensor, the ignition coil, the battery, etc. (arrows point to electrical connector and mounting bolt locations)

7.40 Look for control solenoids in arrays, such as this one on a Dodge Dakota – this computer-controlled switching solenoid (arrow) controls the vacuum signal to the switch/relief valve on the air injection system

7.41 This computer-controlled Auxiliary Air Control (AAC) valve is located on the intake manifold on a Nissan Maxima – as coolant temperature rises, the valve gradually closes, restricting auxiliary air flow

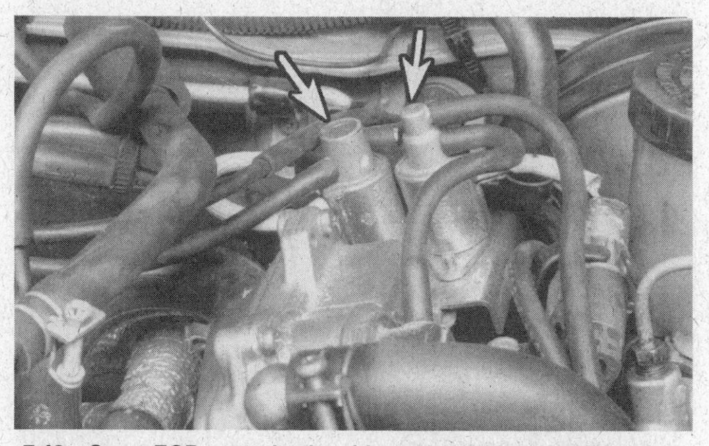

7.42 Some EGR control solenoids are located on small brackets which can be mounted anywhere in the engine compartment – the EGR control solenoid, or vacuum cut solenoid, is on the left on this Nissan Maxima (the control solenoid on the right is for the Air Injection Valve (AIV)

7.43 This EGR control solenoid on a Ford Tempo is bolted to a bracket mounted on the left strut tower, another popular mounting point for solenoids

group of devices known as *actuators,* or *controls* **(see illustrations)**, which alter the operating conditions of the engine (richen or lean the fuel/air mixture, advance or retard the ignition, open or close the EGR valve, open or close the EVAP canister purge valve, etc.).

And that's it! The details vary somewhat from system to system, but not much. All engine management systems use the same three types of components – a bunch of sensors, one computer and several controls. So don't make engine management systems more complicated than they really are.

8 Basic vacuum troubleshooting

What is vacuum?

First, let's look at what vacuum is. In science, the term "vacuum" refers to a total absence of air; in automotive mechanics, vacuum refers a pressure level that's lower than the earth's atmospheric pressure at any given altitude. The higher the altitude, the lower the atmospheric pressure.

You can measure vacuum pressure in relation to atmospheric pressure. Atmospheric pressure is the pressure exerted on every object on earth and is caused by the weight of the surrounding air. At sea level, the pressure exerted by the atmosphere is 14.7 "pounds per square inch" (psi). We call this measurement system "pounds per square inch absolute" (psia).

The gauge on a hand-operated vacuum pump ignores atmospheric pressure at sea level and reads zero instead. In other words, on a gauge, atmospheric pressure, or zero psi, is the starting point of our measurements. We call this measurement system "pounds per square inch gauge" (psig).

But vacuum gauges don't measure vacuum in psia or psig; instead, they measure it in "inches of Mercury" (in-Hg). Once in a while, you'll see another unit of measurement on some gauges; it's expressed in "kilopascals" (kPa). Another unit of measurement, used on manometers, is expressed in "inches of water" (in-H2O). The relationship of these confusing units of measurement is shown in the accompanying table **(see illustration)**. Note that when atmospheric pressure (14.7 psi) is present, automotive gauges indicate zero.

How is vacuum created in an internal combustion engine?

Positive pressure always flows to an area with a "less positive," i.e. relatively negative, pressure. This is a basic law of physics. Viewed from this perspective, an engine is really nothing more than an air pump. As the crankshaft rotates through two full revolutions, the engine cycles through its intake, compression, power and exhaust strokes. The first and last of these strokes – the intake and exhaust strokes – are identical to the action of the intake and exhaust strokes of any air pump: The intake "pulls" in air; the exhaust expels it.

During the intake stroke, the piston moves downward from its top dead center position. At the same time the exhaust valve closes and the intake valve opens. This downward movement of the piston in the cylinder creates a relative vacuum, drawing the air-fuel mixture into the cylinder through the open intake valve.

After the engine compression and power strokes are completed, the intake valve is still closed but the exhaust valve opens as the piston begins moving upward on its exhaust stroke. The rising piston forces the spent exhaust gases out through the open port.

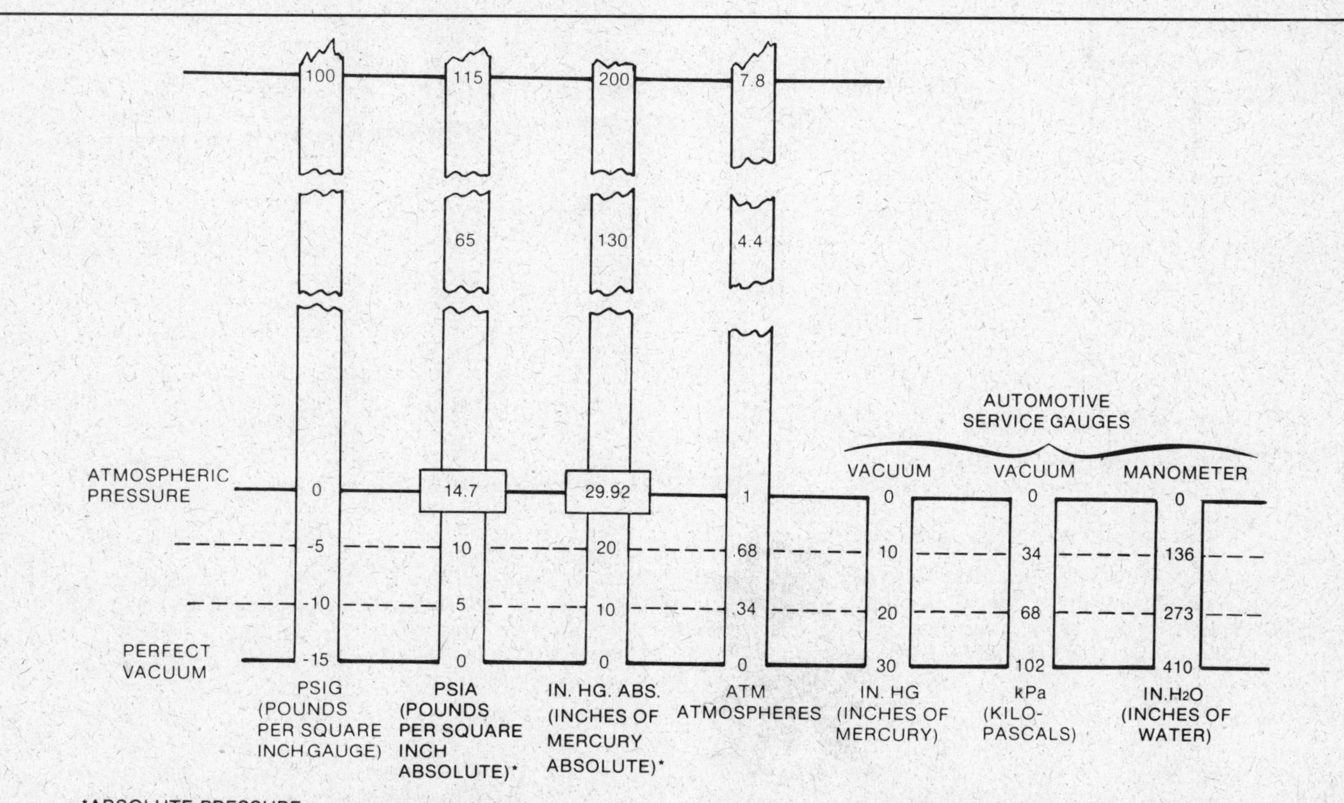

8.1 This table shows the relationship of common units of measurement used for measuring vacuum

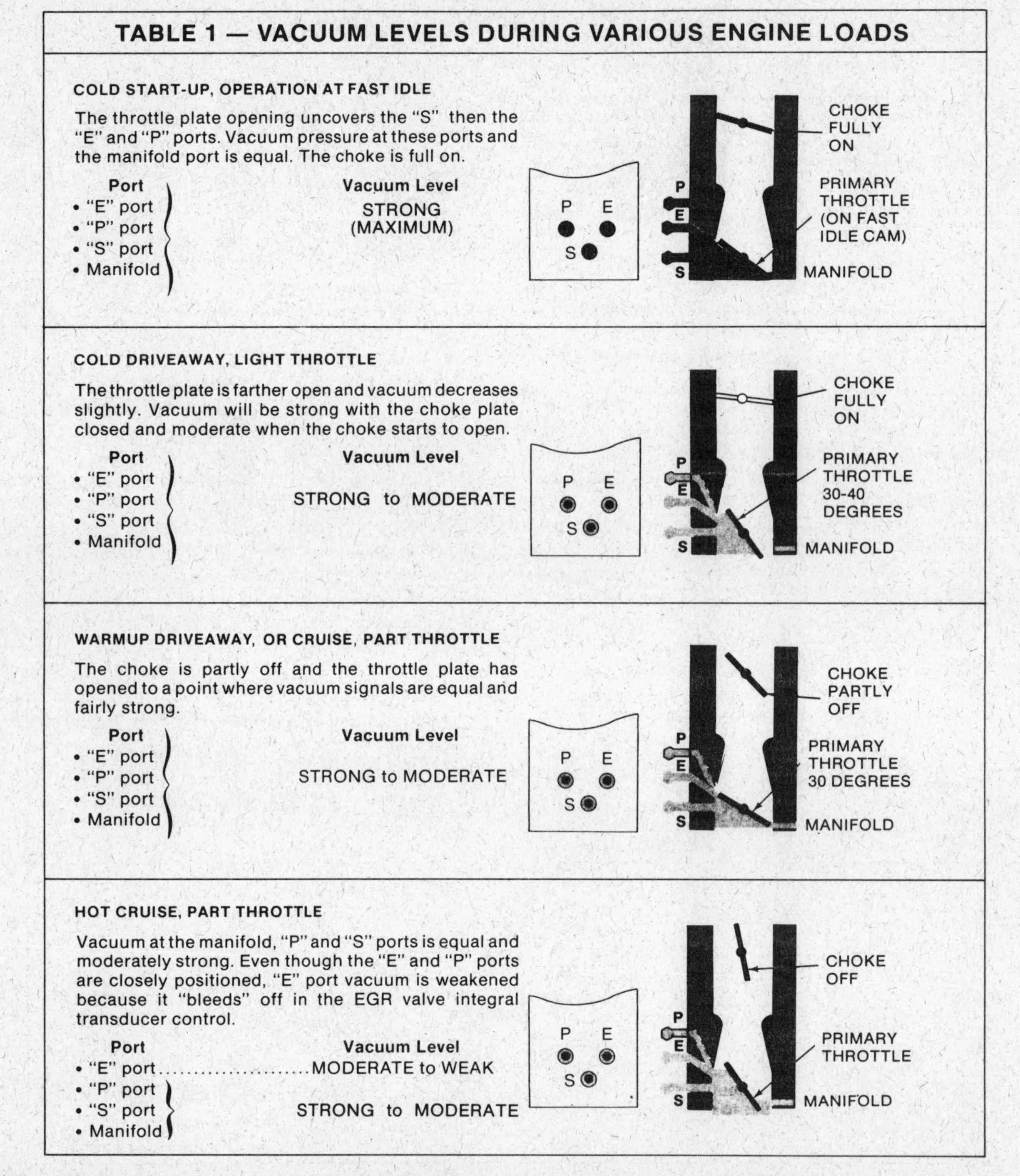

TABLE 1 — VACUUM LEVELS DURING VARIOUS ENGINE LOADS

COLD START-UP, OPERATION AT FAST IDLE

The throttle plate opening uncovers the "S" then the "E" and "P" ports. Vacuum pressure at these ports and the manifold port is equal. The choke is full on.

Port	Vacuum Level
• "E" port • "P" port • "S" port • Manifold	STRONG (MAXIMUM)

COLD DRIVEAWAY, LIGHT THROTTLE

The throttle plate is farther open and vacuum decreases slightly. Vacuum will be strong with the choke plate closed and moderate when the choke starts to open.

Port	Vacuum Level
• "E" port • "P" port • "S" port • Manifold	STRONG to MODERATE

WARMUP DRIVEAWAY, OR CRUISE, PART THROTTLE

The choke is partly off and the throttle plate has opened to a point where vacuum signals are equal and fairly strong.

Port	Vacuum Level
• "E" port • "P" port • "S" port • Manifold	STRONG to MODERATE

HOT CRUISE, PART THROTTLE

Vacuum at the manifold, "P" and "S" ports is equal and moderately strong. Even though the "E" and "P" ports are closely positioned, "E" port vacuum is weakened because it "bleeds" off in the EGR valve integral transducer control.

Port	Vacuum Level
• "E" port	MODERATE to WEAK
• "P" port • "S" port • Manifold	STRONG to MODERATE

8.2a These cutaways (this and facing page) of a typical carburetor (this one's a Motorcraft 740-2V for a Ford) show how vacuum levels change during various engine loads; of course, a throttle body on a fuel-injected vehicle has no choke, but the principles shown here apply to what happens inside a throttle body as well (note the three ports – S for spark vacuum, P for evaporative canister purge and E for EGR valve – on the 740-2V; all carbs/throttle bodies have vacuum ports – find them on your carb or throttle body, and learn where they go)

The partial vacuum created by the engine's intake stroke is relatively continuous, because one cylinder is always at some stage of its intake stroke in a four, six or eight-cylinder engine. On carbureted engines, this intake vacuum is regulated, to some extent, by the position of the choke plate and the throttle valve; on fuel-injected engines, it's regulated strictly by throttle valve position, since there is no choke plate. When the choke plate or throttle valve is in its closed position, air flow is reduced and intake vacuum is higher; as the plate or valve opens, air flow increases and vacuum decreases. The accompanying cutaways show how vacuum levels change during various engine loads **(see illustration)**.

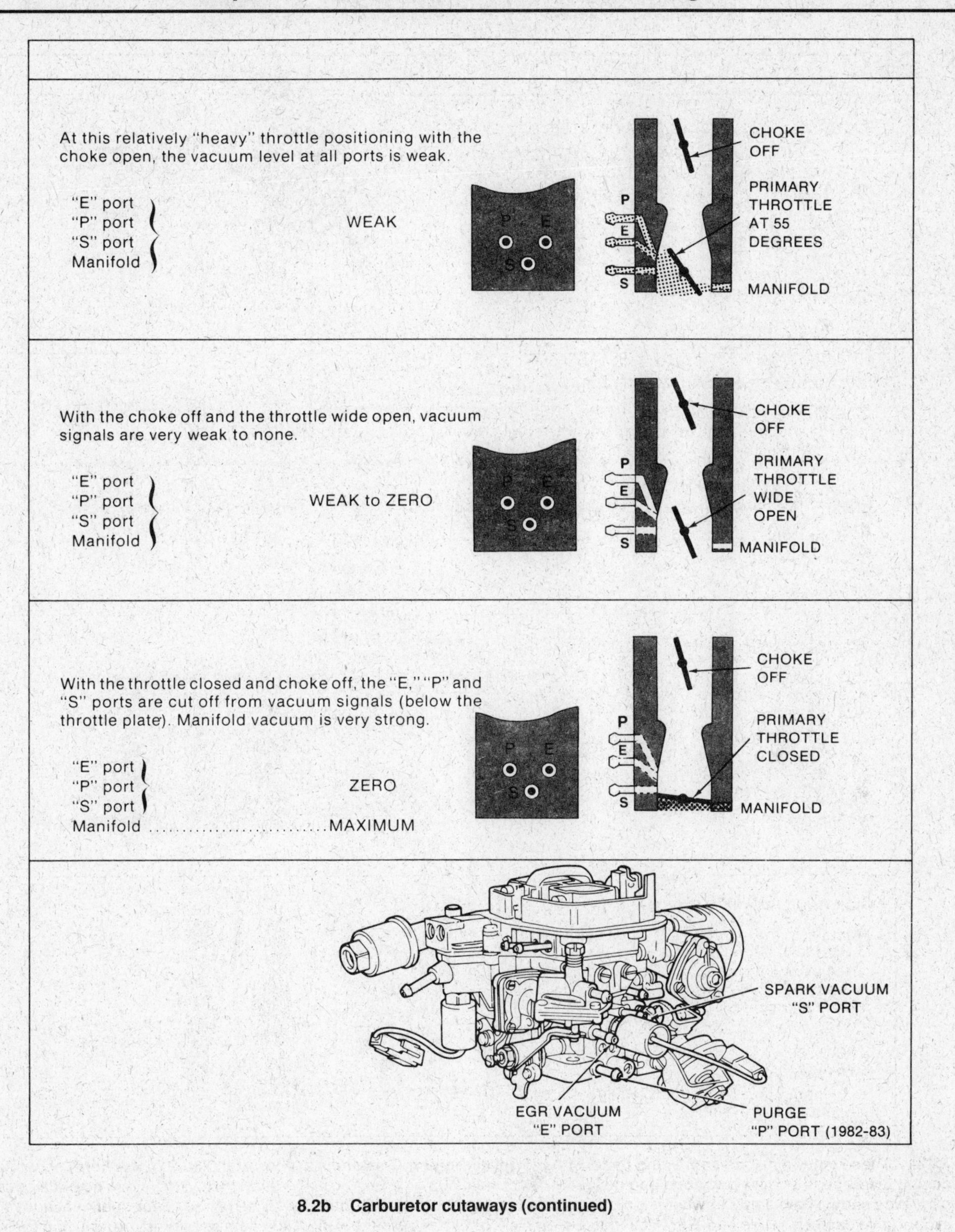

8.2b Carburetor cutaways (continued)

How is vacuum used in automotive emission control systems?

Today, vacuum applications go way beyond fuel metering. Intake vacuum is now used to operate all sorts of devices and systems on automobiles. Vacuum-operated devices are used in the air intake system to control air temperature to improve fuel vaporization and combustion. Vacuum is used to control some automatic chokes and most throttle kickers to ease engine startups, warmups and cold drive-aways. Spark advance on many Seventies and Eighties vehicles is controlled by a vacuum-operated dia-

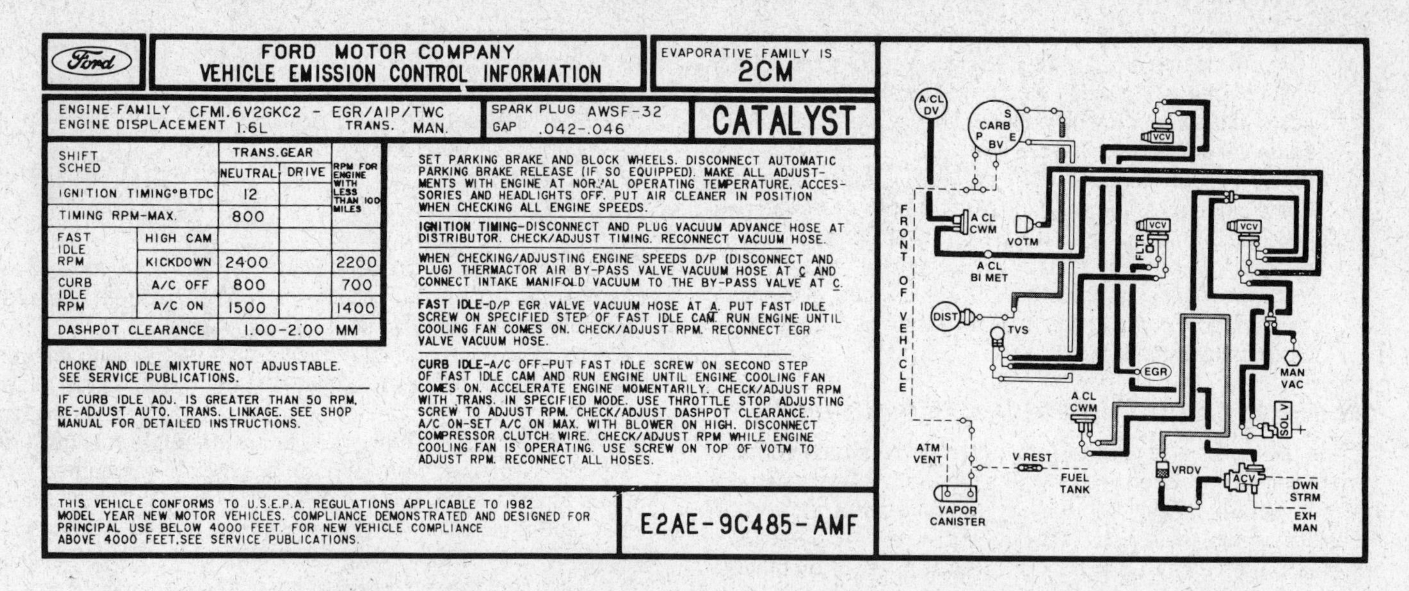

FORD MOTOR COMPANY
VEHICLE EMISSION CONTROL INFORMATION

EVAPORATIVE FAMILY IS 2CM

ENGINE FAMILY CFMI.6V2GKC2 - EGR/AIP/TWC
ENGINE DISPLACEMENT 1.6L TRANS. MAN.

SPARK PLUG AWSF-32
GAP .042-.046

CATALYST

SHIFT SCHED		TRANS.GEAR NEUTRAL	DRIVE	RPM FOR ENGINE WITH LESS THAN 100 MILES
IGNITION TIMING°BTDC		12		
TIMING RPM-MAX.		800		
FAST IDLE RPM	HIGH CAM			
	KICKDOWN	2400		2200
CURB IDLE RPM	A/C OFF	800		700
	A/C ON	1500		1400
DASHPOT CLEARANCE		1.00-2.00 MM		

CHOKE AND IDLE MIXTURE NOT ADJUSTABLE. SEE SERVICE PUBLICATIONS.

IF CURB IDLE ADJ. IS GREATER THAN 50 RPM, RE-ADJUST AUTO. TRANS. LINKAGE. SEE SHOP MANUAL FOR DETAILED INSTRUCTIONS.

SET PARKING BRAKE AND BLOCK WHEELS. DISCONNECT AUTOMATIC PARKING BRAKE RELEASE (IF SO EQUIPPED). MAKE ALL ADJUSTMENTS WITH ENGINE AT NORMAL OPERATING TEMPERATURE, ACCESSORIES AND HEADLIGHTS OFF. PUT AIR CLEANER IN POSITION WHEN CHECKING ALL ENGINE SPEEDS.

IGNITION TIMING-DISCONNECT AND PLUG VACUUM ADVANCE HOSE AT DISTRIBUTOR. CHECK/ADJUST TIMING. RECONNECT VACUUM HOSE.

WHEN CHECKING/ADJUSTING ENGINE SPEEDS D/P (DISCONNECT AND PLUG) THERMACTOR AIR BY-PASS VALVE VACUUM HOSE AT C AND CONNECT INTAKE MANIFOLD VACUUM TO THE BY-PASS VALVE AT C.

FAST IDLE-D/P EGR VALVE VACUUM HOSE AT A. PUT FAST IDLE SCREW ON SPECIFIED STEP OF FAST IDLE CAM. RUN ENGINE UNTIL COOLING FAN COMES ON. CHECK/ADJUST RPM. RECONNECT EGR VALVE VACUUM HOSE.

CURB IDLE-A/C OFF-PUT FAST IDLE SCREW ON SECOND STEP OF FAST IDLE CAM AND RUN ENGINE UNTIL ENGINE COOLING FAN COMES ON. ACCELERATE ENGINE MOMENTARILY. CHECK/ADJUST RPM WITH TRANS. IN SPECIFIED MODE. USE THROTTLE STOP ADJUSTING SCREW TO ADJUST RPM. CHECK/ADJUST DASHPOT CLEARANCE.
A/C ON-SET A/C ON MAX. WITH BLOWER ON HIGH. DISCONNECT COMPRESSOR CLUTCH WIRE. CHECK/ADJUST RPM WHILE ENGINE COOLING FAN IS OPERATING. USE SCREW ON TOP OF VOTM TO ADJUST RPM. RECONNECT ALL HOSES.

THIS VEHICLE CONFORMS TO U.S.E.P.A. REGULATIONS APPLICABLE TO 1982 MODEL YEAR NEW MOTOR VEHICLES. COMPLIANCE DEMONSTRATED AND DESIGNED FOR PRINCIPAL USE BELOW 4000 FEET. FOR NEW VEHICLE COMPLIANCE ABOVE 4000 FEET, SEE SERVICE PUBLICATIONS.

E2AE-9C485-AMF

8.3 **Raise the hood and find your VECI label – many manufacturers place the vacuum diagram right on the VECI label**

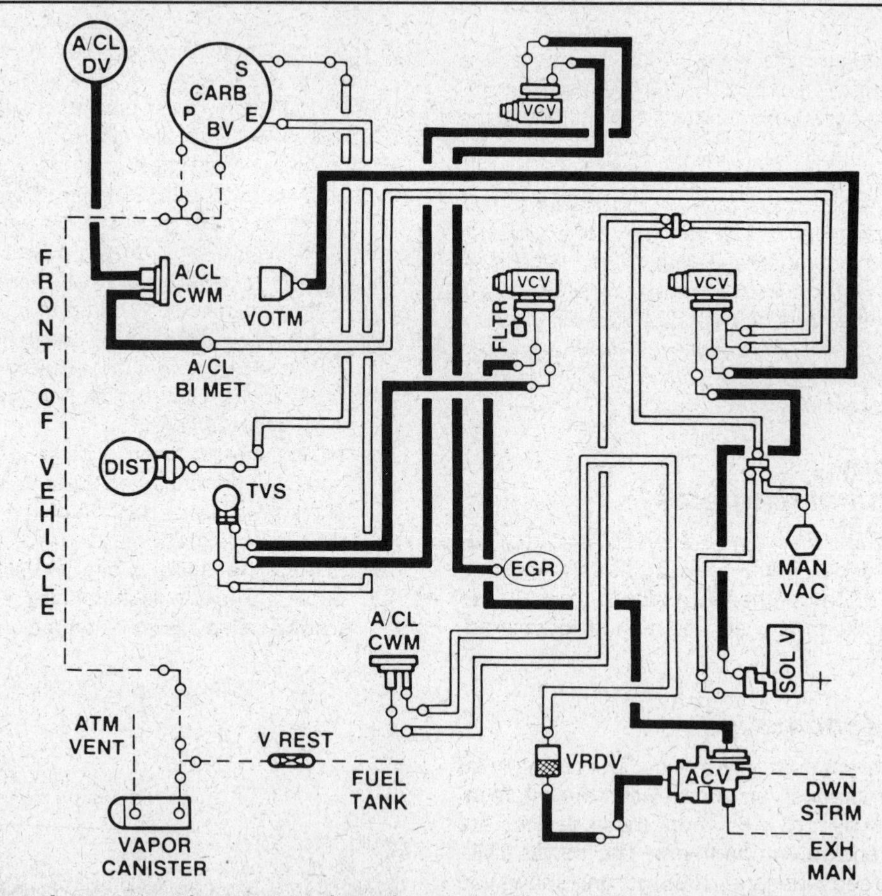

8.4 **Some manufacturers put the vacuum schematic on a decal by itself, such as this Ford diagram, and place it near the VECI label**

phragm at the distributor. Engine vacuum controls operation of accessories such as power brake boosters, automatic transmission vacuum modulators, cruise control systems, air distribution system doors and heater/air conditioner systems.

Most emission control systems also depend on vacuum for proper operation. These systems use numerous vacuum-operated devices that respond to vacuum, electronic, temperature or electronic-temperature inputs to activate and deactivate output actuators which control emissions by altering engine operation in accordance with changing loads and operating temperatures.

There are three basic types of vacuum controls. They are:

1) Vacuum delay valves – Also known as restrictors, these valves are used to delay vacuum flow. They're usually located in the vacuum line between the vacuum source and some vacuum-controlled device.
2) Vacuum diaphragms – These controls, usually referred to in this book as actuators, are used to operate various vehicle mechanical parts, such as throttle linkages on carburetors and throttle bodies, or EGR valves.
3) Vacuum switches and valves – These devices, also known as solenoids, control vacuum flow and are usually operated electrically or thermostatically.

Where can you find vacuum diagrams?

The factory service manual for your vehicle normally contains vacuum diagrams. Even your owner's manual may contain a diagram or two. But the quickest way to determine what vacuum devices are used in the emission control systems on your vehicle is to refer to the vacuum diagram located in the engine compartment. Most vehicles have a vacuum diagram (or "schematic") located somewhere in the engine compartment. It's usually affixed to the underside of the hood for convenient reference when working on your vehicle.

Raise the hood and find your VECI label. Most manufacturers place the vacuum diagram right on the VECI label **(see illustration)**. Some put it on a separate label, near the VECI label **(see illustration)**.

Note: *The diagrams in this manual are typical examples of the type you'll find on your vehicle's VECI label, in your owner's manual, in a Haynes Automotive Repair Manual (ARM) or in a factory service manual. But they're instructional – they DON'T necessarily apply to your vehicle. When you're working on emission control systems, if you notice differences between the vacuum diagram affixed to your vehicle and those in the owner's manual, a Haynes ARM or a factory manual, always go with the one on the vehicle. It's always the most accurate diagram.*

Where can you buy replacement vacuum diagrams?

Like VECI labels, vacuum diagrams are available at your authorized dealer service department. (Of course, as mentioned above, sometimes they're part of the VECI label). You can also find vacuum diagrams in the factory service manual for your vehicle.

Typical vacuum schematics

There are two types of vacuum schematics. The currently favored style shows how various vacuum devices are interconnected, but it doesn't always tell you where these devices are actually located in the engine compartment. The newer style, which is just beginning to see widespread use, not only shows you what the devices look like, it also tells you where they are.

Finding vacuum leaks

Vacuum system problems can produce, or contribute to, numerous driveability problems, including:

1 *Deceleration backfiring*
2 *Detonation*
3 *Hard to start*
4 *Knocking or pinging*
5 *Overheating*
6 *Poor acceleration*
7 *Poor fuel economy*
8 *Rich or lean stumbling*
9 *Rough idling*
10 *Stalling*
11 *Won't start when cold*

The major cause of vacuum-related problems is damaged or disconnected vacuum hoses, lines or tubing. Vacuum leaks can cause problems such as erratic running and rough idling.

For instance, a rough idle often indicates a leaking vacuum hose. A broken vacuum line allows a vacuum leak, which allows more air into the intake manifold than the engine is calibrated for. Then the engine runs roughly due to the leaner air/fuel mixture.

Another example: Spark knock or pinging sometimes indicates a kinked vacuum hose to the EGR valve. If this hose is kinked, the EGR valve won't open when it should. The engine, which requires a certain amount of exhaust gas in the combustion chamber to cool it down, pings or knocks.

Here's another: A misfire at idle may indicate a torn or ruptured diaphragm in some vacuum-activated unit (a dashpot or EGR valve, for instance). The torn diaphragm permits air movement into the intake manifold below the carburetor or throttle body. This air thins out the already lean air/fuel mixture at idle and causes a misfire. A misfire may also indicate a leaking intake manifold gasket or a leaking carburetor or throttle body base gasket. If a leak develops between the mating surfaces of the intake manifold and the cylinder head, or between the carburetor or throttle body base gasket and the intake manifold, the extra air getting into the engine below the gasket causes a misfire.

If you suspect a vacuum problem because one or more of the above symptoms occurs, the following visual inspection may get you to the source of the problem with no further testing.

1) Make sure everything is routed correctly – kinked lines block vacuum flow at first, then cause a vacuum leak when they crack and break.
2) Make sure all connections are tight. Look for loose connections and disconnected lines. Vacuum hoses and lines are sometimes accidentally knocked loose by an errant elbow during an oil change or some other maintenance.
3) Inspect the entire length of every hose, line and tube for breaks, cracks, cuts, hardening, kinks and tears **(see illustration)**. Replace all damaged lines and hoses.

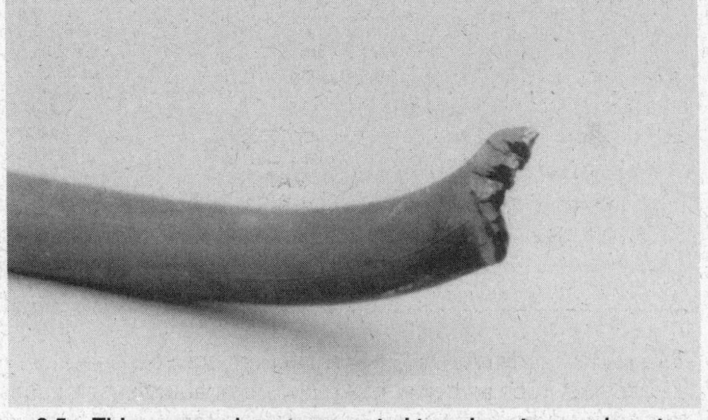

8.5 This vacuum hose was routed too close to an exhaust manifold – after being overheated repeatedly, it finally cracked and broke

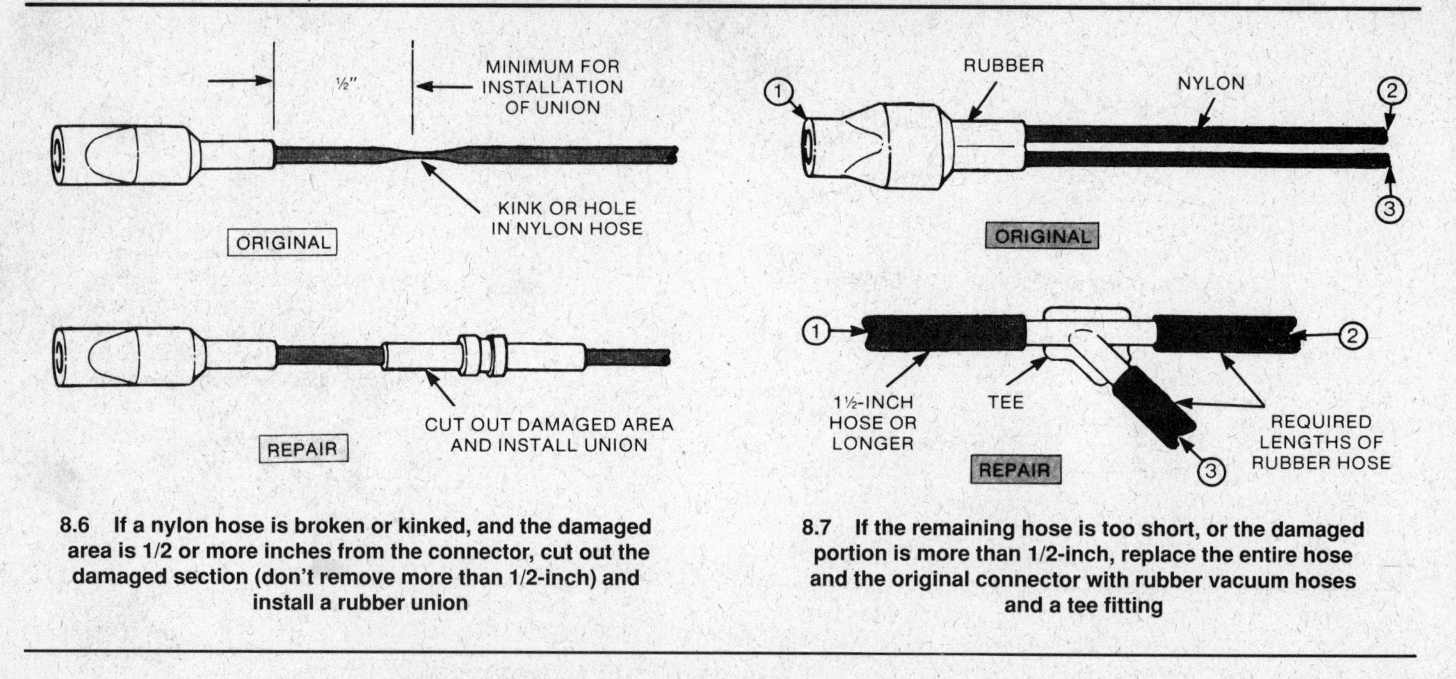

8.6 If a nylon hose is broken or kinked, and the damaged area is 1/2 or more inches from the connector, cut out the damaged section (don't remove more than 1/2-inch) and install a rubber union

8.7 If the remaining hose is too short, or the damaged portion is more than 1/2-inch, replace the entire hose and the original connector with rubber vacuum hoses and a tee fitting

4) When subjected to the high underhood temperatures of a running engine, hoses become brittle (hardened). Once they're brittle, they crack more easily when subjected to engine vibrations. When you inspect the vacuum hoses and lines, pay particularly close attention to those that are routed near hot areas such as exhaust manifolds, EGR systems, reduction catalysts (often right below the exhaust manifold on modern FWD vehicles with transverse engines), etc.
5) Inspect all vacuum devices for visible damage (dents, broken pipes or ports, broken tees in vacuum lines, etc.
6) Make sure none of the lines are coated with coolant, fuel, oil or transmission fluid. Many vacuum devices will malfunction if any of these fluids get inside them.
7) What if none of the above steps eliminates the leak? Grab your vacuum pump and apply vacuum to each suspect area, then watch the gauge for any loss of vacuum.
8) And if you still can't find the leak? Well, maybe it's not in the emissions system; maybe it's right at the source, at the intake manifold or the base gasket between the carburetor or throttle body. To test for leaks in this area, squirt a noncombustible cleaning solvent such as carburetor cleaner or point cleaner (or WD-40, but it's messier) along the gasket joints with the engine running at idle. If the idle speed smooths out momentarily, you've located your leak. Tighten the intake manifold or the carb/throttle body fasteners to the specified torque and recheck. If the leak persists, you may have to replace the gasket.

An alternative to spraying solvent is to use a short length of vacuum hose as a sort of "stethoscope," listening for the high-pitched hissing noise that characterizes vacuum leaks. Hold one end of the hose to your ear and probe close to possible sources of vacuum leakage with the other end. **Warning:** *Stay clear of rotating engine components when probing with the hose.*

Repairing and replacing vacuum hose

Replace defective sections one at a time to avoid confusion or misrouting. If you discover more than one disconnected line during an inspection of the lines, refer to the vehicle vacuum schematic to make sure you reattach the lines correctly. Route rubber hoses and nylon lines away from hot components, such as EGR tubes and exhaust manifolds, and away from rough surfaces which may wear holes in them.

Most factory-installed vacuum lines are rubber, but some are nylon. Connectors can be bonded nylon or rubber. Nylon connectors usually have rubber inserts to provide a seal between the connector and the component connection.

Replacing nylon vacuum lines can be expensive and tricky. Using rubber hose may not be as aesthetically pleasing as the OEM nylon tubing, but it's perfectly acceptable, as long as the hoses and fittings are tightly connected and correctly routed (away from rough surfaces and hot EGR tubes, exhaust manifolds, etc.).

Here are some tips for repairing nylon vacuum hoses and lines:

1) If a nylon hose is broken or kinked, and the damaged area is 1/2-inch or more from a connector, cut out the damaged section (don't remove more than 1/2-inch) and install a rubber union **(see illustration)**.
2) If the remaining hose is too short, or the damage exceeds 1/2-inch in length, replace the entire hose and the original connector with rubber vacuum hoses and a tee fitting **(see illustration)**.
3) If only part of a nylon connector is damaged or broken, cut it apart **(see illustration)** and discard the damaged half of the harness. Then replace it with rubber vacuum hoses and a tee.
4) Suppose you want to repair something a little more complicated **(see illustration)**. Cut the hoses to the required lengths (at least 1-1/2 inches in length) and assemble as shown, using a 4-way connector and an elbow. Just don't forget which line goes to which connector!

Note: *The identification numbers on the accompanying illustrations refer to the old fittings and hoses – and their corresponding replacements – so you don't get mixed up*

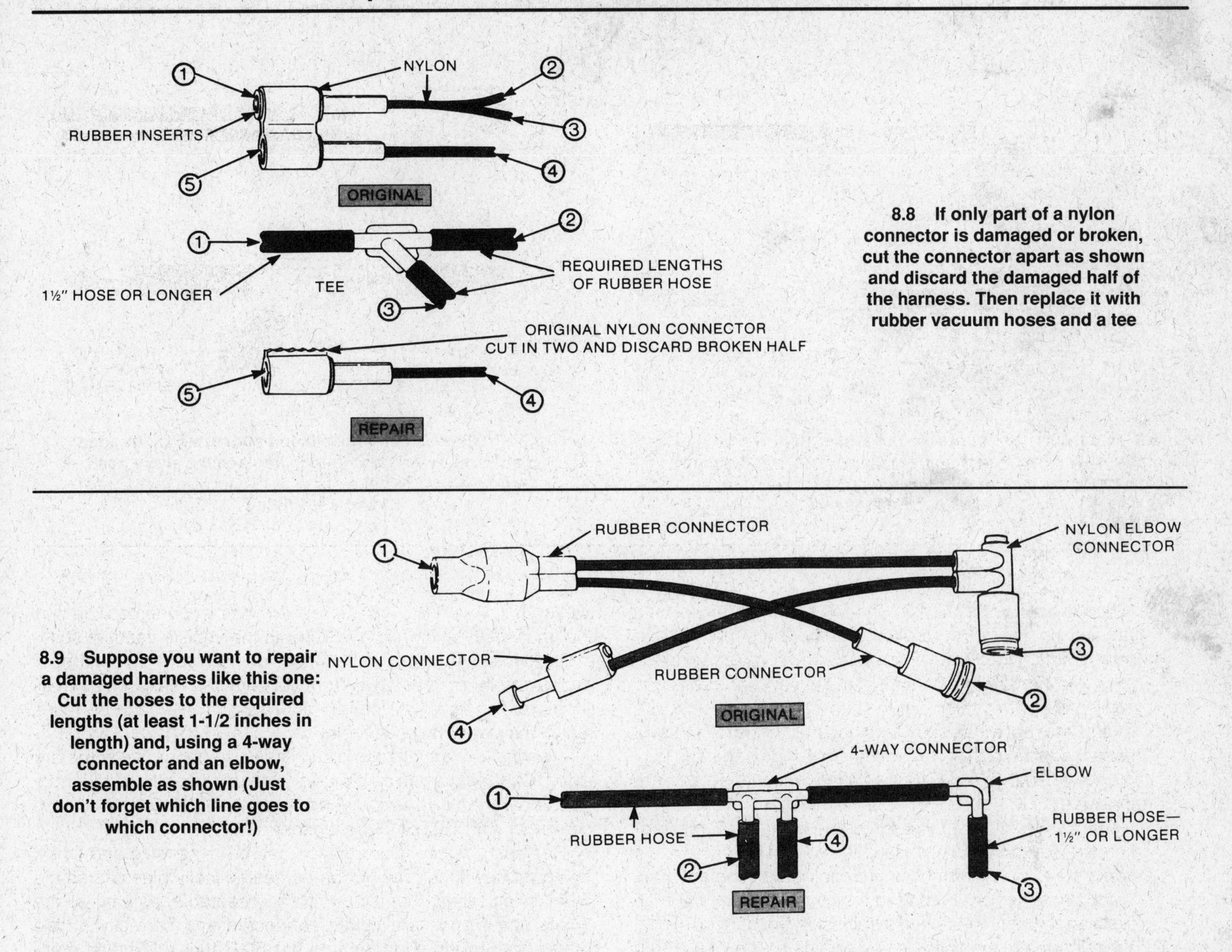

8.8 If only part of a nylon connector is damaged or broken, cut the connector apart as shown and discard the damaged half of the harness. Then replace it with rubber vacuum hoses and a tee

8.9 Suppose you want to repair a damaged harness like this one: Cut the hoses to the required lengths (at least 1-1/2 inches in length) and, using a 4-way connector and an elbow, assemble as shown (Just don't forget which line goes to which connector!)

9 Checking electrical connections

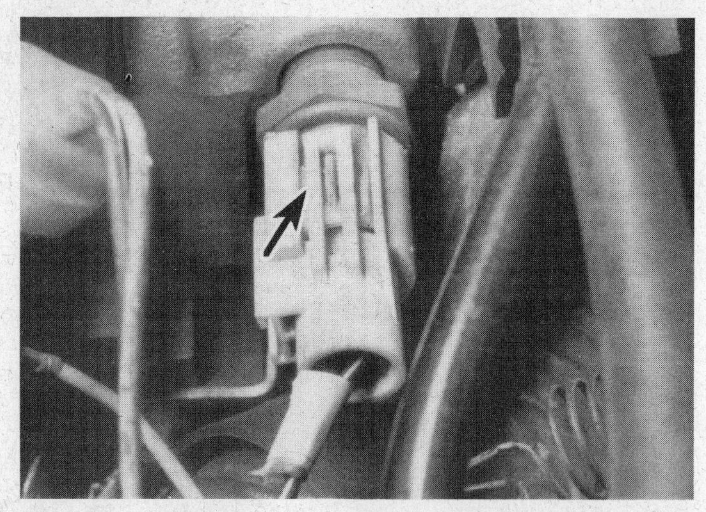

9.1 Most connectors have one or more tabs like this (arrow) that must be lifted before the halves can be separated

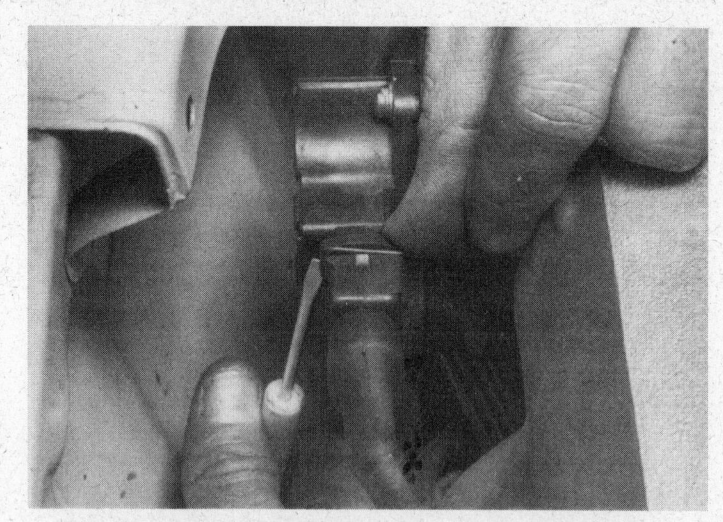

9.2 Some connectors, such as this one on a Toyota throttle position sensor, have a spring clip that must be pried up before the connector can be unplugged

9.3 This Ford SPOUT connector is a bit unusual – a plastic plug must be pulled out of the connector housing before it can be disconnected – don't lose the plug!

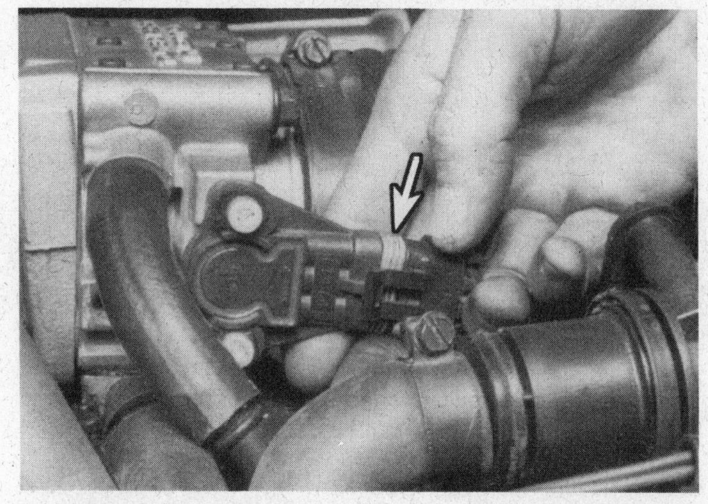

9.4 Many modern engine-management system connectors have flexible seals (arrow) to keep moisture off the terminals and prevent corrosion – make sure the seal isn't damaged in any way

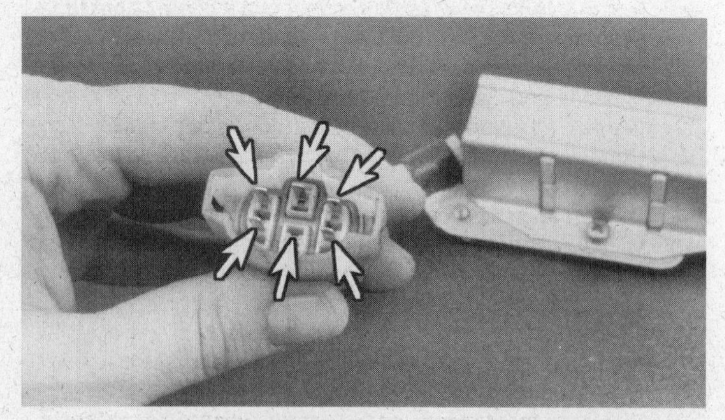

9.5 Check the terminals (arrows) in each connector for corrosion that will cause excessive resistance in the circuit, or even an open circuit

Check the electrical connections to the computer, all sensors and actuators and all other emissions devices. Make sure they're properly mated and tightly coupled. Wiggle and shake the connectors to ensure they're tight. Loose connectors should be unplugged and inspected for corrosion **(see illustrations)**. Look closely at the connector pins and tabs. If corrosion is present, clean it off with a small wire brush and electrical contact cleaner (available in aerosol form). Some connectors might require the use of a special conductive grease to prevent corrosion.

10 Preparing for emissions certification testing

Over half of all states now require regular – annual or biannual – emission certification tests. Many also require smog inspections when vehicle ownership changes hands. Obviously, if you've "modified" (tampered with) your vehicle's smog equipment, this periodic ritual of having your vehicle "smogged" can give you an anxiety attack. And even if you're the type of owner who would never dream of disconnecting a single vacuum line – but you're also the type who neglects scheduled vehicle maintenance, including emission-related components – the prospect of a smog test can be disconcerting. If you fall into either of these two general groups, you're not alone. Recent statistics from some states indicate that nearly one-third of all vehicles fail to meet emissions standards their first time through.

It doesn't have to be that way. With regular engine maintenance and a straight-forward check of emissions-related components, you can catch nearly all the potential failures which might turn up in a state inspection. And you might as well get used to it, because state-certified smog testing is going to become stricter and more frequent as time goes on. Specific testing procedures and standards for various emissions levels vary from state to state, but the idea – lowering the HC, CO and NOx levels – is the same everywhere. So the following information should save you time and money. Time because you won't have to go back several times to pass the test; money because you won't have to shell out extra dollars for the repairs needed to enable your vehicle to pass the test.

Generally, a stock (untampered with) engine will pass the state smog test as long as it's been recently tuned and all the emissions-related hardware is intact, hooked up and working properly. If your vehicle hasn't had a tune-up recently, now is the time to do it.

The following items should be checked carefully before a smog test:

1) Make a quick visual check of all emissions control systems to be sure all components are in place and hooked up correctly. If you have reason to suspect a system is not functioning correctly, check it, as described in Chapter 3.
2) Inspect all underhood vacuum hoses for cracks, loose connections and disconnected hoses.
3) Inspect all underhood electrical wiring for cracks, torn wires, loose or corroded connections and unplugged connectors.
4) Check the air filter carefully, since a dirty, restricted air filter will cause a rich fuel/air mixture, increasing emissions. Also check the PCV filter, if equipped (see Chapter 3).
5) On carbureted models, check the choke to be sure it's opening all the way when the engine is warmed up (here again, the rich fuel/air mixture caused by a closed choke will increase emissions).
6) Finally, before having vehicle car tested, make sure the engine and exhaust (catalytic converter) system are up to normal temperature (10 to 15 minutes of driving time).

How do you pick a shop for an emissions test?

Where do you take your vehicle for a smog test? Well, if you already have a good working relationship with a local shop – and it's certified to do smog testing – by all means stick with that shop. But if you don't have a regular shop, there are several factors worth considering: First, look at what the shop charges. This isn't as important as you might think; the inspection fees charged by private garages are generally regulated by the state, so there may not be that much difference in fees from one shop to another. The important thing to keep in mind is that the fee charged by most shops is usually a lot lower than its normal hourly labor rate. In other words, doing emissions testing can be a marginally profitable – or even a losing proposition – for shops. So how does a shop turn a smog test into a money maker? Basically, by charging you their normal labor rate for fixing problems if your vehicle fails its smog test.

Of course, you're certainly not obligated to have your vehicle repaired by the same shop that inspects it for compliance with state smog laws. But the shop owner hopes you will do just that, to avoid the inconvenience of moving the vehicle to another shop. Whether you decide to have the vehicle repaired by the same shop that inspects it, or take it somewhere else, depends on how much you trust the shop's ethics.

Also, try to find a shop that does not charge for re-testing your vehicle if it should fail the test. Re-testing fees – particularly if your vehicle must be re-tested more than once – can really add up.

The emissions test

Most state smog certification inspections consist of two parts:

1) An under-hood visual inspection – to make sure everything is installed and connected.
2) An analysis of the composition of the exhaust gases coming out the tailpipe, both at idle and at median on-the-road engine speeds.

You might survive the first part of the test even if something is missing or everything isn't hooked up, because the mechanic may or may not be familiar with the emissions devices and systems that are supposed to be fitted to your engine. But don't bet on it. The software programs employed by most state-approved emissions-testing analyzers display this information – component identification and location of all emission components for your specific model – on the video screen of the analyzer.

But even if your vehicle passes the visual inspection phase in spite of a missing or disconnected device or system, the engine will probably fail the second part of the test, which is performed with an infrared gas analyzer.

A probe, which is shoved up the tailpipe, detects the amount of hydrocarbons (HC), carbon monoxide (CO) and oxides of nitrogen (NOx) in the exhaust stream, and transmits this information to a computer which measures these levels to a high degree of accuracy (expressed in "parts per million," or ppm). This analysis of the exhaust gases is usually conducted both at idle speed and at about 2500 to 2800 rpm. If the engine fails either part, the vehicle fails the test.

The analyzer computer prints out two hard copies of the HC and CO readings, one for you and one for the shop's inspection records. How do you know whether the analyzer's conclusions are accurate? Most analyzers are self-calibrating: Every time they're turned on to perform a test, they verify the validity of their calibration against reference gases contained inside their own apparatus.

If your vehicle passes its emissions test, you're allowed to drive it for another year or two. If it fails, you have a "grace period," usually a month, to bring the vehicle up to specification so that it can pass (unless, of course, you have waited until the last minute to submit your vehicle to the test and your registration is about to expire, in which case you have considerably less time to fix it!).

What if your vehicle fails the emissions test?

If your vehicle fails its emissions test, see the information on the Performance Warranty in Section 3 of this Chapter.

11 Resetting emissions maintenance reminder timers

Until recently, the oxygen sensors and EGR systems on most vehicles required periodic inspection or replacement, so they were equipped with emissions maintenance reminder lights which came on periodically to remind you to perform the specified maintenance. After you've completed the service, the timer for the emissions maintenance reminder light must be reset. The following information will help you locate the timer on your vehicle.

You'll note that later models aren't included here. That's because the oxygen sensor and EGR valve on these models don't need to be regularly inspected or replaced; they've been designed to last a lot longer to comply with the Federally mandated extended warranty. If the oxygen sensor or EGR valve fail on a newer vehicle, you'll usually get a trouble code (see Chapter 2).

Also be aware that the following information does not apply to service interval reminder lights that come on periodically to remind you that it's time to perform routine non-emissions service, such as oil and filter changes.

11.1 On American Motors vehicles, you'll find the emissions maintenance E-cell timer under the dash, in the wire harness for the computer; don't try to reset the timer (you can't) – replace it

Alfa Romeo

1 Every 30,000 miles on the Spider 2.OL, Graduate and Quadrifoglio, and every 60,000 miles on the GTV-6 2.5L, a light on the dash will come on as a reminder to check the system and replace the oxygen sensor.

2 To reset the mileage counter after replacing the sensor on the Spider 2.OL, Graduate and Quadrifoglio, locate the counter on the left side of the engine compartment. Remove the plastic cover by drilling through the shank of the attaching screws. Remove the cover, then rotate and press the button. Using new screws, reinstall the cover.

3 On the GTV-6 2.5L, remove the ECU cover panel and the lower right side parcel shelf. Locate the mileage counter under the right side of the dash. Push the white reset button. Make sure the emission reminder light is out.

American Motors

1980 and 1981 models

An emissions maintenance reminder light on the instrument panel, which indicates that the oxygen sensor needs to be serviced, comes on every 30,000 miles. If the sensor is faulty, it must be replaced. After servicing the sensor, reset the light activating switch.

You'll find the switch in the engine compartment, between the upper and lower speedometer cables, next to the firewall. Slide up the rubber boot. With a small screwdriver, turn the reset screw clockwise 1/4-turn until the detent resets in the switch.

1982 through 1984 models (except Alliance and Encore)

The emission maintenance light comes on every 1000 hours of engine operation to let you know that the oxygen sensor must be serviced. After you've serviced the sensor, replace the emission maintenance E-cell timer. You'll find the E-cell timer under the dash, in the wire harness leading to the computer **(see illustration)**. Remove the timer from its enclosure and insert a replacement unit.

1987 Eagle wagon

An emission light timer will start flashing the Oxygen Sensor Service light at 82,500 miles. At this time, both the sensor and timer should be replaced. You'll find the timer under the dash panel (to the right of steering column). Remove the mounting screws and disconnect the wiring. Installation is the reverse of removal.

Audi

The mileage counter is located on the firewall, under the seat, under the dash or behind the instrument cluster, depending on the model. On models which require regular service for both the EGR system and the oxygen sensor, the counter has separate reset buttons for each system. The EGR light comes on at 15,000 mile intervals. Check the system for defects, then reset the counter (see below). The oxygen sensor light comes on at 30,000 mile intervals (60,000 mile intervals on 100/200 models). Replace the sensor, then reset the counter (see below).

Pre-1984 Audi 4000 and 5000S

The inline mileage counter is mounted under the dash, to the left of the steering column near the pedal assembly. Remove the lower dash cover on the driver's side. Once you know where the counter is and you've developed the right "touch," you can dispense with this step and reach the reset button with a bent rod. The reset button for the reminder light is white.

1984 and later Audi 4000 and non-turbo 5000 and 5000S

When the light comes on, replace the oxygen sensor. To reset the mileage counter, you must remove the instrument cluster, disconnect the speedometer cable at the transaxle end and remove the steering wheel to give yourself room to work. Remove the instrument cluster cover screws. Pull the instrument cluster out of the dash as far as the attached wiring and speedometer cable allow.

Find the small plastic cover at the top of the cluster, near the imprinted OXS and break it off. Press the switch to reset the counter. Once you know where the switch is located, you can reach it next time without totally removing the cluster. Installation is the reverse of removal.

1984 through 1988 Audi 5000 Turbo and 100/200 Turbo

The mileage counter is located under the rear seat. Push the seat cushion toward the rear of the vehicle, then lift the front of the cushion to release the cushion retainers. Move the cushion out of the way. The counter is located on the left side of the vehicle. Push the button marked OXS on the reset box. Cycle the ignition to the On position and make sure the reminder light goes off. Installation is the reverse of removal. Replace the oxygen sensor.

All other Audi models

To turn off the reminder light on all other models, trace the speedometer cable to the reset box, which is installed inline with the cable **(see illustration 13.19)**. To reset the counter, press the white button on the box and make sure the reminder light goes out.

BMW

Pre-1983 models

The OXYGEN light in the dash will light up every 30,000 miles (every 25,000 miles on 528i models) as a reminder to replace the sensor. The inline mileage counter is located above the left frame rail, near the transmission. Make sure the white reset button makes an audible "click" when you press it (If it doesn't, the reminder light will remain on). Replace the mileage counter if the button won't click. 528e and 1983 633CSi models have no reset switch. On these models, remove the instrument cluster, then remove and discard the bulb for the OXYGEN light.

1983 and later models

On models through January 1985, when the oxygen sensor light comes on, service the sensor and remove the bulb from the indicator. No reset switch is provided. On February 1985 and later models, there's a reset button on the rear of the light control assembly that's located near the pedal assembly. Press this button to reset the light after you have serviced the oxygen sensor.

Chrysler Motors

1980 passenger cars and 1980 through 1987 light-duty trucks and vans

A mileage counter activates the emissions reminder light. Two types are used. If your vehicle is equipped with the mechanical type, see the reset procedure above for 1980 and 1981 American Motors models.

The electronic type uses a 9-volt battery which supplies power to the electronic counter, preventing memory loss when the vehicle battery is disconnected. On 1987 Dakota models, the mileage counter in the odometer will illuminate the reminder light at 52,500, 82,500 and 105,000 miles.

On all other models, the reminder light will illuminate between 12,000 and 30,000-mile intervals. **Caution:** *During the resetting procedure, the vehicle battery MUST be connected to prevent power loss to the computer memory.*

To reset the electronic type, locate the Green, Red, White or

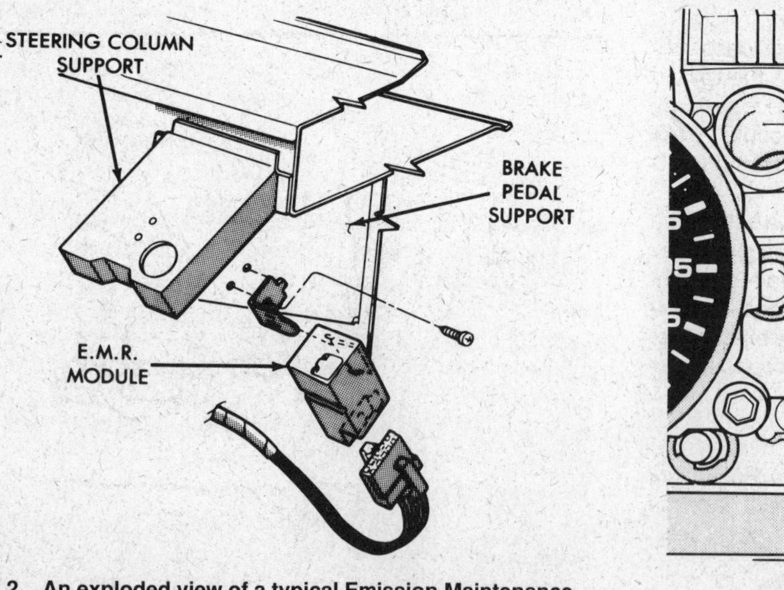

11.2 An exploded view of a typical Emission Maintenance Reminder (EMR) module on the steering column behind the instrument panel (1988 Dodge Ram van shown)

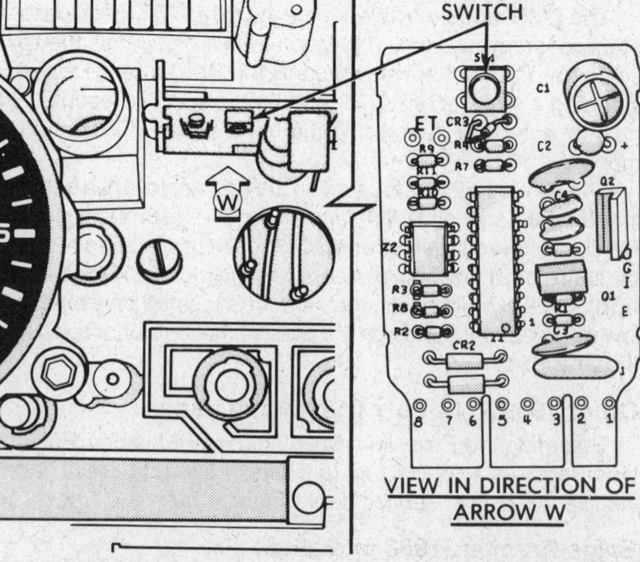

11.3 A typical module reset switch on the back of the instrument cluster (1988 Dodge/Plymouth mini-van)

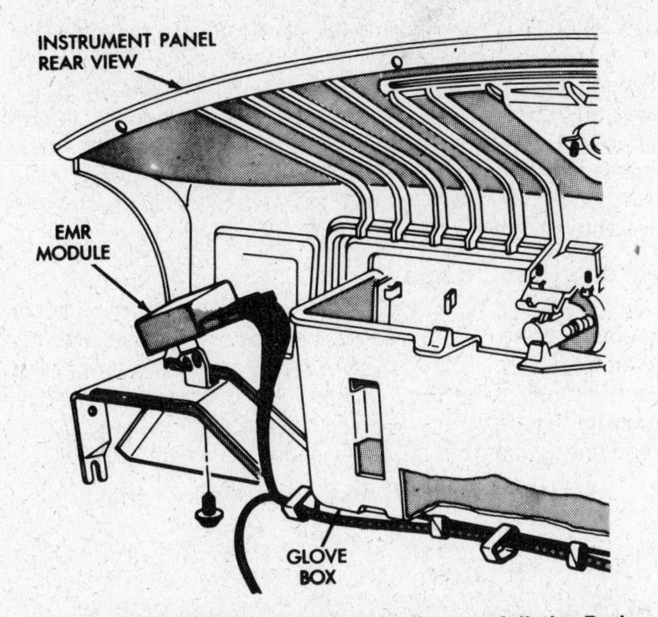

11.4 A Typical EMR module installation on a full-size Dodge Pick-up truck, located behind the far right end of the dash (1988 model shown)

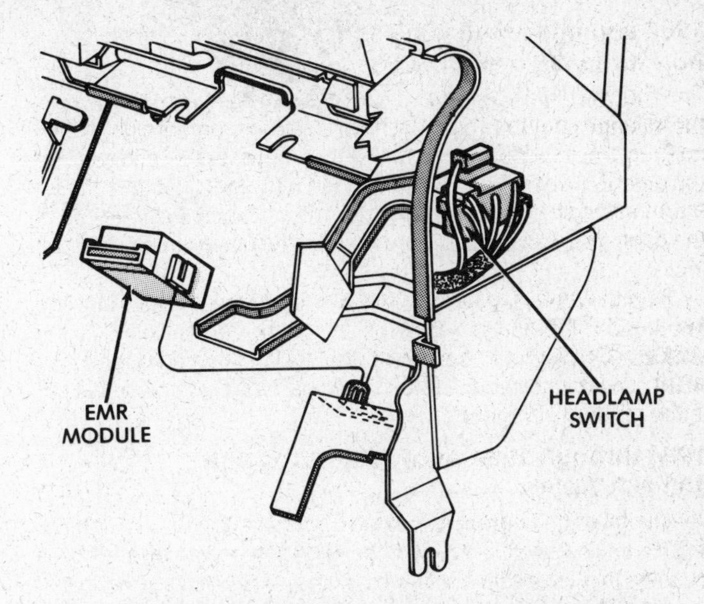

11.5 A typical EMR module installation on a Dodge Dakota Pick-up truck, located on a bracket below the headlight switch on the back of the instrument panel (1988 model shown)

Tan plastic case behind the instrument panel in the lower left instrument cluster area. Slide the case from the bracket and open the cover. Remove the 9-volt battery and insert a small rod or screwdriver into the hole in the switch to close the contacts. Replace the battery with a new 9-volt alkaline type. Close the case switch back into bracket.

1988 and some 1989 light-duty trucks and vans

The Emission Maintenance Reminder (EMR) module's purpose it to remind you to service the vehicle emissions control system. It's not an emissions warning system. The EGR system, PCV valve, oxygen sensor, delay valves and purge valve should all be checked and, if necessary, replaced.

The EMR module will illuminate the "MAINT REQD" dash light at a predetermined time. Mileage alone won't cause the light to come on. The light will stay on until the EMR module is reset by inserting a small screwdriver into the hole in the module (RWD models only) and depressing the reset switch (FWD and RWD models).

The EMR module is located on the steering column behind the instrument panel on RWD vans and in the instrument cluster on FWD vans **(see illustrations)**. On trucks, it's located behind the far right side of the dash panel next to the glove box **(see illustration)**. On Dakota models, the module is located on a bracket below the headlight switch on the back of the instrument panel **(see illustration)**.

Other 1989 light-duty trucks and vans

Resetting the Emissions Maintenance Reminder (EMR) light timer requires a special tool (a Chrysler DRB-II tester). Take the vehicle to a dealer service department to have the timer reset.

Eagle Premier (1988 and 1989)

Every 7500 miles, a Vehicle Maintenance Monitor (VMM) will illuminate a SERVICE interval reminder light to indicate regular maintenance is due. After the required service is performed, press the RESET button on the dash below the VMM display. Hold the button until you hear a beep. The VMM display is cleared.

1988 and 1989 Jeep

Canadian and 49-State models are equipped with an emission maintenance indicator light on the instrument cluster. This light will come on one time at 82,500 miles to alert you that emission service is required. At this time, the oxygen sensor and PCV valve must be replaced and all other emission components should be inspected and serviced or replaced as necessary.

The indicator timer is located under the dash, near the accelerator pedal or to the right of the steering column **(see illustra-**

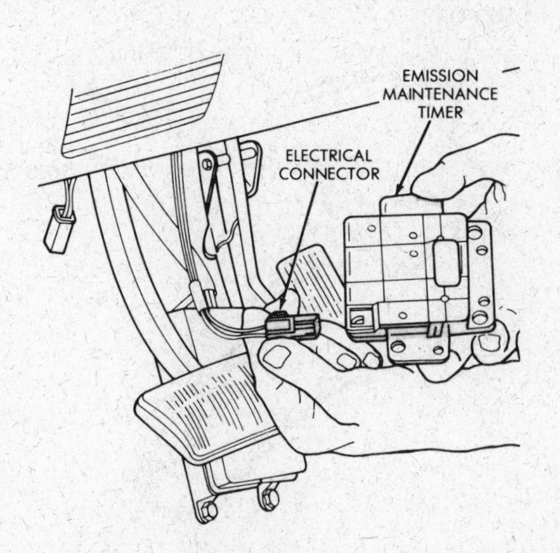

11.6 A typical emissions indicator light timer on a Jeep Wrangler, located near the accelerator pedal

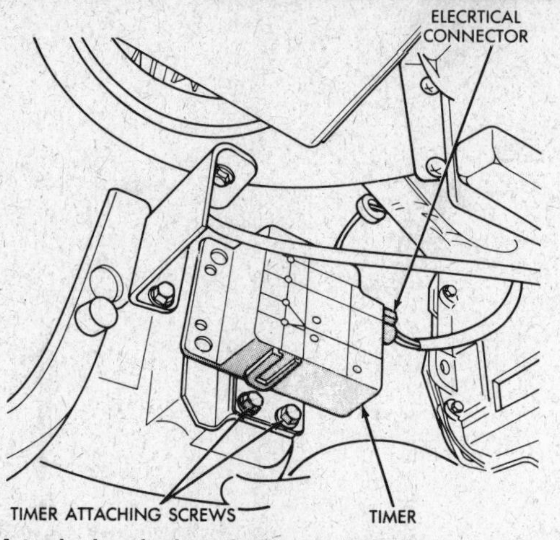

11.7 A typical emissions indicator light timer on a Jeep Cherokee, located to the right of the steering column

11.8 On some carbureted Mitsubishi models (including those sold by Chrysler under the Dodge or Plymouth name), the mileage counter is at the lower right corner of the instrument cluster

tions). The timer can't be reset. To turn off the light, the timer must be replaced or disconnected. Since the timer and the sensor are interdependent, if the timer prematurely fails, the oxygen sensor should be replaced at the same time to preserve the correct replacement interval. Some models are equipped with a computer malfunction indicator light. If it comes on and remains on while you're driving, the vehicle requires service.

After you've repaired the fault(s) and cleared the fault code(s), the malfunction indicator light should go out. Some models may use a dual-function indicator light, which is also used to indicate that emission component service is due. After performing the required service, reset the indicator light.

Chrysler imports

Arrow, Colt, Colt Vista, Champ, Challenger, Conquest and Sapporo cars and D 50/Ram 50 and Arrow Pick-ups and Raider

1 On some carbureted models, an EGR or MAINTENANCE REQUIRED warning light in the dash will come on as a reminder to have the EGR system serviced every 50,000 miles and/or the oxygen sensor replaced every 80,000 miles.

2 After servicing or replacing the components, reset the mileage counter. On some models, the reset switch is located on the back of the instrument cluster, near the speedometer cable junction. Slide the switch to the other side to reset the indicator light. On other models, the reset switch is on the lower right-hand corner of the instrument cluster, behind the instrument cluster face trim **(see illustration)**.

Datsun

See Nissan

Fiat

1 Fuel-injected models are equipped with an oxygen sensor. After 30,000 miles of operation, a warning light in the dash will come on to indicate that the oxygen sensor must be replaced.

2 To reset the mileage counter after replacing the sensor, locate the reset switch. On Brava models, the switch is located behind the left side of the dash.

3 On Strada models, the switch is located under the center of the dash (between the glove box and the radio). On Spider 2000 models, the switch is located under the left side of the dash, above the accelerator pedal. On X1/9 models, the switch is located behind the center console.

4 To reset the switch, cut the retaining wire and remove the screw. Insert a small screwdriver through the housing and press the switch contact. This resets the switch and turns out the EX GAS SENSOR light. Install the cap screw and secure with wire.

Ford Motor Company

1985 through 1987 light-duty trucks, 1988 non-EEC light-duty trucks and 1989 heavy-duty trucks

These vehicles use a maintenance reminder light to indicate emission system maintenance is required. The control unit (timer) for the maintenance light is located under the dash near the steering column or behind the glove box. The control unit may be hidden behind a bracket on some models. The maintenance light is triggered after 2000 key starts (about 60,000 miles). After servicing the emission system, reset the light on models with a resettable timer.

1 Turn off the ignition. Remove the tape from the reset hole in the timer. Lightly push a small Phillips screwdriver into the hole in the timer marked RESET and turn the ignition switch to the RUN position.

2 The light should stay on while the screwdriver is pressed down. Hold down the screwdriver for about five seconds. Remove the screwdriver. The light should go out within two to five seconds. If it doesn't, repeat Steps 1 and 2.

3 Cycle the ignition from the OFF to the RUN position. The light should glow for two to five seconds. This verifies that the maintenance reminder light is reset. **Note:** *Some non-EEC models, such as the 1988 2.0L Ranger and 6.1L and 7.OL gasoline trucks use a non-resettable control unit. Replace it with a resettable type.*

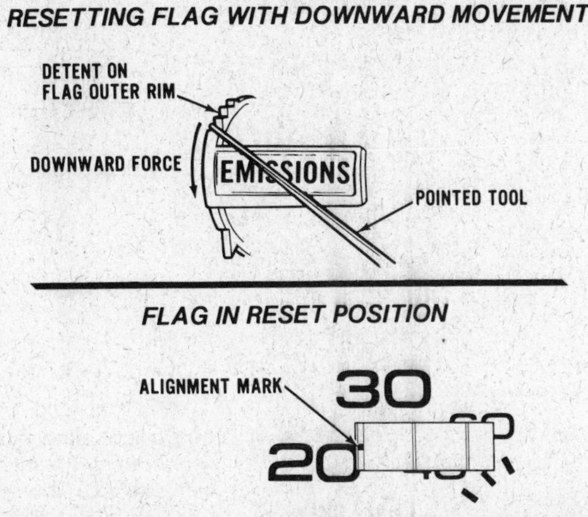

11.9 Using a pointed tool, apply a light downward pressure on the detents of the flag until it's reset – an alignment mark will appear in the left center of the odometer window when the flag is fully reset

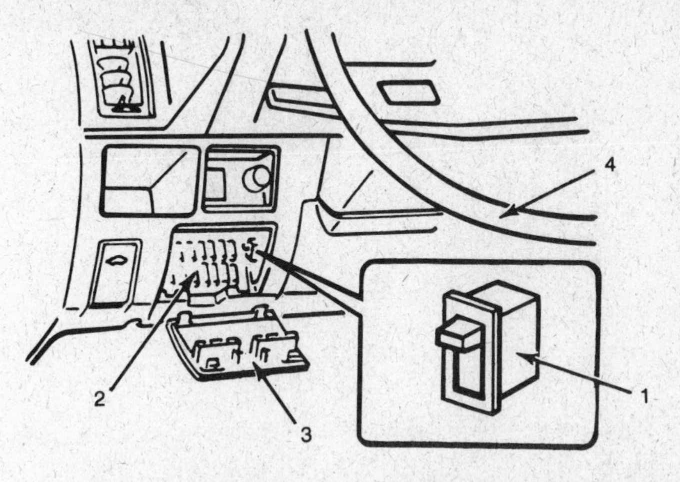

11.10 The SENSOR light cancel switch is on the right side of the fuse box on Chevy Sprints

1	*Cancel switch*	3	*Fuse box cover*
2	*Fuses*	4	*Steering wheel*

General motors

Every 30,000 miles (15,000 for Cadillac), a reminder flag appears in the speedometer face to remind you to service the oxygen sensor.

1980 models (except Cadillac)

Remove the instrument panel trim plate. Remove the instrument cluster lens. Using a pointed tool, apply a light downward pressure on the notches of the flag until it's reset. An alignment mark will appear in the left center of the odometer window when the flag is fully reset **(see illustration)**.

1980 Cadillac

Remove the lower steering column cover. The sensor reset cable is located to the left of the speedometer cluster. Pull the cable lightly (no more than two lbs. force). Reinstall the lower steering column cover.

General Motors imports

GEO

On 1989 Federal Tracker models, the CHECK ENGINE light (the computer system malfunction indicator light) comes on every 50,000 miles for the PCV and EGR systems, every 80,000 miles for the oxygen sensor and every 100,000 miles for the charcoal canister, to remind you to service and/or replace these system components. After servicing and/or replacing components, reset the CHECK ENGINE light by sliding the cancel switch to its opposite position. The three-wire cancel switch is located on the main wiring harness, behind the instrument panel.

1984 through 1986 Sprint

The SENSOR light will start flashing on the dash at 30,000 mile intervals. This indicates the ECM is in good condition and the oxygen sensor needs to be replaced. To reset the SENSOR light, locate the SENSOR light cancel switch on the right side of the fuse box **(see illustration)**. Return the cancel switch to the OFF position. Start the engine to verify that the light is now off.

Isuzu

On Isuzu Pick-up and Trooper II models, the oxygen sensor must be replaced every 90,000 miles. When this mileage has elapsed, a O2 indicator light on the dash will come on. To turn off the indicator light on the Trooper II, slide the reset switch **(see illustration)** on the rear of the instrument cluster. To reset on Pick-up models, you'll need to remove the instrument cluster. Remove

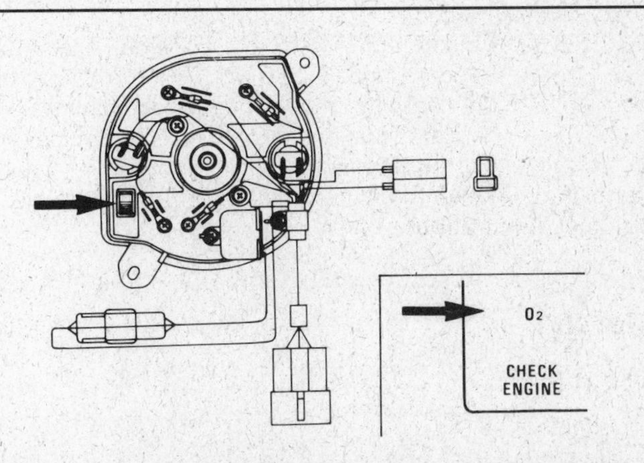

11.11 To turn off the O2 indicator light on an Isuzu Trooper, remove the instrument cluster, locate the reset switch (arrow) on the rear of the instrument cluster and slide the switch to its opposite position

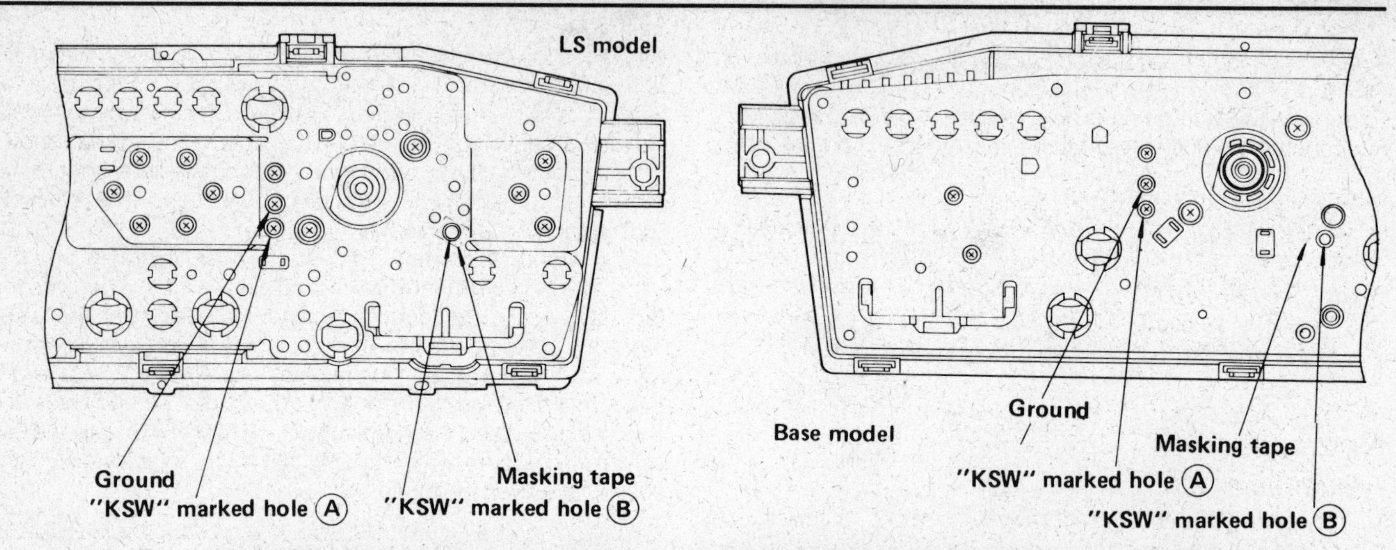

11.12 To turn off the O2 indicator light on an Isuzu Pick-up, remove the instrument cluster, remove the masking tape from hole "B", remove the screw from hole "A" and screw it into hole "B"

the masking tape from hole "B" **(see illustration)**. Remove the screw from hole "A" and insert it into hole "B".

Jaguar

1 Fuel-injected Jaguars are equipped with an oxygen sensor. When the oxygen sensor warning light comes on at 30,000-mile intervals, replace the sensor.

2 To reset the counter after replacing the sensor, locate the interval counter – it's installed inline with the speedometer cable. Reset the counter using Key (BLT-5007) supplied in the kit with the new sensor.

3 Models with electronic speedometers have a mileage counter in the trunk. The counter is behind the left side trim, next to the wheel well, behind the panel at the rear of the seat. To reset the counter, push the white button.

Mazda

All Federal 1988 and 1989 Pick-ups

1 Besides displaying computer system malfunction codes on California models, the CHECK engine warning light on Federal B2200 and B2600 Pick-ups does double duty as an emission maintenance reminder light. It comes on at 60,000-mile intervals for the EGR system and 80,000-mile intervals for the oxygen sensor.

2 To reset the warning light after servicing the indicated component, locate the brown and white, black and green wires under left side of the dash – they are taped to the wiring harness above the fuse/relay block **(see illustration)**.

3 At 60,000-mile intervals, unplug the black wire connector from the brown and white wire connector and plug it into the green wire connector. At 80,000-mile intervals, return the black wire connector to the brown and white wire connector **(see illustration)**.

All Federal 1989 MPVs

1 Besides displaying computer system malfunction codes on California models, the CHECK engine warning light comes on every 80,000 miles to indicate oxygen sensor service is needed.

2 To reset the warning light after servicing the oxygen sensor,

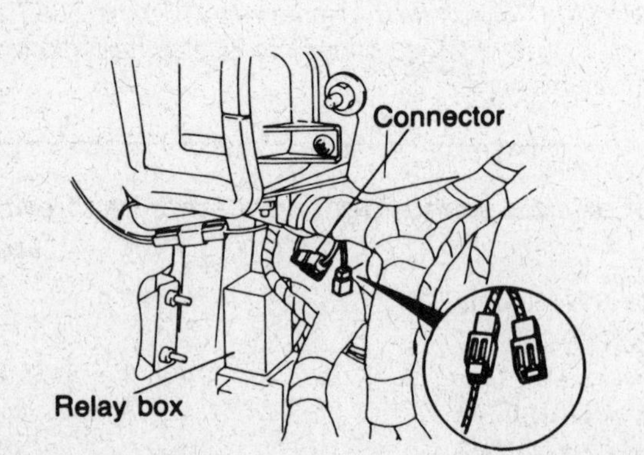

11.13 To reset the CHECK engine warning light on 1988 and 1989 Mazda Federal Pick-ups, locate the brown-and-white, black and green wires under the left side of the dash . . .

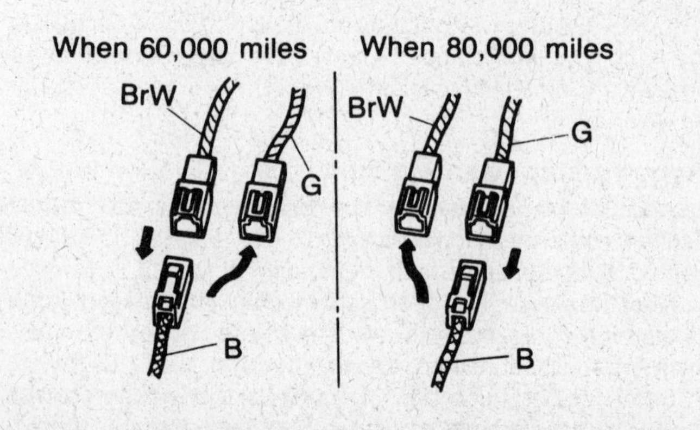

11.14 . . . and, at 60,000-mile intervals, unplug the black connector from the brown-and-white wire connector and plug it into the green wire connector; at 80,000-mile intervals, return the black wire connector to the brown-and-white wire connector

remove the instrument cluster. Locate the reset holes labelled "NC and "NO" on the rear of the instrument cluster. Remove the screw from the lettered reset hole and install it in the other hole. After another 80,000 miles, return the screw to the original hole.

Federal B2000 Pick-up

The warning light comes on every 60,000 miles to indicate that it's time to service the EGR system. To reset the warning light after service, locate the black wire and green wire connector under the dash (it's above, and to left of, the fuse box). Unplug the connector and leave it disconnected. The light won't come on again.

Mercedes Benz

1980 through 1985 Models

1 When the OX-SENSOR light comes on at 30,000 miles, you must replace the oxygen sensor and reset the mileage counter.
2 To reset the counter on 280 series vehicles, locate the mileage counter – it's inline with the speedometer cable, under the dash. Unplug the wiring plug from the counter and leave it disconnected. No reset switch is provided.
3 To reset the counter on all other models, you'll have to partially remove the instrument cluster. Insert a hooked steel wire between the right side of the cluster and the dash. Turn the hook to engage the cluster and pull the cluster out of the spring retaining clips. Remove the bulb from the lower corner of the cluster. Press the cluster into position. No reset switch is provided.

1986 through 1989 Models

The O2 SENSOR light is used as a malfunction indicator for the oxygen sensor circuit. There is no mileage counter. No reset procedure is required. Servicing and repairing the oxygen sensor circuit should turn off the light.

Mitsubishi

See Chrysler imports.

Nissan

CHECK engine light

Some 1988 and 1989 models are equipped with a CHECK engine light. The light comes on when the computer control unit senses a system malfunction. It will reset itself only when the system is repaired.

Oxygen sensor warning light

1 After 30,000 miles of operation, the oxygen sensor light in the dash will come on to indicate that oxygen sensor should be inspected. If the sensor is faulty, it must be replaced.
2 After the sensor has been inspected and/or replaced, turn off the warning light. Most models use a wire harness connector which, when disconnected, turns off the light.
3 Some 1985 through 1987 Nissan models use a reset relay located behind the left or right kick panel or the glove box, under the center console or a seat. To reset the relay, go to Step 6.
4 On 1985 and 1986 Federal Pick-ups, disconnect the yellow and white harness at 50,000 miles, and the yellow and black harness at 100,000 miles. On 1985 and 1986 California Pick-ups, disconnect the yellow wire harness at 90,000 miles. The harnesses are located above the hood release cable, under the dash.
5 To locate the single warning light harness connector on models without sensor light relays, see the accompanying chart showing the location of oxygen sensor light connectors on various models. Unplug connector and leave it unplugged. The reminder light will no longer illuminate.
6 On models with sensor light relays, locate the relay. See the accompanying chart showing the location of the oxygen sensor light relay on various models. To reset the relay, push the reset button on the relay or remove the tape over the reset hole and insert a small screwdriver into the reset hole. Push lightly to reset. Reset the relay at 30,000 and 60,000 miles. At 90,000 miles, locate and disconnect the warning light wire connector. See the accompanying chart showing the location of the oxygen sensor light connector on various models.

Nissan oxygen sensor light relay locations

Model	*Location*
Maxima	
1985 and 1986	Left kick panel
1987	Right kick panel
Pick-up (California)	
1986	Right kick panel
Pulsar/Pulsar NX	
1986	Right kick panel
1987	Left kick panel
Sentra	
1986 and 1987	Right kick panel
Stanza	
1986 and 1987	Right kick panel
Stanza Wagon	
1986 and 1987	Under right seat
200SX	
1985 1/2 through 1987	Behind grille, left of console
300 ZX	
1986 and 1987	Near glove box

Note: *The 1987 digital dash uses a relay. On the 1987 analog dash, unplug one of three connectors located near glove box at 30,000 mile intervals.*

Nissan oxygen sensor light connector locations

Model	*Location*
Maxima and Maxima/810	
1980 through 1984	
Yellow and blue wire	Near the hood release
1985 through 1987	
Green-and-red and green-and-white wires	Near the hood release
Pick-up	
1980 through 1984	
Yellow-and-white wire	Near the hood release

Nissan oxygen sensor light connector locations (con't)

Model	*Location*
Pulsar/Pulsar NX	
1983 through 1986	
Light green-and-black and light green wires or Two black-and-white wires	Near the fuse box
1987	
Red-and-black and red-and-blue wires	Above the fuse box
Sentra	
1982 and 1983	
Green-and-yellow or green-and-black wire	Above the fuse box
1984	
Light green-and-black and light green wires	Near the hood release
1985 and 1986	
Light green-and-black wire	Above the fuse box
1987	
Red-and-blue and red-and-black wires	Above the fuse box
Stanza	
1984 through 1986	
Yellow-and-red wire or yellow and yellow-and-green wires	Behind the left kick panel
1987	
Green and brown wires	Above the fuse box
Stanza Wagon	
1986 and 1987	
Red-and-yellow or red-and-blue wire	Behind the instrument panel
200SX	
1980 through 1984	
Green-and-white wire	Under far right side of dash
1985 through 1987	
Pink and purple wires	Behind the fuse box
280ZX	
1982 and 1983	
Green-and-yellow wire	Under right side dash
300ZX	
1984	
White connector	Behind the left kick panel
1985 and 1986	
White connector	Above the hood release
1987*	
Gray-and-red and gray-and-blue wires or yellow wire	Above the hood release

** Digital dash only. On analog dash, at 30,000-mile intervals, unplug one of the three connectors located behind the glove box*

Peugeot

EGR warning light

1 All 1980 604 models are equipped with an EGR warning light which comes on at 12,500-mile intervals. When the light comes on, service the EGR system, then reset the mileage counter.

2 To reset the mileage counter, unbolt the maintenance switch from the inside of the left front wheelwell. Pull down the switch without disconnecting the speedometer cables.

3 Remove the outer and inner covers from the counter and turn the reset button counterclockwise until it reaches the stop point. This resets the counter to zero. Make sure the warning light is out. Replace the covers and reinstall the counter.

Oxygen sensor warning light

1 All 505 models with gasoline engines are equipped with an oxygen sensor warning light which comes on at 30,000-mile intervals. When the warning light comes on, replace the oxygen sensor and reset the mileage counter.

2 To reset the mileage counter, locate the switch below the brake master cylinder and remove the plastic cover and plug. Use a small punch or rod to press the reset button. Replace the plug and cover.

Porsche

All 1980 and later models are equipped with an oxygen sensor. On all models except the 928S with LH-Jetronic fuel injection and the 944, an OXS light comes on at 30,000-mile intervals as a reminder to replace the sensor. (On 928S models with LH-Jetronic and on 944 models, there is no warning light – replace the oxygen sensor every 60,000 miles.) After replacing the sensor, reset the warning light as follows:

1 On 911 models, disconnect the battery negative cable and remove the speedometer. The counter will be visible through the speedometer mounting hole.

2 Use a thick piece of wire or a thin rod to press the white reset button. Push the reset button all the way in against its stop. Make sure the warning light is out.

3 On 924 models, after replacing the oxygen sensor, and with the vehicle still raised, locate the mileage counter on the left engine mount and use a thick wire or thin metal rod to push in the reset button. Be sure to push the button in all the way to the stop. Make sure the OXS light is out.

4 On 928 and 928S models with CIS fuel injection, the counter is located at the right of the passenger's seat floor. Remove the counter cover retaining screw and cover. Press the reset button all the way in against the stop. Make sure the warning light is out.

Renault

1 On models equipped with an oxygen sensor, a dash-mounted warning light comes on at 30,000-mile intervals as a reminder to replace the sensor.

2 To reset the mileage counter after replacing the sensor, locate the counter inline with the speedometer cable. Cut the retaining wires and remove the cover by disengaging the clips.

3 Turn the reset button 1/4-turn counterclockwise towards the "O" mark to reset the counter. Make sure the OXYGEN SENSOR light is out. Replace the cover and secure it with new wires.

Saab

1 All 1980 and later models use an oxygen sensor. On all models except 1985 through 1987 Turbo models, the EXH maintenance light on the dash comes on every 30,000 miles. The oxygen sensor should be replaced at this time. The above-mentioned tur-

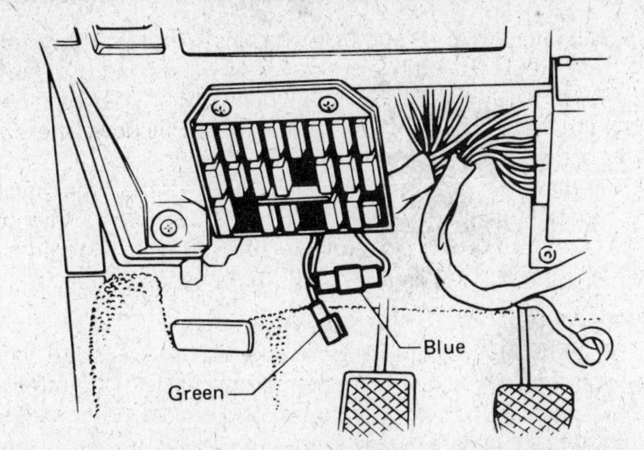

11.15 To reset the EGR warning light on 1985 through 1987 Subaru models with the 1.8L engine, remove the cover from underneath the left side of the dash, pull down the connectors from behind the fuse panel, unplug the blue connector and plug it into the green connector (1985 Subaru shown)

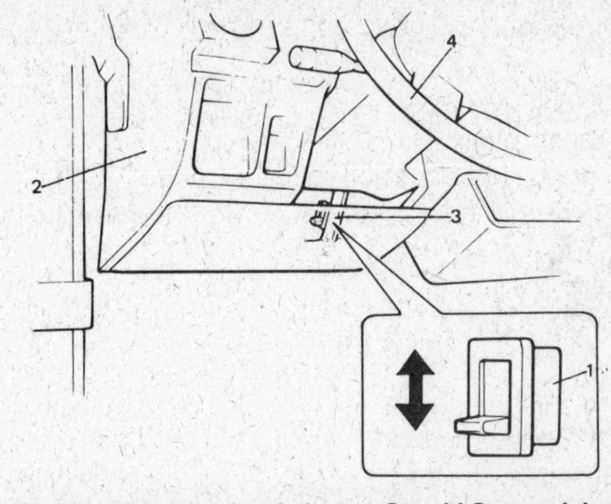

11.16 To reset the cancel switch on a Suzuki Samurai, locate the switch under the dash near the steering column, and flip the switch to its opposite position

1 *Cancel switch*
2 *Dash*
3 *Steering column mounting bracket*
4 *Steering wheel*

bo models don't use an EXH maintenance light. On these models, replace the oxygen sensor every 60,000 miles

2 To reset the mileage counter for the EXH light, press the reset button on the mileage counter. The counter is located under the instrument cluster, next to the flasher relay. Reach under the knee panel, find the counter and press the reset button.

Subaru

On 1985 through 1987 models with a 1.8L engine, an EGR warning light will turn on. To reset the light after servicing the EGR system, remove the left cover under the instrument panel. Pull down the connectors behind the fuse panel **(see illustration)**. Unplug the connector from the blue connector and plug it into the green connector.

Suzuki

1 On Samurai models, a SENSOR light will start flashing every 60,000 miles. The light will only flash with a warm engine at 1500 to 2000 RPM, indicating the ECM is in good condition and the oxygen sensor needs replacement.

2 When all emission system service procedures have been completed, locate the SENSOR light cancel switch under the dash, near the steering column **(see illustration)** and flip the switch to its opposite position. Start the engine and drive the vehicle to verify that the light doesn't flash.

Toyota

1 All 1980 Celica (six-cylinder), Supra and Cressida models and all 1981 models with gasoline engines (except Starlet models) are equipped with an oxygen sensor. At 30,000-mile intervals, a mileage counter activates a warning light on the dash. The oxygen sensor must be serviced at this time. After servicing the sensor, reset the warning light.

2 To reset the warning light, remove the white cancel switch **(see illustration)** from the top of the left kick panel, except on Cressida and 1980 Celica Supra models. On Cressida models, remove the small panel next to the steering column. On the 1980 Celica Supra, the black cancel switch is located on the bracket above the brake pedal.

3 On all models, open the switch cover and move the switch to the opposite position **(see illustration)**.

11.17 The cancel switch for the oxygen sensor warning light is located behind the left kick panel on most 1980 and 1981 Toyota models (on Cressida models, it's behind the small panel next to the steering column; on Celica Supra models, it's on the bracket above the brake pedal)

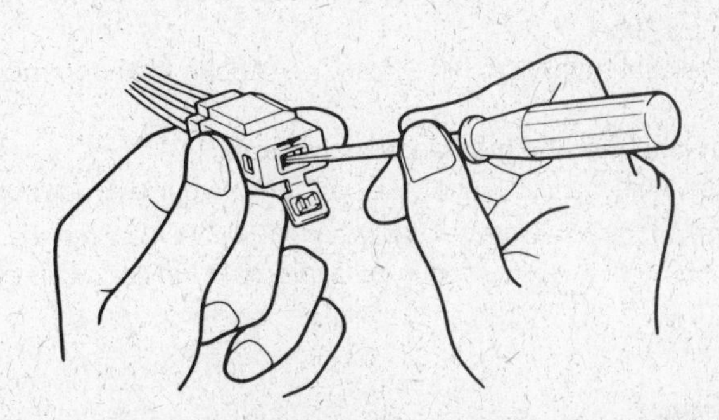

11.18 To reset the cancel switch on 1980 and 1981 Toyotas, open the switch cover and move the switch to its opposite position with a small screwdriver

Triumph

1 All fuel-injected models are equipped with an oxygen sensor. When the light on the dash comes on at 30,000-mile intervals, replace the oxygen sensor.

2 After replacing the sensor, locate the interval counter inline with the speedometer cable. Reset the counter with the key (BLT-5007) supplied in the kit with the new sensor.

Volkswagen

EGR maintenance light

Check EGR system operation when the EGR light comes on every 15,000 miles. After checking the system, reset the mileage counter (see below).

Oxygen sensor warning light

Replace the oxygen sensor when the OXS light comes on at 30,000 miles. After the new sensor is installed, reset the mileage counter (see below).

Resetting the mileage counter

1 On some Rabbit and Pick-up models, remove the instrument cluster cover plate. Using a hooked rod, reach into the opening at the upper left corner by the speedometer and pull the release arm to reset the counter. The left arm resets the EGR, the right arm resets the oxygen sensor.

2 On Vanagan models, locate the mileage counter under the spare tire or under the driver's floorboard inline with the speedometer cable. Using a pointed instrument, depress the reset button. Make sure the light is out.

3 On all other models, locate the mileage counter on the firewall, inline with the speedometer cable, and push the white reset button **(see illustration)**. Make sure the light is out.

Volvo

1 On 1980 through 1985 models (and 1986 760 GLE models) equipped with an oxygen sensor system, a warning light in the dash will come on every 30,000 miles as a reminder to replace the sensor.

2 To reset the warning light after replacing sensor on 1980 through 1984 models, locate the mileage counter inline with the speedometer cable. Press the reset button. Make sure the reminder light is out.

3 On 1985 models and 1986 760 GLE models, locate the unit under the dash (follow the wires or the small cable from the back of the speedometer to the unit). Remove the retaining screw and switch cover. Press the reset button. Make sure the reminder light is out. Install the switch cover.

4 On 1986 models (except 760 GLE) and 1987 and 1989 models, there's no warning light for oxygen sensor service intervals.

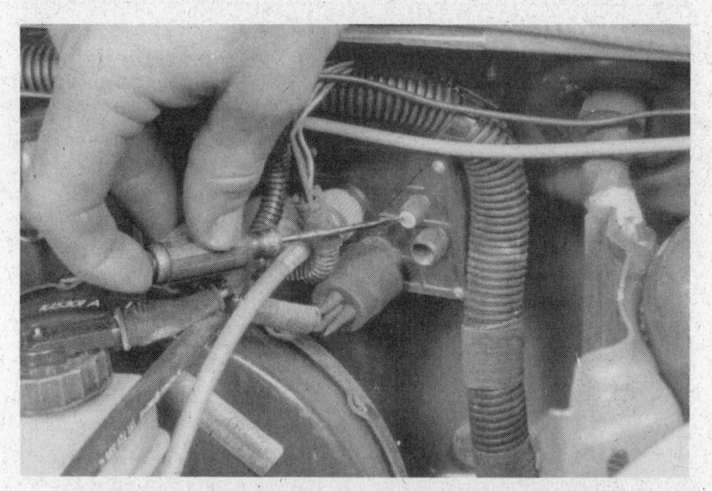

11.19 On most 1980 through 1984 VW models (except Vanagons), you'll find the mileage counter on the firewall, inline with the speedometer cable – to reset it, simply push in the oxygen sensor reset button (1980 VW Pick-up shown)

12 Can you modify an emissions-controlled vehicle?

Basically, yes!, so long as you leave all emissions systems intact and the EPA (or, if you live in California, the California Air Resources Board [CARB]) has certified all the components you're planning to install. **Note:** *The components must be certified for use on your particular vehicle. Also, nothing you do on the vehicle can be considered "tampering."*

What is tampering?

EPA regulations and many state laws prohibit tampering with the emissions components originally installed on the vehicle. They also prohibit replacing an emissions-related component with a non-original component, unless that component has been specifically certified for use on the vehicle. "Emissions-related" components are any components that have an effect on emissions. Although their primary function is not emissions control, authorities consider many engine, fuel and exhaust system components to be emissions related. These include camshafts, intake and exhaust manifolds, air cleaners and carburetors or fuel-injection systems. Even aftermarket replacement computer chips must be certified for use in the vehicle.

Tampering with emissions system components (for example, removing an air pump, disconnecting an EGR vacuum hose or even replacing an air cleaner with an aftermarket unit that doesn't have the original emissions provisions) is not only illegal, it can *decrease* your engine's performance. Vehicles designed for operation with emissions control devices often run better when the systems are hooked up and working properly. This is particularly true of newer, computer-controlled vehicles in which the computer uses information from the emissions systems to control the engine's operation.

Since most smog inspections involve a visual inspection as well as exhaust gas analysis, you'll be caught if you tamper with your emissions systems or emissions-related components. And there's usually no limit on how much you have to spend to correct components that have been tampered with; you'll have to put everything back the way it was, regardless of cost, or you won't get registered.

And don't figure you can fool the inspectors. They have on-line information or books that tell what equipment must be installed on the vehicle, and inspectors are usually pretty good at spotting non-original equipment, such as a non-certified high-performance carburetor.

Don't fall into the trap of figuring you can change a camshaft without worry, since it's inside the engine and won't be found on a visual inspection. Many long-duration camshafts cause the engine to emit excessive pollutants at idle. You may pass the visual inspection, but you might fail the exhaust analysis. Considering the amount of work and expense involved in replacing a camshaft, it pays to make sure it is certified for use in your vehicle.

What modifications can you make?

Aftermarket equipment

Many aftermarket manufacturers offer equipment that is designed for use on emission-controlled vehicles. These components include carburetors, intake manifolds, exhaust headers and camshafts **(see illustrations)**. Avoid components that are for racing use only. These often say "for off-highway use only" or "not for use in pollution-controlled vehicles" in their product literature. If you doubt any piece of equipment, check with the manufacturer to see if the component is EPA or CARB certified.

12.1 This Holley high-performance carburetor is certified for use on many Chevrolets through the mid 1970's – certified replacement carburetors like this are available for many earlier models that don't have a computer

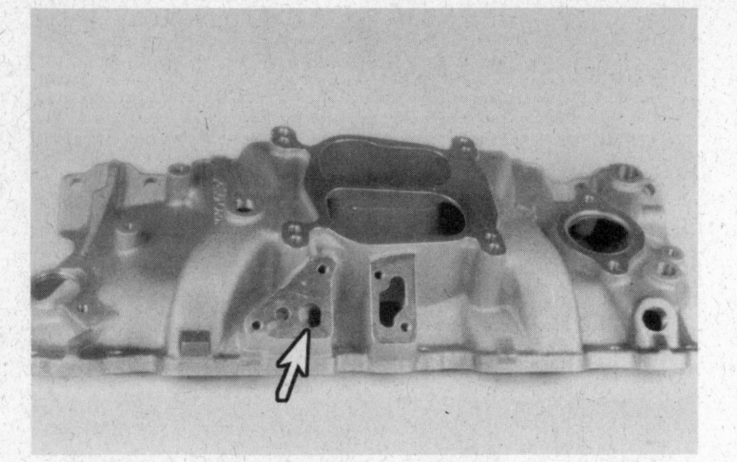

12.2 If your vehicle is equipped with EGR, make sure the certified aftermarket intake manifold you select has a mounting point for the EGR valve (arrow)

12.3 Computer-controlled ignition systems like this one can give a slight increase in performance and fuel economy to vehicles that don't already have a computer – check to be sure they're certified for use on your vehicle

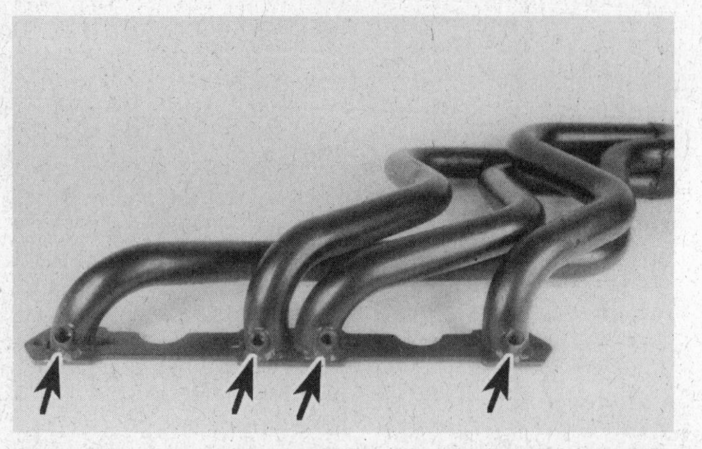

12.4 If you have an air injection system on your vehicle and the injection tubes are threaded into the exhaust manifold, make sure the certified exhaust headers you select have provisions for mounting the tubes in the same place (arrows), and . . .

12.5 . . . if your vehicle is equipped with an oxygen sensor, be sure to use a collector on the end of the header that has a mounting hole for the sensor (arrow)

Engine swaps

Most states permit swapping similar engines. However, If you are planning to swap an engine into your vehicle from a different-year vehicle, check the laws in your state to determine whether the emissions requirements apply to the vehicle or the engine. For example, if you have an older vehicle and are swapping in a newer engine, you may only be required to meet the emissions standards the vehicle has always had to meet. But, in some states, you may be required to meet new, possibly tougher, standards based on the emissions controls the vehicle from which the engine is coming originally had installed on it.

Conversely, installing an older engine in a newer vehicle does not necessarily mean your emissions requirements will be lenient. In fact, you'll often be required to update the older engine with the same level of emissions equipment originally installed on the newer vehicle.

Troubleshooting

1 General information

A malfunctioning emissions control component or system can cause a variety of problems, ranging from obvious symptoms like excessive noise, engine overheating, backfiring and visible exhaust smoke to problems that are harder to track down, such as an occasional fuel smell, poor driveability or decreased fuel mileage. On the other hand, sometimes the only symptom will be a failed emissions test.

This Chapter provides a reference guide to the more common problems which may occur during the operation of your vehicle. The first part of this Chapter deals with symptom-based diagnosis, which applies to vehicles equipped with computer-controlled fuel and emissions control systems as well as older, pre-computer vehicles. Under each symptom is a list of possible causes, with references to the appropriate Chapter and Section which will cover the system or component more thoroughly.

The second part of this Chapter deals exclusively with vehicles having computer-controlled fuel and emissions control systems. Most vehicles falling into this category have the ability to store "trouble codes" within the memory of their computers when a problem with the fuel or emissions control system occurs. On most vehicles these codes can be retrieved and, while they can't indicate the exact problem, they can direct you to the malfunctioning circuit or system, which will reduce your diagnosis time. If the vehicle you are working on is equipped with a computer-controlled fuel and emissions system, consult this part first to see if any trouble codes are stored. If so, refer to the indicated Chapter and Section that covers that particular system or component for a more complete diagnosis procedure. If there aren't any stored trouble codes, then refer to the symptom-based diagnosis.

Take note that in certain instances when a problem arises in the fuel or emissions control system of a computer-controlled vehicle, no code will be stored. And, under some circumstances, a code may set with no noticeable driveability symptom.

2 Symptom-based troubleshooting

Note: *The causes listed here are primarily related to the emissions and engine management systems. For other possible causes of the listed symptoms, refer to the* Haynes Automotive Repair Manual *for your specific vehicle.*

1 Engine noise

Grinding or rumbling – Air injection pump defective (see Chapter 3, Section 4)
Groaning – leak in Air injection system (see Chapter 3, Section 4)
Clatter – heat control valve defective (see Chapter 3, Section 5)
Hiss – vacuum leaks (see Chapter 1, Section 8)

2 Engine cranks but won't start

Carbon (charcoal) canister full of fuel (see Chapter 3, Section 2)
Faulty MAP, MAF or coolant sensor or circuit (see Chapter 3, Section 8)
EGR valve stuck open (see Chapter 3, Section 1)
Faulty canister vent valve (see Chapter 3, Section 2)
Incorrect fuel pressure

3 Engine cranks but is hard to start

Dirty air filter
PCV valve stuck open (see Chapter 3, Section 3)
Vacuum leak (see Chapter 1, Section 8)
Faulty carburetor bowl vent valve causing flooding (see Chapter 3, Section 2)
Defective coolant sensor or circuit (see Chapter 3, Section 8)
Defective air temperature sensor or circuit (see Chapter 3, Section 8)
Defective MAF sensor or circuit (see Chapter 3, Section 8)
Defective MAP sensor or circuit (see Chapter 3, Section 8)
Faulty TPS or circuit (see Chapter 3, Section 8)
Cold engine: Malfunctioning choke (carbureted models) or fuel injection system

4 Engine starts but won't run

Faulty canister vent valve (see Chapter 3, Section 2)
EGR valve stuck open (see Chapter 3, Section 1)

5 High oil consumption

PCV valve clogged or stuck open (see Chapter 3, Section 3)

6 Rough idle

Clogged air filter
Incorrect ignition timing
Dirty throttle plate or throttle bore (fuel-injected vehicles)
Minimum idle speed adjustment out of specification (fuel-injected vehicles) (refer to the VECI label under the hood)
EGR valve stuck open or leaking (see Chapter 3, Section 1)
Vacuum leak (see Chapter 1, Section 8)
PCV valve stuck open or closed (see Chapter 3, Section 3)
Cold engine:
Heat control valve stuck open (see Chapter 3, Section 5)
EFE heater inoperative (see Chapter 3, Section 5)
Warm engine:
Heat control valve stuck closed (see Chapter 3, Section 5)
Power to EFE heater after engine has warmed up (see Chapter 3, Section 5)
TPS or circuit malfunctioning or out of adjustment (see Chapter 3, Section 8)
MAF sensor or circuit out of adjustment or malfunctioning (see Chapter 3, Section 8)

7 Hesitation or stumble on acceleration

Accelerator pump in carburetor defective
Faulty TPS or circuit (see Chapter 3, Section 8)
Malfunctioning air temperature sensor or circuit (see Chapter 3, Section 8)
MAP sensor or circuit faulty (see Chapter 3, Section 8)
Leak in air intake duct or faulty MAF sensor or circuit (see Chapter 3, Section 8)
Ignition timing incorrect
Dirty throttle plate or throttle bore (fuel-injected vehicles)

8 Sluggish performance

Restricted exhaust system (most likely the catalytic converter) (see Chapter 3, Section 7)
Vacuum leak (see Chapter 1, Section 8)
EGR valve stuck open (see Chapter 3, Section 1)
EFE heater inoperative (cold engine) or restricted (see Chapter 3, Section 5)
Heat control valve stuck open (during cold engine operation) (see Chapter 3, Section 5)
Heat control valve stuck shut (during warm engine operation) (see Chapter 3, Section 5)
Incorrect ignition timing
Low or uneven cylinder compression pressures
Choke plate not opening fully
MAP sensor or circuit malfunctioning (see Chapter 3, Section 8)

9 Stalls on deceleration or when coming to a quick stop

EGR valve stuck open (see Chapter 3, Section 1)
Leak at base of EGR valve (see Chapter 3, Section 1)
Idle speed too low
TPS misadjusted or defective (see Chapter 3, Section 8)
Idle Speed Control or Electronic Air Control Valve misadjusted or malfunctioning (see Chapter 3, Section 8)

10 Surging at steady speed

Dirty air filter
Vacuum leak (see Chapter 1, Section 8)
Carburetor bowl vent valve stuck open (see Chapter 3, Section 2)
EGR valve stuck or leakage around base (see Chapter 3, Section 1)
Problem with oxygen sensor or circuit (see Chapter 3, Section 8)
Misadjusted or defective TPS or circuit (see Chapter 3, Section 8)
Defective Mass Air Flow (MAF) sensor or circuit (see Chapter 3, Section 8)
Defective MAP sensor or circuit (see Chapter 3, Section 8)
Misadjusted or malfunctioning mixture control solenoid in carburetor (General Motors carbureted models with computer control [see Chapter 3, Section 6])
Fuel pressure incorrect (fuel-injected vehicles)
Defective Vehicle speed sensor or circuit (see Chapter 3, Section 8)
Torque Converter Clutch (TCC) engaging/disengaging (see Chapter 3, Section 8)

11 Engine diesels (runs on) when shut-off or idles too fast

Vacuum leak (see Chapter 1, Section 8)
EGR valve stuck closed, causing overheating (see Chapter 3, Section 1)
Heat control valve stuck closed (see Chapter 3, Section 5)
Idle speed too high – check for correct minimum idle speed (fuel-injected vehicles [refer to the VECI label under the hood]), Idle Speed Control (ISC) motor or solenoid (see Chapter 3, Section 8) or fuel cutoff solenoid (carbureted models [see Chapter 3, Section 6])
Excessive engine operating temperature

12 Backfiring (through the intake or exhaust)

Vacuum leak in the PCV or canister purge line (see Chapter 1, Section 8 and Chapter 3, Sections 2 or 3)
Faulty air injection valve (see Chapter 3, Section 4)
Incorrect ignition timing

13 Poor fuel economy

Dirty air filter
EFE heater inoperative (see Chapter 3, Section 5)
Heat control valve stuck open or closed (see Chapter 3, Section 5)
PCV problem – valve stuck open or closed, or dirty PCV filter (see Chapter 3, Section 3)
Carburetor bowl vent valve stuck open, or faulty canister purge valve (see Chapter 3, Section 2)
Heated air intake flap stuck shut (see Chapter 3, Section 5)
Defective oxygen sensor (see Chapter 3, Section 8)

14 Pinging (spark knock)

Ignition timing incorrect
Heated air intake flap stuck closed (engine warm) (see Chapter 3, Section 5)
Bi-metal sensor in air cleaner housing malfunctioning (engine warm) (see Chapter 3, Section 5)
Power to EFE heater when the engine is warm (see Chapter 3, Section 5)
EGR valve inoperative (see Chapter 3, Section 1)

15 Engine runs hot

Heated air intake flap stuck closed (see Chapter 3, Section 5)
Bi-metal sensor in air cleaner housing malfunctioning (engine warm) (see Chapter 3, Section 5)
Power to EFE heater when the engine is warm (see Chapter 3, Section 5)
EGR valve inoperative or restricted EGR passage (see Chapter 3, Section 1)

16 Exhaust smoke

Black (overly rich fuel mixture) – Dirty air filter or restricted intake duct
Blue (burning oil) – PCV valve stuck open or PCV filter dirty (see Chapter 3, Section 3)

17 Fuel smell

Fuel tank overfilled
Leaking canister (see Chapter 3, Section 2)
Fuel vapor line or return line leaking or disconnected (see Chapter 3, Section 2)
Fuel feed line leaking

3 Computer trouble codes

General information

On-board computer systems not only control engine functions to give you better driveability and emissions, but on most systems they also have a built-in diagnostic feature. The computer detects faults and stores them as trouble codes, which can then generally be easily retrieved. A trouble code doesn't necessarily indicate the exact cause of a problem, but it will direct you to a particular circuit or system, which will simplify diagnosis. While it may not be possible for the home mechanic to repair all of these faults, the codes can allow you to be better informed when explaining a problem to the mechanic.

Retrieving codes

There are a variety of ways to retrieve trouble codes, depending on the manufacturer. Most systems work in conjunction with a light on the dash which comes on when a fault is stored. The light is marked "CHECK ENGINE", "POWER LOSS" or the like and is used to blink the codes stored in the computer. On other models, the code can be accessed by connecting a voltmeter to the diagnostic connector and counting the needle sweeps or in an LED readout on the computer itself. Once the codes are retrieved, check them against the chart for your vehicle. **Note:** *Because engine management systems may differ from year-to-year, certain trouble codes may indicate different problems from one year to the next. Since this is the case, it would be a good idea to consult your dealer or other qualified repair shop before replacing any electrical component, as they are usually expensive and can't be returned once they are purchased.*

Some models require a special readout meter or tool to retrieve the codes. These readout meters or tools are designed to be used by dealer mechanics and are usually quite expensive. Consequently, in this book code retrieval procedures will be limited to those which don't require such meters or tools.

Trouble code charts by manufacturer

Acura

3.1 The ECU on all Legend Sedans and Integra models through 1989 is located under the passenger front seat

Note 1: *Because engine management systems may differ from year-to-year, certain trouble codes may indicate different problems from one year to the next. Since this is the case, it would be a good idea to consult your dealer or other qualified repair shop before replacing any electrical component, as they are usually expensive and can't be returned once they are purchased.*

Note 2: *This information applies to 1986 through 1990 Integra and Legend models only.*

On these models the Engine Control Unit (ECU) stores the codes which are accessed by reading the flashing Light Emitting Diode (LED) on the unit. If the ECU has two LEDs, the red one is for codes. The ECU on Legend sedans and Integras through 1989 is located under the front passenger seat **(see illustration)**. On Legend coupes and 1990 Integras, the ECU is found under the dashboard on the passenger's side behind the carpet; Legends incorporate a flip-out mirror so the LED can be seen.

When the ECU sets a code, the Engine or S light on the dashboard will come on. To view the codes, turn the ignition switch On, then count and record the number of flashes. On 1986 through 1989 models, the light will blink a sequence the sum total representing the code number (for example, 14 short blinks is code 14). On 1990 models, the light will hold a longer blink to represent the first digit of a two-digit number and then will blink short for the second digit (1 long and 8 short blinks is 18). If the system has more than one problem, the codes will be displayed in sequence, pause, then repeat.

To erase the codes after making repairs, remove the Hazard fuse at the battery positive terminal (Integra) or Alternator Sense fuse in the underhood relay box (Legend) for at least ten seconds.

Code	Probable cause
1	Oxygen sensor or circuit (Integra)
1	Front oxygen sensor (Legend)
2	Rear oxygen sensor (Legend)
3	Manifold absolute pressure sensor or circuit
4	Crank angle sensor or circuit (Integra)
5	Manifold absolute pressure sensor or circuit
6	Coolant temperature sensor or circuit
7	Throttle angle sensor or circuit
8	TDC sensor or circuit
9	Crank angle sensor or circuit
10	Intake air temperature sensor or circuit
12	EGR control system
13	Atmospheric pressure sensor or circuit
14	Idle control system
15	Ignition output signal
16	Fuel injector (Integra)
17	Vehicle speed sensor or circuit (Integra)
18	Ignition timing adjustment (Legend)
19	Lock-up control solenoid valve (Integra)
20	Electric load (Integra)
21	Front spool solenoid valve (Legend)
22	Front valve timing oil pressure switch (Legend)
23	Front knock sensor (Legend)
30	AT FL signal A (Legend)
31	AT FL signal B (Legend)
35	TC standby signal (Legend)
36	TC FC signal (Legend)
41	Front oxygen sensor heater (Legend)
42	Rear oxygen sensor heater (Legend)
43	Fuel supply system (Integra)
43	Front fuel supply system (Legend)
44	Rear fuel supply system (Legend)
45	Front fuel metering (Legend)
46	Rear fuel metering (Legend)
47	Fuel pump (Legend)
51	Rear spool solenoid valve (Legend)
52	Rear valve timing oil pressure switch (Legend)
53	Rear knock sensor (Legend)
54	Crank angle B (Legend)
59	No. 1 cylinder position (Legend)

1989 and later 325i, 325ix

Code	Probable cause
1	Air flow sensor
2	Oxygen sensor
3	Coolant temperature
4	Idle speed controller

1989 and later 735i, 750i, and 5-series models

Code	Probable cause
1000, 2000	End of diagnosis
1211, 2211	Control unit
1215, 2215	Air flow sensor
1221, 2221	Oxygen sensor
1222, 2222	Oxygen sensor, regulation
1223, 2223	Coolant temperature sensor
1224, 2224	Air temperature sensor
1231, 2231	Battery voltage out of range
1232, 2232	Idle switch
1233, 2233	Full throttle switch
1251, 2251	Fuel injectors, final stage 1
1252, 2252	Fuel injectors, final stage 2
1261, 2261	Fuel pump relay
1262, 2262	Idle speed controller
1263, 2263	Tank vent
1264, 2264	Oxygen sensor heating relay
1444, 2444	No faults in memory

Note 1: *This information applies to 1989 and later models only.*

Note 2: *Because engine management systems may differ from year-to-year, certain trouble codes may indicate different problems from one year to the next. Since this is the case, it would be a good idea to consult your dealer or other qualified repair shop before replacing any electrical component, as they are usually expensive and can't be returned once they are purchased.*

On 1989 through 1991 325i, 325ix models, turn the ignition switch On (engine Off). The Check Engine light will blink the codes and then stay on once all codes have displayed. After making repairs, the memory is cleared by starting the engine five to ten times.

On 1989 through 1991 735i, 750i models, turn ignition switch On (engine Off) and depress accelerator pedal five times. The Check Engine light will then blink to display the codes. Clear the memory after repairs by momentarily unplugging the engine control unit. On 12-cylinder engines, codes beginning with 1 indicate a fault on the right bank (cylinders 1 through 6) and codes beginning with 2 denote a fault on the left bank (cylinders 7 through 12).

On 1989 and later 5-Series models, turn the ignition switch to ON (engine not running) and depress the accelerator pedal 5 times (be sure to reach wide open throttle each time). The check engine light will blink to display the codes. Clear the memory by disconnecting the negative battery terminal for 5 seconds or longer.

Chrysler

Dodge, Plymouth and Chrysler domestic cars and light trucks

Note 1: *The following information does not apply to Chrysler Corporation imports.*

Note 2: *Codes can only be retrieved on fuel-injected models. Because engine management systems may differ from year-to-year, certain trouble codes may indicate different problems from one year to the next. Since this is the case, it would be a good idea to consult your dealer or other qualified repair shop before replacing any electrical component, as they are usually expensive and can't be returned once they are purchased.*

Turn the ignition switch On, Off, On, Off, On and watch the flashes of the Power Loss or Check Engine light on the dash. The codes will blink the number of the first digit, then pause and blink the number of the second digit. For example, Code 23 would be 2 blinks, pause, 3 blinks.

Code	Probable cause
11	Engine not cranked since battery was disconnected/no distributor input signal
12	Memory standby power lost
13*	MAP (Manifold Absolute Pressure) sensor vacuum circuit
14*	MAP (Manifold Absolute Pressure) sensor electrical circuit
15**	Vehicle speed/distance sensor circuit
16*	Loss of battery voltage
17	Engine running too cold
21**	Oxygen sensor circuit
22*	Coolant temperature sensor unit
23	Throttle body temperature sensor circuit
24*	Throttle position sensor circuit
25**	ISC (Idle Speed Control) motor driver circuit
25	AIS (Automatic Idle Speed) motor driver circuit
26*	Peak injector current has not been reached or injector circuits have high resistance
27*	Fuel Injector Control circuit/or injector output circuit not responding
31**	Canister purge solenoid circuit failure

Code	Probable cause (continued)
32**	EGR (Exhaust Gas Recirculation) system failure/power loss light circuit (some 1987 models)
33	A/C clutch cutout relay circuit
34	Speed control vacuum or vent control solenoid circuits/an open or shorted circuit at the EGR solenoid (1987 models)
35	Idle switch circuit/cooling fan relay circuit
36*	Air switching solenoid circuit (non-turbo) or wastegate solenoid circuit on turbocharged models
37	Part throttle unlock solenoid driver circuit (automatic transmission only) or shift indicator light circuit (lockup converter)
41	Charging system excess or lack of field current
42	Automatic Shutdown relay driver circuit (ASD)
43	Ignition coil control circuit/or spark interface circuit
44	Loss of FJ2 to logic board/battery temperature out of range (1987) or failure in the SMEC/SBEC
45	Overboost shut-off circuit (1987) on MAP sensor reading above overboost limit detected/overdrive solenoid (A-500 orA-518 automatic transmission)
46*	Battery voltage too high
47	Battery voltage too low
51**	Oxygen sensor indicates lean
52**	Oxygen sensor indicates rich
53	Module internal problem/SMEC/SBEC failure. Internal engine controller fault condition detected
54	Problem with the distributor synchronization circuit
55	End of code output
61*	BARO solenoid failure
62	Emissions reminder light mileage is not being updated
63	EEPROM write denied/controller failure

** These codes light up Check Engine light*

***These codes light up Check Engine light on vehicles with special California emissions controls*

Eagle

Note: *Because engine management systems may differ from year-to-year, certain trouble codes may indicate different problems from one year to the next. Since this is the case, it would be a good idea to consult your dealer or other qualified repair shop before replacing any electrical component, as they are usually expensive and can't be returned once they are purchased.*

1988 and later Medallion, Premier (early), Summit and Talon

Locate the diagnostic connector in or under the glove compartment. Connect an analog voltmeter to the upper right (+) and lower left (-) connector terminals. Turn on the ignition On (engine Off) and watch the voltmeter needle. It will display the codes as sweeps of the needle. For example, two sweeps followed by three sweeps is code 23. Count the number of needle sweeps and write the codes down for reference. After making repairs, disconnect the battery to erase codes from the computer memory.

1988 and later Medallion, Premier (early), Summit and Talon

Code	Probable cause
11	Oxygen sensor
12	Airflow sensor
13	Intake air temperature sensor
14	Throttle position sensor
15	Motor position sensor
21	Coolant temperature sensor
22	Crank angle sensor
23	Top dead center sensor
24	Vehicle speed sensor
25	Barometric pressure sensor
41	Injector
42	Fuel pump
43	EGR

1991 Premier (late)

Turn the ignition switch On, Off, On, Off, On and watch the flashes of the Power Loss or Check Engine light on the dash. The codes will blink the number of the first digit, then pause and blink the number of the second digit. For example, Code 23 would be 2 blinks, pause, 3 blinks.

1991 Premier (late)

Code	Probable cause
11	Ignition reference circuit
13	MAP sensor vacuum circuit
14	MAP sensor electrical circuit
15	Speed/distance sensor circuit
17	Engine running too cool
21	Oxygen sensor circuit
22	Coolant temperature sensor circuit
23	Charge temperature circuit
24	Throttle position sensor circuit
25	Automatic idle speed control circit
26	Peak injector current not reached
26	Injector circuit
27	Fuel injection circuit control
32	EGR system
33	Air conditioner clutch relay
34	Speed control solenoid driver circuit
35	Fan control relay circuit
42	Automatic shutdown relay circuit
43	Ignition coil circuit

Code	Probable cause (continued)
51	Oxygen sensor lean
52	Oxygen sensor rich
53	Internal engine controller fault
54	Fuel sync pick-up circuit
55	End of message
63	EEPROM write denied
77	Speed control power relay

Ford

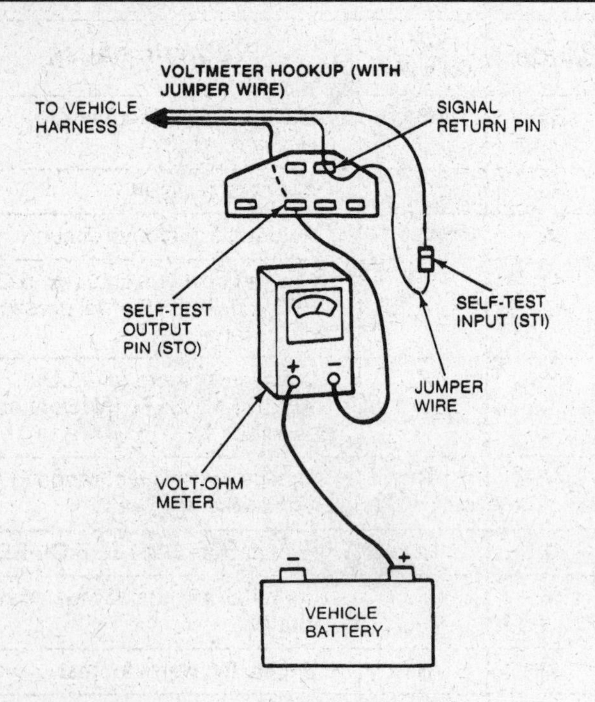

3.2a To output codes on a Ford with the EEC IV system, connect a voltmeter as shown and, using a jumper wire, bridge the self-test input connector to the signal return pin (terminal number 2)

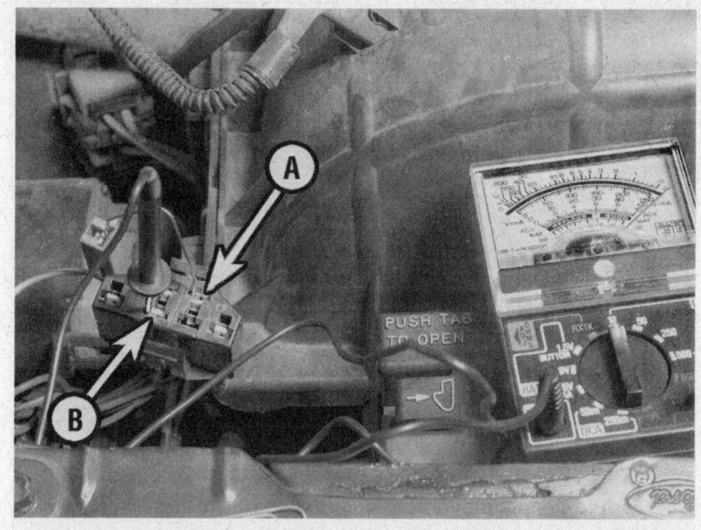

3.2b This is how it looks on a real vehicle – insert a jumper wire from terminal number 2 (A) to the self-test input connector, then install the negative probe of the voltmeter into terminal number 4 (B) and the positive probe to the battery positive terminal

Note: *Because engine management systems may differ from year-to-year, certain trouble codes may indicate different problems from one year to the next. Since this is the case, it would be a good idea to consult your dealer or other qualified repair shop before replacing any electrical component, as they are usually expensive and can't be returned once they are purchased.*

The EEC-IV system is the only Ford engine management system covered by this manual, as retrieving the codes from the other systems is beyond the scope of the home mechanic.

EEC IV system

In the engine compartment, find the "Self-Test" connector. Usually, the connector has two parts: a large one with six output terminals and the single input terminal. The connector is located at different locations on different models. For example, on most front wheel drive models it's located on the right side of the firewall, near the strut tower. On unibody rear drive models and full-size vans, it's located near the battery. On conventional frame rear wheel drive models it's located on the left side of the firewall, near the hood hinge. On the Continental/Mark VII it's located near the voltage regulator. On F-series trucks and Ranger/Bronco II models it's located near the carbon canister. On Explorer models it's located in the right-hand side of the engine compartment, near the blower motor.

With the engine off, connect the positive probe of an analog voltmeter to the battery positive post. Unplug the "Self-Test" connector. Connect a jumper wire between the input to the pin 2 on the larger connector and connect the voltmeter negative probe to pin 4 **(see illustrations)**. Set the voltmeter on a 15 or 20-volt scale, then connect a timing light to the engine. The three types of codes this test will provide are:

O – Key On Engine OFF (KOEO) (on-demand codes with the engine off)

C – Continuous Memory (codes stored when the engine was running)

R – Engine Running (ER) (codes produced as the engine is running)

O (KOEO)

Turn on the ignition on and watch the voltmeter needle. It will display the codes as sweeps of the needle. For example, two sweeps followed by three sweeps is code 23, with a four second delay between codes. Write the codes down for reference. The codes will appear in numerical order, repeating once.

C (Continuous Memory)

After the KOEO codes are reported, there will be a short pause and any stored Continuous Memory codes will appear in order. Remember that the "Pass" code is 11, or sweep, two second pause, sweep.

R (Engine running)

Start the engine. The first part of this test makes sure the system can advance the timing. Check the ignition timing. It should be advanced about 20-degrees above base timing (check the VECI label for the base timing specification).

Shut off the engine, restart it and run it for two minutes, then turn it off for ten seconds before restarting it. The voltmeter needle should make some quick sweeps, then show an engine code (two sweeps for a four cylinder engine, three for a six, four for a V8). After another pause will be one sweep, the signal to tap the accelerator so the system can check throttle component operation. After this there will be a pause, followed by the Engine Running codes which will appear in the same manner as before, repeating twice.

Ford

1983 through 1990 Car models

Code	Test condition	Probable cause
11	O,R,C	System OK, testing complete
12	R	Idle speed control out of specified range
13	O,R,C	Normal idle not within specified range
14	O,C	Ignition profile pickup erratic
15	O	ROM test failure
15	C	Power interrupt to computer memory
16	R	Erratic idle, oxygen sensor out of range or throttle not closing
17	R	Curb idle out of specified range
18	R	SPOUT circuit open
19	O	No power to processor
19	R	Erratic idle speed or signal
19	C	CID sensor failure
21	O,R,	Coolant temperature out of specified range
21	O,R,C	Coolant temperature sensor out of specified range
22	O,R,C	MAP, BARO out of specified range
23	O,R,C	Throttle position signal out of specified range
24	O,R	Air charge temperature low
25	R	Knock not sensed in test
26	O,R	Mass Air Flow sensor or circuit
27	C	Vehicle Speed Sensor or circuit
28	O,R	Vane air temperature sensor or circuit
29	C	No continuity in Vehicle Speed Sensor circuit
31	O,R,C	Canister or EGR valve control system
32	O,R,C	Canister or EGR valve control system
33	R,C	Canister or EGR valve not operating properly
34	O,R,C	Canister or EGR valve control circuit

Code	Test condition	Probable cause
35	O,R,C	EGR pressure feedback, regulator circuit
38	C	Idle control circuit
39	C	Automatic overdrive circuit
41	C	Oxygen sensor signal (ex. 5.OL SEFI); 5.OL SEFI, fuel pressure out of range
41	R	Lean fuel mixture (ex. 5.OL SEFI); 5.OL SEFI, injectors out of balance
42	R	Fuel pressure out of range (5.OL SEFI only)
42	R,C	Fuel mixture rich (ex. 5.OL SEFI)
43	C	Lean fuel mixture at wide open throttle
43	R	Engine too warm for test
44	R	Air management system inoperative
45	R	Thermactor air diverter circuit
45	C	Distributorless Ignition System (DIS) coil pack, 1 circuit failure
46	R	Thermactor air bypass circuit
46	C	Distributorless Ignition System (DIS) coil pack, 2 circuit failure
47	R	Low flow of unmetered air at idle
48	R	High flow of unmetered air at idle
48	C	Distributorless Ignition System (DIS) coil pack, 3 circuit failure
49	C	SPOUT signal defaulted to 10-degrees
51	O,C	Coolant temperature sensor out of specified range
52	O,R	Power steering pressure switch out of specified range
53	O,C	Throttle Position Sensor input out of specified range
54	O,C	Vane air flow sensor or air charge temperature sensor

1983 through 1990 Car models (con't)

Code	Test condition	Probable cause
55	R	Charging system under specified voltage (1983 through 1988 ex. 3.8L TBI)
55	R	Open ignition key power circuit (1983 through 1988 3.8L TBI only, 1989 through 1990 all)
56	O,R,C	Mass Air Flow sensor or circuit
57	C	Transmission neutral pressure switch circuit
58	O	CFI – idle control circuit; EFI – vane air flow circuit
58	R	Idle speed control motor or circuit
58	C	Vane air temperature sensor or circuit
59	O,C	Transmission throttle pressure switch circuit (ex. 3.OL SHO models)
59	O,C	Low speed fuel pump circuit (3.OL SHO, 3.8L Supercharged)
61	O,C	Coolant temperature switch out of specified range
62	O	Transmission circuit fault
63	O,C	Throttle Position Sensor or circuit
64	O,C	Vane air temperature or air charge temperature sensor
65	C	Fuel control system not switching to closed loop
66	O,C	No Mass Air Flow sensor signal
67	O,R,C	Neutral drive switch or circuit
67	C	Air conditioner clutch switch circuit
68	O,R,C	TBI – idle tracking switch; EFI – vane air temperature circuit
69	O,C	Vehicle Speed Sensor or circuit
71	C	TBI – idle tracking switch; EFI – electrical interference
72	C	System power circuit, electrical interference
72	R	No Manifold Absolute Pressure or Mass Air Flow sensor signal fluctuation
73	O,R	Throttle Position Sensor or circuit
74	R	Brake on/off ground circuit fault
75	R	Brake on/off power circuit fault
76	R	No vane airflow change
77	R	Throttle "goose" test not performed
78	C	Power circuit
79	O	Air conditioner clutch circuit
81	O	Thermactor air circuit, turbo boost circuit
82	O	Thermactor air circuit, integrated controller circuit
82	O	Supercharger bypass circuit
83	O	EGR control circuit (3.8L CFI, 2.3L EFI ex. turbo)
83	O	Cooling fan circuit (2.3L Turbo, 2.5L, 3.0L, 3.8L EFI only)
83	O,C	Low speed fuel pump relay (all 1983 through 1988, 1989 and 1990 SHO models)
83	O,C	EGR solenoid or circuit (1989 and 1990 ex. SHO)
84	O,R	EGR control circuit
85	O,R	Canister purge circuit (ex. 2.3L Turbo) or transmission shift control circuit
85	C	Excessive fuel pressure or flow
85	O	Canister purge circuit
86	C	Low fuel pressure or flow
87	O,R, C	Fuel pump circuit
88	O	2.3L Turbo – clutch converter circuit; others – integrated controller
89	O	Lock-up solenoid
91	R,C	Oxygen sensor problem, fuel pressure out of specified range or injectors out of balance
92	R	Fuel mixture rich, fuel pressure high
93	O	Throttle Position Sensor or circuit
94	R	Secondary air system inoperative
95	O,C	Fuel pump circuit problem
95	R	Thermactor air diverter circuit
96	O,C	Fuel pump circuit
96	R	Thermactor air bypass circuit

1983 through 1990 Car models (con't)

Code	Test condition	Probable cause
98	R	Repeat test sequence (1983 through 1988 models)
99	R	Repeat test sequence (1983 through 1988 models)
99	R	System hasn't learned to control idle speed

Ford

1983 and later Truck models

Code	Test condition	Probable cause
11	O,R,C	System OK, test sequence complete
12	R	Idle speed control out of specified range
13	R,C	Controlled idle out of specified range
14	C	Ignition profile pickup erratic
15	O	ROM test failure
15	C	Power interrupt to keep alive memory
16	R	Erratic idle, oxygen sensor out of specified range
18	C	No tach signal to ECM
18	R	Spout circuit open (ex. 4.OL)
18	C	Erratic IDM, input to ECM or spout circuit open
19	O	No power to processor
21	O,R,C	Coolant temperature sensor out of specified range
22	O,R,C	MAP or BARO sensor out of specified range
23	O,R	Throttle Position Sensor out of specified range
24	O,R	Air temperature low
25	R	Knock not sensed in test
26	O,R	MAF sensor or circuit (4.OL models)
26	O,R	Transmission oil temperature sensor, ex. 4.OL
28	C	Loss of primary tach – right side

Code	Test condition	Probable cause
29	C	Vehicle Speed Sensor circuit
31	O,R,C	EGR valve control, feedback sensor out of specified range (ex. V8 models)
31	O,R	EVAP control system (V8 models)
32	R,C	EGR, pressure feedback not controlling properly (1985 through 1989 models)
33	R,C	EGR valve not closing
34	O,R,C	EGR control circuit (ex. V8 models)
34	O,R,C	EVAP control system (V8 models)
35	O,R,C	No EGR position signal, RPM too low
35	O,R,C	EVAP control system (V8 models)
41	C	Oxygen sensor signal
42	R,C,	Fuel mixture rich
43	R	Oxygen sensor cooled
44	R	Air management system not functioning properly
45	C	Coil 1, 2 or 3 failure
46	R	Thermactor air bypass circuit
47	O	4WD switch closed
51	O,C	Coolant temperature sensor out of specified range
52	O,R	Power steering pressure switch out of specified range
53	O,C	Throttle sensor input out of specified range
54	O,C	Vane air flow sensor, air charge temperature sensor
55	R	Charging system
56	C	Mass Air Flow sensor above voltage (4.OL models)
56	O,C	Transmission oil temperature sensor (ex. 4.OL models)
58	R	Idle tracking switch circuit

Ford

1983 and later Truck models (con't)

Code	Test condition	Probable cause
59	C	Transmission throttle pressure switch circuit (1985 through 1988 models)
59	C	2-3 shift error (1989 models)
61	O,C	Coolant temperature switch out of specified range
62	O,R	Transmission 4/3 circuit fault
63	O,C	Throttle Position Sensor or circuit
64	O,C	Vane air temperature, air charge temperature sensor
65	R	Charging system over voltage (1985 through 1988 models)
65	R	Overdrive cancel switch not charging
66	C	MAF circuit below voltage (4.OL models)
66	O,C	Transmission oil temperature sensor grounded (ex. 4.OL models)
67	O	Neutral drive switch
67	C	Air conditioning compressor clutch switch circuit
68	O	Idle tracking switch (1985 through 1988 models)
69	C	Vehicle Speed Sensor circuit
69	O	3-4 shift error
72	R	No Manifold Absolute Pressure or Mass Air Flow signal change
73	O,R	No Throttle Position Sensor signal fluctuation
74	R	Brake on/off ground circuit fault
75	R	Brake on/off power circuit fault
77	R	Throttle “blip” test not performed
78	C	Time delay relay or circuit
79	-	Air conditioner on during test
81	O	Thermactor air bypass circuit
82	O	Thermactor air system circuit
83	O	EGR control circuit
84	O	EGR vent circuit
85	O	Canister purge circuit
86	O	Shift solenoid circuit
87	O	Fuel pump circuit
88	O	Choke relay out of specified range
88	C	Loss of dual plug input control
89	O	Clutch converter override circuit
91,92	O	Shift solenoid circuit
93	O	Coast clutch solenoid
94	O	Converter clutch solenoid
95	O,C	Fuel pump circuit
96	O,C	Fuel pump circuit
97	O	Overdrive cancel indicator light circuit
98	R	Test failure, repeat sequence (1985 through 1988)
98	O	Electronic pressure control driver
99	O,C	Electronic pressure control circuit

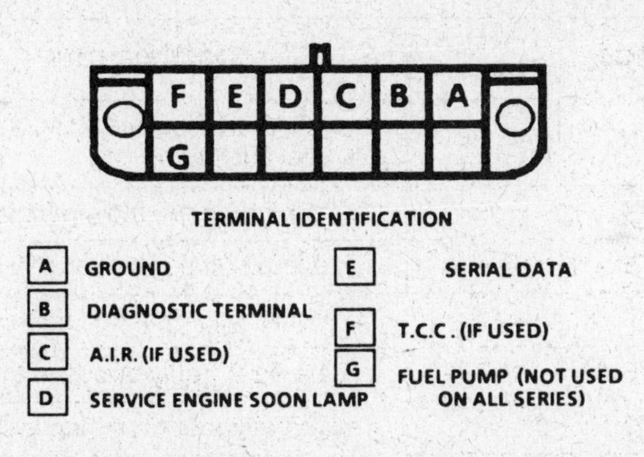

3.3 On most GM models (domestic) the ALDL connector is located under the dash, usually on the driver's side – to output trouble codes, jump terminals A and B with the ignition On

General Motors (domestics)

(except Geo, Nova and Sprint models)

Note: *Because engine management systems may differ from year-to-year, certain trouble codes may indicate different problems from one year to the next. Not all of the codes listed apply to all models. Since this is the case, it would be a good idea to consult your dealer or other qualified repair shop before replacing any electrical component, as they are usually expensive and can't be returned once they are purchased.*

All, except Cadillac with 4.1L, 4.5L, 4.6L, 4.9L and 6.0L engines AND 1988 to 1990 Oldsmobile Toronado with CRT display

The "CHECK ENGINE" light on the instrument panel will come on whenever a fault in the system has been detected, indicating that one or more codes pertaining to this fault are set in the Electronic Control Module (ECM). To retrieve the codes, you must use a short jumper wire to ground a diagnostic terminal. This terminal is part of an electrical connector known as the Assembly Line Diagnostic Link (ALDL) **(see illustration)**. On most models the ALDL is located under the dashboard on the the driver's side. If the ALDL has a cover, slide it toward you to remove it. Push one end of the jumper wire into the ALDL diagnostic terminal and the other into the ground terminal. **Caution:** *Don't crank the engine with the diagnostic terminal grounded – the ECM could be damaged.*

When the diagnostic terminal is grounded with the ignition On and the engine stopped, the system will enter Diagnostic Mode and the "CHECK ENGINE" light will display a Code 12 (one flash, pause, two flashes). The code will flash three times, display any stored codes, then flash three more times, continuing until the jumper is removed.

After checking the system, clear the codes from the ECM memory by interrupting battery power. Turn off the ignition switch (otherwise the expensive ECM will be damaged) disconnect the negative battery cable for at least ten seconds, then reconnect it.

Cadillac 4.1L, 4.5L, 4.6L, 4.9L and 6.0L engines

To retrieve codes on these models, turn the ignition switch to ON (engine not running), then press the OFF and WARMER buttons on the climate control panel at the same time. Codes will be displayed on the climate control indicator with either an E or EO preceding them. Press the RESET and RECALL buttons at the same time on 1984 to 1986 models, or AUTO on 1987 and later models to exit the diagnostic mode. Clear codes by pressing the OFF and HIGH buttons at the same time.

1988 to 1990 Oldsmobile Toronado with CRT display

To retrieve codes on these models, turn the ignition switch to ON (engine not running), then press the OFF and WARMER buttons on the climate control panel at the same time. Any stored codes will be displayed. To clear codes, press HIGH to access the ECM system, then press LOW after each message until "CLEAR CODES" is displayed. Press BI-LEVEL to exit the diagnostic mode.

Code	Probable cause
12	Diagnostic mode
13	Oxygen sensor or circuit
14	Coolant sensor or circuit/high temperature indicated
15	Coolant sensor or circuit/low temperature indicated
16	System voltage high/ECM voltage over 17.1 volts (could be alternator problem)
16	DIS (Distributorless Ignition System) circuit (Chevrolet cars only)
17	Crank signal circuit (shorted) or faulty ECM
18	Crank signal circuit (open) or faulty ECM
18	Cam and crank sensor sync error (models with DIS ignition)
19	Fuel pump circuit (shorted)
19	Crankshaft position sensor (1988 to 91)
20	Fuel pump circuit (open)
21	Throttle Position Sensor (TPS) circuit or plunger
22	Throttle Position Sensor (TPS) out of adjustment
23	Mixture Control (M/C) solenoid circuit (carbureted models)

Code	Probable cause
23	Manifold Absolute Temperature (MAT) sensor or circuit (fuel-injected models) (low temperature indicated)
23	Electronic Spark Timing (EST)/bypass circuit problem (Cadillac DFI models)
24	Vehicle Speed Sensor (VSS) or circuit
25	Manifold Air Temperature (MAT) sensor or circuit (high temperature indicated)
25	Modulated displacement failure (1981 Cadillac V8-6-4 only)
25	Electronic Spark Timing (EST) (Cadillac HT4100 only)
26	Quad Driver Circuit (dealer serviced)
26	Throttle switch circuit shorted
27	Throttle switch circuit open
27	Gear Switch Diagnosis (dealer serviced)
28	Pressure Switch Manifold check (PSM) vehicles with 4L80-E transmissions
28	Same as Code 27
29	Same as Code 27
30	ISC circuit problem (Cadillac TBI)
30	RPM error (Cadillac MFI)
31	Turbo over boost (Turbo models only)
31	Park/Neutral Switch (3.3L V6)
31	Manifold Air Temperature (MAT) sensor or circuit (Cadillac DFI models)
31	Canister purge solenoid circuit
31	Camshaft sensor or circuit
31	EGR circuit (1988 to 1990 TBI)
31	Shorted MAP sensor circuit
32	BARO sensor or circuit (carbureted models)
32	EGR circuit (fuel-injected models)
32	Digital EGR circuit (3.1L V6)
32	Open MAP sensor circuit
33	Manifold Absolute Pressure (MAP) sensor or circuit (low vacuum)
33	MAF (Mass Air Flow) sensor or circuit
34	MAF (Mass Air Flow) sensor or circuit
34	Vacuum sensor or Manifold Absolute Pressure (MAP) sensor (high vacuum) or circuit
35	Idle Air Control (IAC) valve or circuit

Code	Probable cause
35	Idle Speed Control (ISC) switch or circuit (shorted)
35	BARO sensor or circuit (shorted) (Cadillac DFI models)
36	BARO sensor or circuit (open) (Cadillac DFI models)
36	Mass Air Flow (MAF) sensor burn-off circuit
36	Distributorless Ignition System (DIS) (Quad-4)
36	Transaxle shift control (1991)
36	Closed throttle shift control (1991)
36	DIS ignition circuit (Corvette)
37	Manifold Absolute Temperature (MAT) sensor or circuit (shorted) (Cadillac HT4100)
37	MAT sensor temperature high (1984-86)
38	Manifold Absolute Temperature (MAT) sensor or circuit (open) (Cadillac HT4100)
38	Brake Input Circuit (brake light switch)
38	MAT sensor temperature low (1984-86)
39	Torque Converter Clutch (TCC)
40	Power steering pressure switch circuit
41	No distributor signals to ECM, or faulty ignition module
41	Cam sensor or circuit
41	Cylinder select error
41	Quad 4 engine 1X reference (check ignition module/ECM wiring)
42	Electronic Spark Timing (EST) circuit
42	Front oxygen sensor lean (Cadillac MFI)
43	Electronic Spark Control unit (ESC)
43	Throttle Position Sensor (TPS) out of adjustment
43	Front oxygen sensor rich (Cadillac MFI)
43	Knock sensor signal
44	Oxygen sensor or circuit – lean exhaust
45	Oxygen sensor or circuit – rich exhaust
46	Power steering pressure switch (4 cylinder – air conditioned models)
46	Vehicle Anti-Theft System (VATS)
46	Right to left fueling imbalance (Cadillac)

Code	Probable cause
47	A/C clutch and cruise circuit
48	Misfire diagnosis
48	EGR system fault (Cadillac)
51	PROM, MEM-CAL or ECM problem
52	CALPAK or ECM problem
53	System over-voltage (ECM over 17.7 volts)
53	EGR system (carbureted models)
53	Distributor signal interrupt (1983 and later Cadillac HT4100)
53	Alternator voltage out of range
53	Vehicle anti-theft circuit (5.0L TBI)
54	Mixture control (M/C) solenoid or ECM (carbureted models)
54	Fuel pump circuit (fuel-injected models)
55	Oxygen sensor circuit or ECM
55	TPS out of range (Cadillac)
55	Fuel lean monitor (Corvette)
56	Vacuum sensor circuit
56	Quad driver B circuit (3.8L)
56	Anti-theft system (Cadillac)
58	PASS key fuel enable circuit
60	Transmission not in drive (Cadillac)
61	Oxygen sensor signal faulty
61	Cruise vent solenoid (3.8L)
61	Secondary part throttle valve (Corvette)
62	Transaxle gear switch signal circuits (3.1L V6/Quad-4 engines)
62	Engine oil temperature sensor (Corvette)
62	Cruise vacuum circuit (3.8L)
62	Engine oil temperature sensor (Corvette)
63	EGR flow check (3.8L)
63	MAP sensor voltage high
63	Right side oxygen sensor circuit open (Corvette)
64	Same as Code 63 (3.8L)
64	MAP sensor voltage low
64	Right side oxygen sensor lean (Corvette)
65	Same as Code 63 (3.8L)
65	Right side oxygen sensor rich (Corvette)

General Motors (domestics)

Code	Probable cause
65	Cruise servo position sensor (3.8L)
65	Fuel Injection Circuit (Quad-4 engines)
66	Air conditioning pressure sensor circuit
67	A/C pressure sensor or clutch circuit (Chevrolet)
67	Cruise switch circuit
68	A/C relay circuit (Chevrolet)
68	Cruise system problem
69	A/C clutch circuit (Chevrolet)
69	A/C head pressure switch circuit
70	Intermittent TPS (Cadillac)
71	Intermittent MAP (Cadillac)
72	Gear selector switch (Chevrolet)
73	Intermittent coolant sensor (Cadillac)
74	Intermittent MAT (Cadillac)
75	Intermittent speed sensor (Cadillac)
80	TPS idle learn (Cadillac 4.6L)
80	Fuel system rich (Cadillac)
81	Cam reference problem (Cadillac)
82	Reference signal high (Cadillac)
85	Idle throttle angle high (Cadillac 4.6L)
85	Throttle body service required (Cadillac)
95	Engine stall detected (Cadillac)
99	Power management, cruise control system
107	PCM/BCM data link problem
108	PROM checksum mismatch
109	PCM memory reset (Cadillac)
110	Generator L-terminal circuit (Cadillac)
112	EEPROM failure (Cadillac)
131	Knock sensor failure (Cadillac)
132	Same as 131

General Motors (imports)

1989 through 1991 Geo Metro, Prizm, Storm, Tracker, 1985 through 1988 Sprint, 1988 Nova, 1985 through 1989 Spectrum and 1990 and 1991 Storm

Code retrieval

Note: *Because engine management systems may differ from year-to-year, certain trouble codes may indicate different problems from one year to the next. Since this is the case, it would be a good idea to consult your dealer or other qualified repair shop before replacing any electrical component, as they are usually expensive and can't be returned once they are purchased.*

1985 through 1988 Spectrum

A three terminal connector, located near the ECM, is used to access the diagnostic system. This terminal is known as the Assembly Line Diagnostic Link (ALDL). To activate the system, turn the ignition switch to ON (engine not running), and insert a jumper wire across the two outside cavities of the three terminal connector. The "Check Engine" light should flash code twelve (one flash followed by a pause and two flashes), this indicates the diagnostic system is working. Without turning the key to OFF remove the jumper wire from the ALDL and start the engine. **Note:** *Remove the jumper wire from the test terminal before starting the engine.* Run the engine until the check engine light comes back on, indicating a problem and a stored trouble code. Then insert the jumper wire back across the two outer terminals of the ALDL and the "Check Engine" light will flash the trouble code. To clear the trouble code memory, disconnect the battery for ten seconds or more.

1989 Spectrum and 1990 and later Storm

You must use a short jumper wire to ground the diagnostic connector. This terminal is part of the Assembly Line Diagnostic Link (ALDL), located under the dash near the ECM. Turn the ignition switch to ON (engine not running). Jumper the two outer cavities of the three-terminal connector. The "CHECK ENGINE" light will flash a Code 12 three times, then display the stored codes. After making repairs, clear the memory by removing the ECM fuse for at least ten seconds.

1985 and 1986 Sprint

Move the "Cancel" switch next to the fuse block to the On position. Turn the ignition switch On (engine Off). The "Sensor" light in the instrument cluster should light, but not flash. Start the engine and run it until normal operating temperature is reached, then rev it up to around 1500 to 2000 rpm. The "Sensor" light should now flash. If it doesn't, check these components:

Oxygen sensor
Mixture control solenoid
Idle mixture and carburetor
Thermal switch
Wire connections for the emission control systems
ECM
Idle and Wide Open Throttle (WOT) switches

After diagnosis and repairs, clear the memory by turning off the "Cancel" switch with engine running.

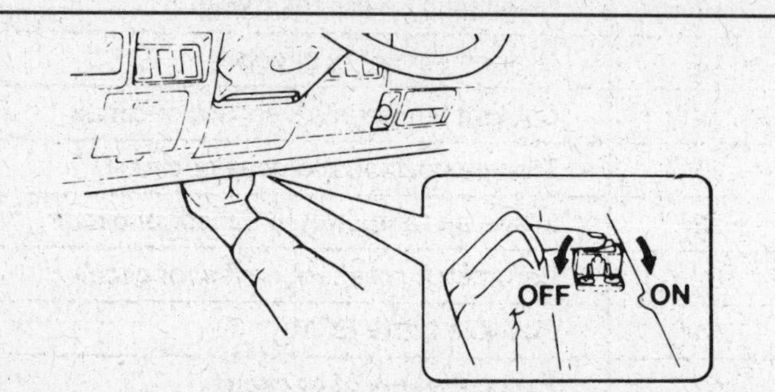

3.4 On 1987 and 1988 Sprint models, turn the Diagnostic switch to the On position (with the ignition switch in the On position) to retrieve the trouble codes

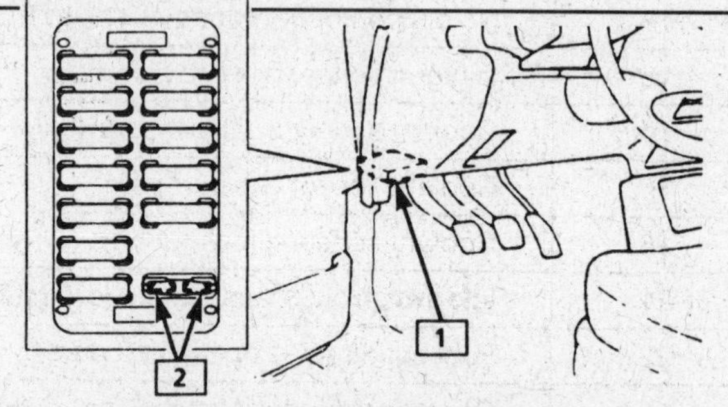

3.5 On 1989 and later Geo Metro models, insert the spare fuse into the fuse block (1) Diagnostic terminal (2) to retrieve the codes

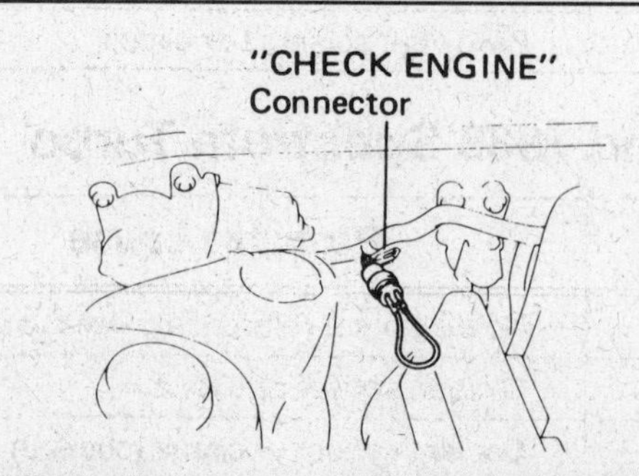

3.6 To display the codes on 1988 Nova (fuel-injected) models, insert a jumper wire into the Check Engine connector with the ignition switch in the On position

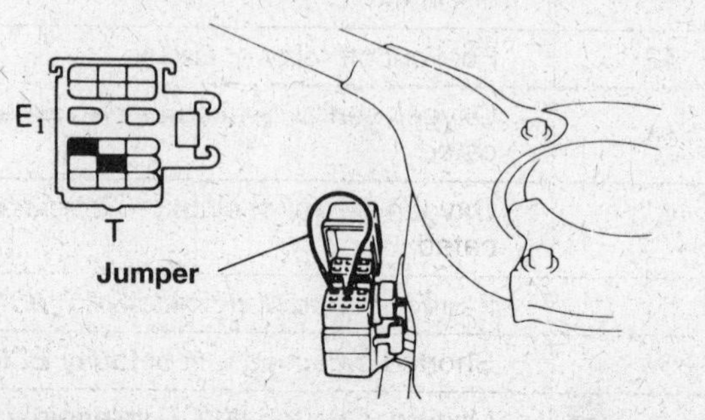

3.7 On 1989 and later Geo Prizm models, insert a jumper wire between terminals T and E1 of the Diagnostic connector to retrieve the codes

1987 and 1988 Sprint

With the engine at normal operating temperature, turn the "Diagnostic" switch located under the steering column to the On position **(see illustration)**. The codes will then be flashed by the "CHECK ENGINE" light on the dashboard. After checking the system, clear the codes from the ECM memory by turning the "Diagnostic" switch Off.

1989 and later Geo Metro

Insert the spare fuse into the "Diagnostic" terminal of the fuse block **(see illustration)**. Turn the ignition switch On (engine Off). Read the diagnostic codes as indicated by the number of flashes of the "CHECK ENGINE" light on the dashboard. Normal system operation is indicated by Code 12. If there are any malfunctions, the light will flash the requisite number of times to display the codes in numerical order, lowest to highest. After testing, remove the fuse from the "Diagnostic" terminal and clear the codes by removing the tail light fuse (otherwise the clock and radio will have to be reset).

1988 Nova fuel-injected models

With the ignition switch On, use a jumper wire to bridge both the terminals of the "CHECK ENGINE" connector located near the wiper motor **(see illustration)**. The "CHECK ENGINE" light will flash any stored codes. After checking, clear the codes by removing the ECM fuse (with the engine Off) for at least ten seconds.

1985 through 1989 Spectrum

Code	Probable cause
12	No distributor reference pulses to ECM
13	Oxygen sensor or circuit
14	Coolant sensor or circuit (shorted)
15	Coolant sensor circuit (open)
16	Coolant sensor circuit (open)
21	Idle switch out of adjustment (or circuit open)
22	Fuel cut off relay or circuit (open)
23	Open or grounded Mixture Control (M/C) solenoid or circuit
25	Open or grounded vacuum switching valve or circuit
42	Fuel cut off relay or circuit
44	Oxygen sensor or circuit – lean exhaust indicated
45	Oxygen sensor or circuit – rich exhaust indicated
51	Faulty or improperly installed PROM
53	Shorted switching unit or faulty ECM
54	Mixture Control (M/C) solenoid or circuit shorted, or faulty ECM
55	Faulty ECM

General Motors *(imports)*

1989 and later Geo Prizm models

With the ignition On (engine Off), use a jumper wire to bridge terminals T and E1 of the "Diagnostic" connector in the engine compartment **(see illustration)**. Start the engine; the "CHECK ENGINE" light will then flash any stored codes. After checking, clear the codes by removing the ECM fuse (with the engine Off) for at least ten seconds.

1989 and later Geo Tracker

On 1989 and 1990 models, insert the spare fuse into the diagnostic terminal of the fuse block. On 1991 and later models, use a jumper wire to bridge the black wire and the blue/yellow wire of the ECM check connector located in the engine compartment near the battery.

Turn the ignition switch On (engine Off). Read the diagnosis codes as indicated by the number of flashes of the "CHECK ENGINE" light on the dashboard. Normal system operation is indicated by Code 12. Code 12 will flash three times, then if there are any malfunctions, the light will flash the requisite number of times to display the codes in numerical order, lowest to highest. After testing, remove the fuse or jumper wire and clear the codes by removing the tail light fuse (otherwise the clock and radio will have to be reset).

1987 and 1988 Sprint

Code	Probable cause
12	Diagnostic function working
13	Oxygen sensor or circuit
14	Coolant temperature sensor or circuit
21	Throttle position switches or circuit
23	Intake air temperature sensor or circuit
32	Barometric pressure sensor or circuit
51	Possible faulty ECM
52	Fuel cut solenoid or circuit
53	Secondary air sensor or circuit
54	Mixture control solenoid or circuit
55	Bowl vent solenoid or circuit

1987 and 1988 Spectrum Turbo

Code	Probable cause
12	No distributor reference pulses to ECM
13	Oxygen sensor or circuit
14	Coolant sensor or circuit (shorted)
15	Coolant sensor or circuit (open)
16	Coolant sensor or circuit (open)

General Motors *(imports)*

1987 and 1988 Spectrum Turbo (continued)

Code	Probable cause
21	Throttle Position Sensor (TPS) voltage high
22	Throttle Position Sensor (TPS) voltage low
23	Manifold Air Temperature (MAT) sensor or circuit
24	Vehicle Speed Sensor or circuit
25	Air Switching Valve (ASV) or circuit
31	Wastegate control
33	Manifold Absolute Pressure (MAP) sensor voltage high
34	Manifold Absolute Pressure (MAP) sensor voltage low
42	Electronic Spark Timing (EST) circuit
43	Detonation (knock) sensor or circuit
45	Oxygen sensor – rich exhaust
51	Faulty PROM or ECM

1989 and later Prizm and 1988 Nova EFI

Code	Probable cause
Continuous Flashing	System normal
12	RPM signal
13	RPM signal
14	Ignition signal
21	Oxygen sensor or circuit
22	Coolant temperature sensor or circuit
24	Manifold Air Temperature sensor or circuit
25	Air/fuel ratio lean
26	Air/fuel ratio rich
27	Sub-oxygen sensor
31	Mass Air Flow (MAF) sensor or circuit
41	Throttle Position Sensor (TPS) or circuit
42	Vehicle Speed Sensor (VSS)
43	Starter signal
51	A/C Switch signal
71	EGR system

1987 and 1988 Sprint Turbo, 1989 and later Metro, Tracker, Storm

Code	Probable cause
12	Diagnostic function working
13	Oxygen sensor or circuit
14	Coolant temperature sensor or circuit (open)
15	Coolant temperature sensor or circuit (shorted)
21	Throttle position sensor or circuit (open)
22	Throttle position sensor or circuit (shorted)
23	Intake air temperature sensor or circuit (open)
24	Vehicle Speed Sensor (VSS) or circuit
25	Intake air temperature sensor or circuit (shorted)
31	High turbocharger pressure (1987 and 1988 models)
31	Barometric pressure sensor or circuit (1989 through 1991 models)
32	Barometric pressure sensor or circuit (1989 through 1991 models)
32	EGR system (1991 through 1993 models)
33	Air flow meter (Turbo models)
33	Manifold Absolute Pressure (MAP) sensor (1990 and 1991 models)
41	Ignition signal problem
42	Crank angle sensor (except Storm)
42	Electronic Spark Timing (EST) (Storm)
44	ECM idle switch circuit (1987 through 1989 models)
44	Oxygen sensor or circuit – lean exhaust
45	Oxygen sensor or circuit – rich exhaust
51	EGR system (except Storm)
51	ECM (Storm)
53	ECM ground circuit
On Steady	ECM fault

Honda

Note: *Because engine management systems may differ from year-to-year, certain trouble codes may indicate different problems from one year to the next. Since this is the case, it would be a good idea to consult your dealer or other qualified repair shop before replacing any electrical component, as they are usually expensive and can't be returned once they are purchased.*

Codes can only be retrieved on 1985 and later PGM-FI (fuel-injected) models. The PGM-FI or "CHECK ENGINE" light on the dash will flash when there are faults in the system and codes are set. The codes are displayed on the ECU by either four lights flashing in combination (1985 through 1987 models) or a single light (1988 and later models). ECU location varies, depending on model and year.

1985 Accord and 1985 through 1987 Civic

The ECU is located under the passenger's seat and displays the codes on four lights numbered, from left to right, 8-4-2-1. With the ignition On (engine Off), the lights will display the codes in ascending order. After making repairs, clear the codes from the ECU by turning the ignition switch Off and disconnecting the negative battery cable for at least ten seconds.

1986 and 1987 Accord and Prelude

The ECU is located under the driver's seat. With the ignition switch On, the red light on the ECU will display the codes by blinking (code 12 would be one blink, pause, two blinks) with a two second pause between codes. After making repairs, clear the ECU memory by removing the number 11 fuse from the fuse box for ten seconds.

1988 through 1991 Accord, Civic and Prelude

Pull back the carpeting on the passenger's side kick panel for access to the ECU. With the ignition On, the light on the ECU will display the codes by flashing. To clear the codes after making repairs, make sure the ignition is Off, then disconnect the negative battery cable for ten seconds.

1985 through 1987 models

LED display	Symptom	Possible cause
1 (Dash warning light on)	Engine will not start	Check for a disconnected control unit ground connector. Also check for a loose connection at the ECU main relay resistor. Possible faulty ECU
2 (Dash warning light on)	Engine will not start	Check for a short circuit in the combination meter or warning light wire. Also check for a disconnected control unit ground wire. Possible faulty ECU
3 1	System does not operate	Faulty ECU
4 2	System does not operate	Faulty ECU
5 2 1	Fuel-fouled spark plugs, engine stalls, or hesitation	Check for a disconnected MAP sensor coupler or an open circuit in the MAP sensor wire. Also check for a faulty MAP sensor
6 4	System does not operate	Faulty ECU
7 4 1	Hesitation, fuel-fouled spark plugs or the engine stalls frequently	Check for disconnected MAP sensor vacuum hose
8 4 2	High idle speed during warm-up, continued high idle or hard starting at low temperature	Check for a disconnected coolant temperature sensor connector or an open circuit in the coolant temperature sensor wire. Also check for a faulty coolant temperature sensor

Honda 1985 through 1987 models (continued)

LED display	Symptom	Possible cause
9 — 4 2 1	Poor engine response when opening the throttle rapidly, high idle speed or engine does not rev-up when cold	Check for a disconnected throttle angle sensor connector. Also check for an open circuit in the throttle angle sensor wire. Possible faulty throttle angle sensor
10 — 8	Engine does not rev-up, high idle speed or erratic idling	Check for a short or open circuit in the crank angle sensor wire. Spark plug wires interfering with the crank angle sensor wire. Also the crank angle sensor could be faulty
11 — 8 1	Same as above	Same as above
12 — 8 2	High idle speed or erratic idling when very cold	Check for a disconnected intake air temperature sensor or an open circuit in the intake air temperature sensor wire. Possible faulty intake air temperature sensor
13 — 8 2 1	Continued high idle speed	Check for a disconnected idle mixture adjuster sensor coupler or an open circuit in the idle mixture adjuster sensor wire. Possible faulty idle mixture adjuster sensor
14 — 8 4	System does not operate at all	Faulty ECU
15 — 8 4 1	Poor acceleration at high altitude when cold	Check for a disconnected atmospheric pressure sensor coupler or an open circuit in the atmospheric pressure sensor wire. Possible faulty atmospheric pressure sensor
16 — 8 4 2	System does not operate at all	Faulty ECU
17 — 8 4 2 1	Same as above	Same as above

1988 and later models

Code	Probable cause
0	Faulty ECU
1	Oxygen content
2	Faulty ECU
3/5	Manifold Absolute Pressure (MAP) sensor or circuit
4	Crank angle sensor or circuit
6	Coolant temperature sensor or circuit
7	Throttle angle sensor or circuit
8	TDC position/crank angle sensor or circuit
9	Crank angle sensor or circuit
10	Intake air temperature sensor or circuit
11	No particular symptom shown or system does not operate – faulty ECU
12	Exhaust Gas Recirculation (EGR) failure
13	Atmospheric pressure sensor circuit
14	Electronic Air Control Valve (EACV)
15	No ignition output signal – possible faulty igniter
16	Fuel injector circuit
17	Vehicle speed sensor or circuit
19	Lock-up control solenoid valve (automatic transmission)
20	Electric load – possible open or grounded circuit in ECU wiring

Hyundai

Note: *Because engine management systems may differ from year-to-year, certain trouble codes may indicate different problems from one year to the next. Since this is the case, it would be a good idea to consult your dealer or other qualified repair shop before replacing any electrical component, as they are usually expensive and can't be returned once they are purchased.*

1988 Stellar

With ignition off, connect an analog voltmeter to the diagnostic connector located in the engine compartment, behind the right strut tower. Turn the ignition On (engine Off) and watch the voltmeter needle. It will display the codes as sweeps of the needle. The needle will sweep in long or short pulses over a ten-second period with each period separated by six-second intervals.

Clear the codes after repairs by disconnecting the negative battery cable for 15 seconds.

Short sweep = 0
Long sweep = 1
10000 = 1
01000 = 2
11000 = 3
00100 = 4
10100 = 5
01100 = 6
11100 = 7
00010 = 8
00000 = 9

1989 and later Sonata, 1990 and later Excel

Locate the diagnostic connector. On Sonata models this is under the dash, to the left of the steering column and on Excel models, under the driver's side kick panel. Connect an analog voltmeter to the diagnostic connector ground terminal (lower left cavity) and MPI diagnostic terminal (upper right cavity). Turn ignition On. Count the voltmeter needle sweeps and write them down for reference. Long sweeps indicate the first digit in two-digit codes. The short sweeps indicate the second digit. For example, two long sweeps followed by one short sweep indicates a code 21.

To clear the codes, disconnect the negative battery cable for 15 seconds.

1988 Stellar

Code	Probable cause
1	Oxygen sensor or circuit
2	Ignition signal
3	Airflow sensor or circuit
4	Atmospheric pressure sensor or circuit
5	Throttle position sensor or circuit
6	Idle Speed Control (ISC) motor position sensor or circuit
7	Coolant temperature sensor or circuit
8	TDC sensor or circuit
9	Normal

1989 and later Sonata, 1990 and later Excel

Code	Probable cause
1	Electronic Control Unit (ECU) (one long needle sweep)
9	ECU normal state
11	Oxygen sensor or circuit
12	Airflow sensor or circuit
13	Intake air temperature sensor or circuit
14	Throttle Position Sensor (TPS) or circuit
15	Motor position sensor or circuit
21	Coolant temperature sensor or circuit
22	Crank angle sensor or circuit
23	TDC sensor or circuit
24	Vehicle Speed Sensor or circuit
25	Barometric pressure sensor or circuit
41	Fuel injector or circuit
42	Fuel pump or circuit
43	EGR system

Isuzu

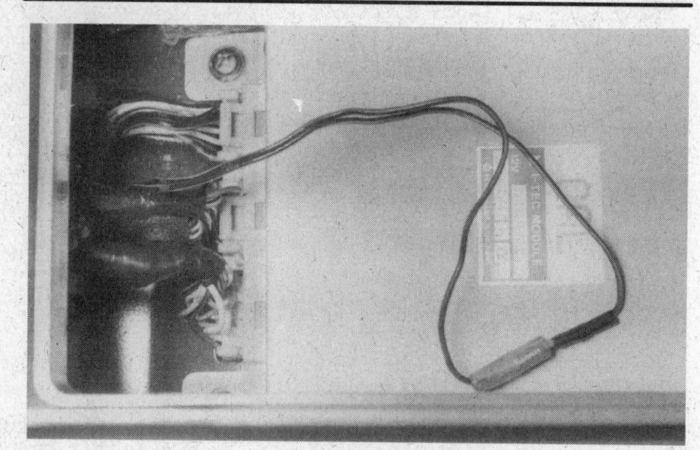

3.8 Typical DIAG hook-up – make sure the ignition switch is On before connecting the terminals

Note: *Because engine management systems may differ from year-to-year, certain trouble codes may indicate different problems from one year to the next. Since this is the case, it would be a good idea to consult your dealer or other qualified repair shop before replacing any electrical component, as they are usually expensive and can't be returned once they are purchased.*

I-Mark (RWD), 1982 through 1987 California pick-up, 1984 and later Amigo, Trooper, Rodeo, Pick-up, 1983 through 1989 non-turbo Impulse

The above models that have a "CHECK ENGINE" light on the dash have the self-diagnostic feature. To retrieve the codes, find the DIAG warning light harness connectors. These can be located in the engine compartment, under the dash or near the ECM. The connectors are usually tucked or taped out of the way in the harness. With the ignition switch On, connect the two DIAG leads together to ground them **(see illustration)**.

Trooper and Rodeo V6 and I-Mark (FWD)

To retrieve the codes, use a short jumper wire to ground the "Diagnostic" terminal. This terminal is part of an electrical connector known as the Assembly Line Diagnostic Link (ALDL). The ALDL is usually located under the dashboard or in the console near the ECM. Push one end of the jumper wire into the ALDL diagnostic terminal and the other into the ground terminal. On I-mark models terminals A and C must be jumpered (the two outer terminals on the three terminal connector). On V6 Trooper and Rodeo models, jumper terminals A and B.

All of the above models

With the diagnostic terminal now grounded and the ignition on with the engine stopped, the system will enter Diagnostic Mode and the "CHECK ENGINE" light will display a Code 12 (one flash, pause, two flashes). The code will flash three times, display any stored codes, then flash three more times, continuing until the jumper is removed.

After checking the system, remove the jumper and clear the codes from the ECM memory by removing the appropriate fuse (ECM on four-cylinder models, ELM on V6) for ten seconds.

Trouble code chart for Isuzu models listed on this page

Code	Probable cause
12	Normal
13	Oxygen sensor or circuit
14	Coolant sensor shorted (high temperature indicated) – V6 models
15	Coolant sensor open (low temperature indicated) – V6 models
16	Same as 15
21/43/65	Throttle valve switch/Wide Open Throttle (WOT) position sensor/1989 Manifold Absolute Pressure (MAP) circuit failure
21	Throttle Position Sensor (TPS) – V6 models
22	Starter signal system/1988 and 1989 fuel cut solenoid circuit failure
23	Mixture control solenoid circuit failure – 1988 and 1989 models
24	Vehicle Speed Sensor (VSS) circuit – V6 models
25	AIR V.S.V. circuit failure
26	Canister VSV (Vacuum Switching Valve) circuit failure – 1988 and 1989 models
31	No ignition reference to ECM – 1988 and 1989 models
32	Exhaust Gas Recirculation (EGR) system failure
33	Injector circuit failure
33	Manifold Absolute Pressure (MAP) sensor voltage high – V6 models
34	Manifold Absolute Pressure (MAP) sensor voltage low – V6 models
34	Exhaust Gas Recirculation (EGR) sensor or circuit failure
35	Power transistor circuit failure
41	Crank angle sensor or circuit
42	Electronic spark timing circuit failure – V6 models
42	Fuel cut-off relay – four-cylinder models
43	Electronic spark control failure – V6 models
44	Oxygen sensor (lean condition indicated)
45	Oxygen sensor (rich condition indicated)
51	Fuel cut-off solenoid shorted – carbureted four-cylinder models

Trouble code chart for Isuzu models listed on previous page (continued)

Isuzu

Code	Probable cause
51/52	Electronic Control Module (ECM) failure
53	Faulty Electronic Control Module (ECM) or VCV (Vacuum Switching Valve)
54	Shorted vacuum control solenoid/Faulty Electronic Control Module (ECM) – 1988 and 1989 models
54	Fuel pump circuit failure – V6 models
55	Faulty Electronic Control Module (ECM)
61/62	Air flow sensor circuit failure
63	Vehicle Speed Sensor (VSS) circuit
27/64	Driver transistor
65	Full-throttle switch
66	Knock sensor failure
71	Throttle position switch signal abnormal
72	V.S.V. for EGR system
73	Same as 72

Jeep

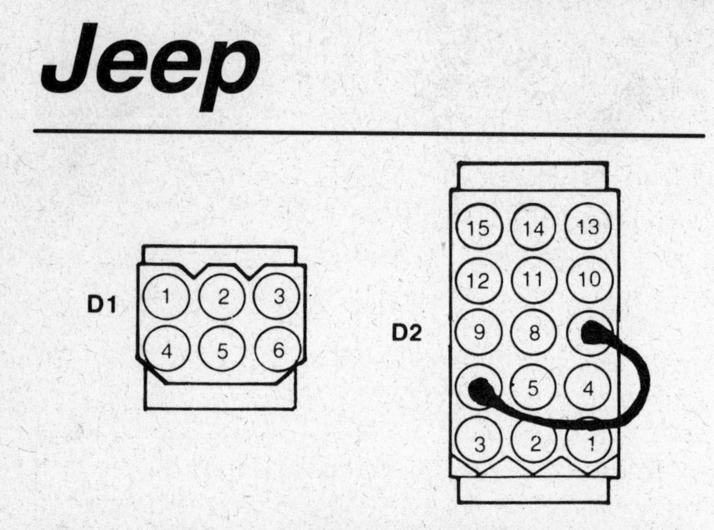

3.9 On 1984 through 1986 four-cylinder and V6 models, jump terminals 6 and 7 of the diagnostic connector to output the codes

Note: *Because engine management systems may differ from year-to-year, certain trouble codes may indicate different problems from one year to the next. Since this is the case, it would be a good idea to consult your dealer or other qualified repair shop before replacing any electrical component, as they are usually expensive and can't be returned once they are purchased.*

1984 through 1986 four-cylinder and V6 models

Insert a jumper between terminals 6 and 7 of the 15 terminal diagnostic connector **(see illustration)**. Turn on the ignition switch (engine Off) and the "CHECK ENGINE" light will flash the codes. Code 12 will flash three times, then other codes will be displayed. Disconnect ECM fuse for ten seconds to clear memory.

1991 and later fuel-injected models

Turn the ignition switch On, Off, On, Off, On and watch the flashes of the Power Loss or Check Engine light on the dash. The codes will blink the number of the first digit, then pause and blink the number of the second digit. For example, Code 23 would be 2 blinks, pause, 3 blinks.

1984 through 1986 four-cylinder and V6 models

Code	Probable cause
12	No tach signal to ECM
13	Oxygen sensor or circuit
14	Coolant temperature sensor or circuit (shorted)
15	Coolant temperature sensor or circuit (open)
21	Throttle Position Sensor or circuit
23	Open or grounded Mixture Control (M/C) solenoid
34	Vacuum sensor or circuit
41	No distributor reference pulses to ECM
42	Electronic Spark Timing (EST) or EST bypass circuit open or grounded
44	Air/fuel mixture lean
44 & 45	Faulty oxygen sensor
45	Air/fuel mixture – rich condition indicated
51	Faulty PROM or installation
54	Shorted Mixture Control (M/C) solenoid or faulty ECM
55	Grounded V-REF, faulty oxygen sensor or ECM

1991 and later fuel-injected models

Code	Probable cause
11	Ignition
13	Manifold Absolute Pressure (MAP) sensor vacuum
14	Manifold Absolute Pressure (MAP) sensor electrical
15	Speed sensor or circuit
17	Engine running too cool
21	Oxygen sensor or circuit
22	Coolant temperature sensor or circuit
23	Air charge temperature
24	Throttle Position Sensor (TPS) sensor or circuit
25	Air Induction System (AIS) control
27	Fuel injector control
33	Air conditioning clutch relay
34	Speed control solenoid driver
35	Fan control relay
41	Alternator field
42	Automatic shutdown relay
44	Battery temperature sensor
46	Battery over voltage
47	Battery under voltage
51	Oxygen sensor – lean condition indicated
52	Oxygen sensor – rich condition indicated
53	Internal engine controller fault
54	Distributor sync pickup
62	Emissions Maintenance Reminder (EMR) mileage accumulator
63	Controller failure EEPROM write denied
76	Fuel pump resistor bypass relay

Note: *Because engine management systems may differ from year-to-year, certain trouble codes may indicate different problems from one year to the next. Since this is the case, it would be a good idea to consult your dealer or other qualified repair shop before replacing any electrical component, as they are usually expensive and can't be returned once they are purchased.*

Locate the diagnostic connector and with the engine Off, connect an analog voltmeter which will be used to display the codes. The connector location and voltmeter hookup details vary with model and year:

1987 and 1988 Galant, Starion, Van, Montero V6 and all 1989 and later models

The diagnostic connector is in or under the glove compartment. Connect the voltmeter to the upper right (+) and lower left (-) terminals.

1983 through 1986 fuel-injected models

The diagnostic connector is located under the battery or on the right side firewall near the ECU, depending on model.

1987 and 1988 Mirage Turbo

Connect the voltmeter to a good ground and the single wire connector on the firewall near the set timing connector.

1987 and 1988 Tredia and Cordia

The voltmeter is connected to the lower right (-) and upper terminals of the diagnostic connector.

All models

Turn on the ignition on and watch the voltmeter needle. It will display the codes as sweeps of the needle. Count the number of needle sweeps and write the codes down for reference.

1987 Galant, 1989 Starion and 1987 and 1988 Van models

The voltmeter needle sweeps will be of long or short duration every two seconds over a ten second period. Use this code to decipher the sweeps:

Short sweep = 0, long sweep = 1

00000 = 0
10000 = 1
01000 = 2
11000 = 3
00100 = 4
10100 = 5
01100 = 6
11100 = 7
00010 = 8

Mitsubishi

All 1989 through 1991 models and 1988 Galant

The voltmeter sweeps will be of short duration to indicate ones and long duration to indicate tens.

Precis (1990 and later fuel-injected models only)

Code	Probable cause
11	Oxygen sensor or circuit
12	Air flow sensor or circuit
13	Intake air temperature sensor or circuit
14	Throttle Position Sensor (TPS) or circuit
15	Motor position sensor or circuit
21	Engine coolant temperature sensor or circuit
22	Crank angle sensor or circuit
23	No. 1 cylinder (TDC) Top Dead Center sensor or circuit
24	Vehicle speed read switch
25	Barometric pressure sensor
41	Fuel injector
42	Fuel pump or circuit
43	EGR (Exhaust Gas Recirculation) temperature sensor or circuit

All 1989 and later fuel-injected cars (except Precis), 1988 Galant

Code	Probable cause
1	Electronic Control Unit (ECU) (one long needle sweep)
9	Normal state (continuous short flashes)
11	Oxygen sensor or circuit
12	Air flow sensor or circuit
13	Intake air temperature sensor or circuit
14	Throttle Position Sensor (TPS) or circuit
15	Motor position sensor or circuit
21	Coolant temperature sensor or circuit
22	Crank angle sensor or circuit

Mitsubishi *(continued)*

All 1989 and later fuel-injected cars (except Precis), 1988 Galant (continued)

Code	Probable cause
23	Top Dead Center sensor or circuit
24	Vehicle Speed Sensor or circuit
25	Barometric pressure sensor or circuit
31	Detonation sensor
36	Ignition timing adjustment
39	Front oxygen sensor
41	Fuel injector failure
42	Fuel pump or circuit
43	Exhaust Gas Recirculation (EGR) system
44	Ignition coil (except DOHC V6)
44	Power transistor for coil (1-4) (DOHC V6)
52	Power transistor for coil (2-5) (DOHC V6)
53	Power transistor for coil (3-6) (DOHC V6)
61	ECM and transmission interlock
62	Induction control valve position sensor

1989 Starion, 1983 through 1988: all except 1988 Galant

Code	Probable cause
1	Exhaust gas sensor and/or ECU
2	Crankshaft angle sensor or ignition signal
3	Air flow sensor
4	Atmospheric pressure sensor
5	Throttle angle sensor
6	Idle Speed Control (ISC) motor position sensor
7	Engine coolant temperature sensor
8	Top Dead Center (TDC) sensor or vehicle speed sensor

1989 and later fuel-injected Pick-ups/Montero

Code	Probable cause
1	Engine Control Unit (ECU) (one long needle sweep)
9	Normal state (continuous short flashes)
11	Oxygen sensor or circuit
12	Air flow sensor or circuit
13	Intake air temperature sensor or circuit
14	Throttle Position Sensor (TPS) or circuit
21	Coolant temperature sensor or circuit
22	Crank angle sensor or circuit
23	Top Dead Center sensor or circuit
24	Vehicle Speed Sensor or circuit
25	Barometric pressure sensor or circuit
41	Fuel injector
42	Fuel pump or circuit
43	Exhaust Gas Recirculation (EGR) system

Nissan/Datsun

Pick-ups (fuel-injected models through 1991), Maxima (1985 through 1991), Sentra (through 1990) and 300ZX (1987 through 1989)

1 On 300ZX models, to start the diagnostic procedure, expose the inspection lamps by removing the dash side panel. On other models, remove the ECU from under the dash (1989 and later Maxima models) or under the passenger's seat (all other models). **Caution:** *Do not disconnect the electrical connector from the ECU or you will erase any stored diagnostic codes.* Turn the ignition switch to ON (throttle body fuel-injected [TBI] pick-up models) or start the engine and warm it to normal operating temperature (all other models). Turn the diagnostic mode selector on the ECU fully clockwise or turn the mode selector to ON **(see illustrations)**. Wait until the inspection lamps flash (the LED-type inspection lamps are located on the side or top of the ECU). After the inspection lamps have flashed three times, turn the diagnostic mode selector fully counterclockwise or turn the mode selector to OFF. The ECU is now in the self-diagnostic mode.

2 Now, count the number of times the inspection lamps flash. First, the red lamp flashes, then the green lamp flashes. The red lamp denotes units of ten, the green lamp denotes units of one. Check the trouble code chart for the particular malfunction.

3 For example, if the red lamp flashes once and the green lamp flashes twice, the ECU is displaying the number 12, which indicates the air flow meter is malfunctioning.

4 If the ignition switch is turned off at any time during a diagnostic readout, the procedure must be re-started.

5 The stored memory or memories will be lost if, for any reason, the battery terminal is disconnected.

6 On TBI-equipped pick-up models, to erase the memory after self-diagnosis codes have been noted or recorded, turn the diagnostic mode selector to ON. After the inspection lamps have flashed four times, turn the diagnostic mode selector to OFF. Turn the ignition switch to OFF.

7 On all other models, erase the memory by turning the diagnostic mode selector on the ECU fully clockwise. After the inspection lamps have flashed four times, turn the mode selector fully counterclockwise. This will erase any signals the ECU has stored concerning a particular component.

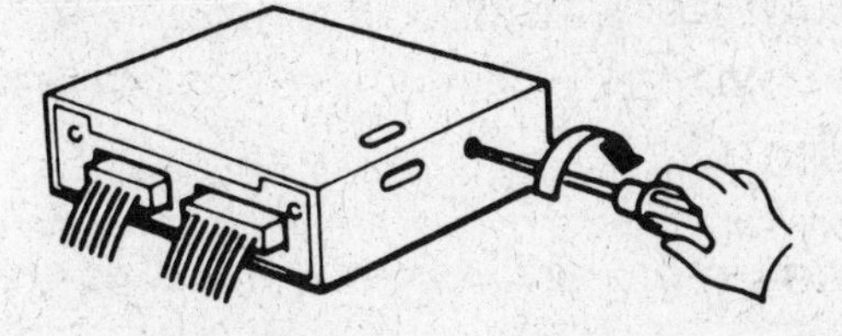

3.10a On all except TBI-equipped pick-ups and 1984 through 1986 300ZX, select the diagnostic mode by turning the ECU mode selector clockwise until it stops

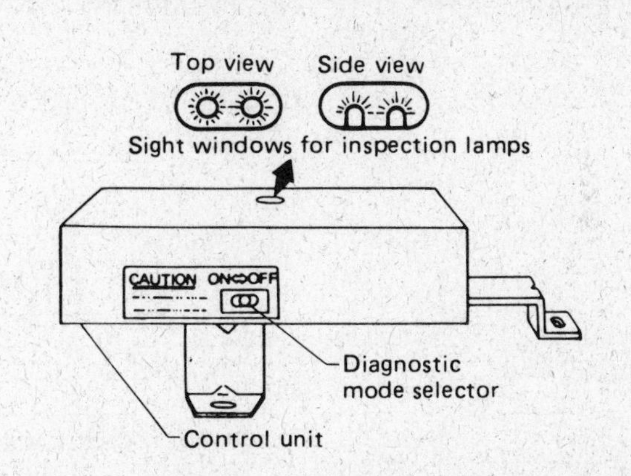

3.10b On TBI-equipped pick-ups, activate the diagnostic mode by pushing the mode switch to the left – the red and green lights should begin flashing

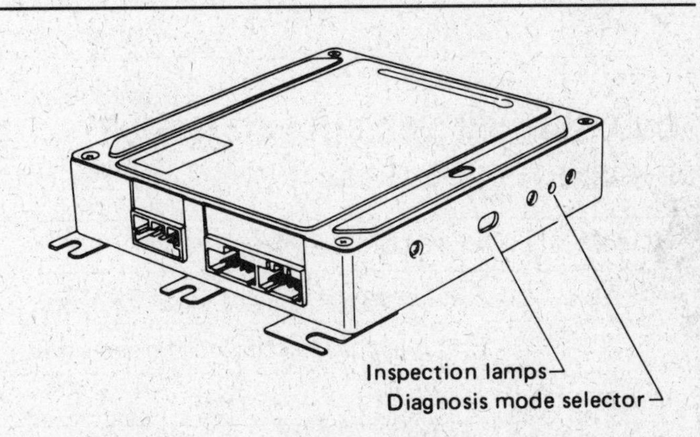

3.11 On 1984 through 1986 300ZX models, verify the diagnostic mode selector is turned fully counterclockwise using a small screwdriver

300ZX (1984 through 1986)

Extracting trouble codes

8 Locate the ECCS control module under the passenger's side kick panel.

9 Remove the module retaining bolts and pull the module out so you can handle it. **Caution:** *Do not disconnect the electrical connector(s) to the module or the trouble codes will be erased.*

10 Verify the diagnosis mode selector **(see illustration)** is turned fully counterclockwise using a small screwdriver.

11 Turn the ignition switch to ON.

12 Check that the inspection lamps stay on to check the bulbs.

13 Turn the mode selector fully clockwise.

14 Now, count the number of times the inspection lamps flash. First, the red lamp flashes, then the green lamp flashes. The red lamp denotes units of ten, the green lamp denotes units of one. For example, code 23 would be indicated by the red lamp flashing twice and the green lamp flashing three times. Check the trouble code chart. Confirm that the lamps are displaying codes 23, 24 and 31 on turbo models. If not, write down the malfunction code.

15 Depress the accelerator one time and release it.

16 On VG30ET only, shift the transmission through the gears, ending in Neutral.

Nissan/Datsun *(continued)*

17 On 1984 and 1985 models, make sure the inspection lamps are displaying codes 24 (VG30ET) and 31. If not, write down the malfunction code.
18 Start the engine, and, on 1986 VG30ET models with automatic transmission, apply the brake and move the selector to "D."
19 Confirm the codes displayed: They should be 14 for VG30ET engines and 31 for others.
20 If a VG30ET model, drive the vehicle at more than 6 mph.
21 Make sure the lamps are flashing code 31. If not, write down the malfunction code.
22 Add a load to the system by turning the air conditioning switch On, then Off.
23 The lamps should display code 44.
24 Turn the diagnosis mode selector fully counterclockwise.
25 Turn the engine off.
26 See the decoding chart for trouble code identification.
27 Check the malfunctioning area, then erase the memory. **Caution:** *The crank angle sensor plays an important part in the electronic computer control system and a malfunctioning sensor is sometimes accompanied by a display which shows malfunctions in other signal systems. If this happens, start with the crank angle sensor.*

Erasing the trouble codes from memory

28 Turn the switch to the On position.
29 Turn the diagnosis mode selector fully clockwise and hold it there for more than two seconds.
30 Turn the ignition switch to Off.
31 After correcting a malfunctioning system, be sure to erase the memory
32 Reverse the removal procedures to install the computer control module.

Nissan trouble code chart

Note: *This chart is only for the models listed on the previous page*

Display code	Probable cause
Code 11 (1 red flash, 1 green flash)	Crank angle sensor/circuit
Code 12 (1 red flash, 2 green flashes)	Air flow meter/circuit open or shorted
Code 13 (1 red flash, 3 green flashes)	Cylinder head temperature sensor (Maxima and 300ZX models); coolant temperature sensor circuit (all other models)
Code 14 (1 red flash, 4 green flashes)	Vehicle speed sensor signal circuit is open
Code 15 (1 red flash, 5 green flashes)	Mixture ratio is too lean despite feedback control. Fuel injector clogged
Code 21 (2 red flashes, 1 green flash)	Ignition signal in the primary circuit is not being entered to the ECU during cranking or running
Code 22 (2 red flashes, 2 green flashes)	Fuel pump circuit (Maxima and 1987 and later 300ZX models); idle speed control valve or circuit (all other models)
Code 23 (2 red flashes, 3 green flashes)	Idle switch (throttle valve switch) signal circuit is open
Code 24 (2 red flashes, 4 green flashes)	Park/Neutral switch malfunctioning
Code 25 (2 red flashes, 5 green flashes)	Idle speed control valve circuit is open or shorted
Code 31 (3 red flashes, 1 green flash)	1984 through 1986 300ZX models: problem in air conditioning system. All other models: ECU control unit problem
Code 32 (3 red flashes, 2 green flashes)	1984 through 1986 300ZX models: check starter system. All other models: EGR function
Code 33 (3 red flashes, 3 green flashes)	Oxygen sensor or circuit (300ZX left side)

Nissan/Datsun

Nissan trouble code chart (continued)	
Display code	**Probable cause**
Code 34 (3 red flashes, 4 green flashes)	Detonation (knock) sensor
Code 35 (3 red flashes, 5 green flashes)	Exhaust gas temperature sensor
Code 41 (4 red flashes, 1 green flash)	Maxima and 1984 through 1987 300ZX models: fuel temp. sensor circuit. All other models: air temperature sensor circuit
Code 42 (4 red flashes, 2 green flashes)	1988 and later 300ZX models: fuel temperature sensor circuit. All other models: throttle sensor circuit open or shorted
Code 43 (1987 Sentra only) (4 red flashes, 3 green flashes)	The mixture ratio is too lean despite feedback control. Fuel injector is clogged
Code 43 (all others) (4 red flashes, 3 green flashes)	Throttle position sensor circuit is open or shorted
Code 44 (4 red flashes, 4 green flashes)	No trouble codes stored in ECU
Code 45 (4 red flashes, 5 green flashes)	Injector fuel leak
Code 51 (5 red flashes, 1 green flash)	Fuel injector circuit open
Code 53 (5 red flashes, 3 green flashes)	Oxygen sensor (300ZX right side)
Code 54 (5 red flashes, 4 green flashes)	Short between A/T control unit and ECU
Code 55 (5 red flashes, 5 green flashes)	Normal engine management system operation is indicated

Subaru

Note: *Because engine management systems may differ from year-to-year, certain trouble codes may indicate different problems from one year to the next. Since this is the case, it would be a good idea to consult your dealer or other qualified repair shop before replacing any electrical component, as they are usually expensive and can't be returned once they are purchased.*

Connect the male and female connectors under steering wheel to the left of module. Turn ignition On (engine Off). The codes are displayed as pulses on Light Emitting Diode (LED) mounted on module. The long pulses indicate tens and the short pulses ones.

1983 carbureted models

Code	Probable cause
11, 12, 21, 22	Ignition pulse system
14, 24, 41, 42	Vacuum switches stay on or off
15, 51, 52	Solenoid valve stays on or off
23	Oxygen sensor or circuit
32	Coolant temperature sensor or circuit
33	Main system in feedback
34,43	Choke power stays on or off
42	Clutch switch or circuit

1984 carbureted models

Code	Probable cause
11, 18	Ignition pulse system
22	Vehicle Speed Sensor (VSS) or circuit
23	Oxygen sensor or circuit
24, 25	Coolant temperature sensor or circuit
31, 32	Duty solenoid valve or circuit
33	Main system in feedback
34, 35	Back up system
42, 45	Vacuum switch stays on or off
52, 53	Solenoid valve control system
54, 55	Choke control system
62	Exhaust Gas Recirculation (EGR) solenoid valve control
63, 64	Canister solenoid valve or circuit
73, 77	Ignition pulse system

1985 and 1986 carbureted models

Code	Probable cause
11	Ignition pulse system
22	Vehicle Speed Sensor (VSS) or circuit
23	Oxygen sensor or circuit
24	Coolant temperature sensor or circuit
25	Manifold vacuum sensor or circuit
32	Duty solenoid valve or circuit
33	Main system in feedback
34	Back up system
42	Clutch switch or circuit
52	Solenoid valve control system
53	Fuel pump or circuit
54	Choke control system
55	Upshift control
62	Exhaust Gas Recirculation (EGR) solenoid valve control
63	Canister solenoid valve or circuit
64	Vacuum line control valve or circuit
65	Float chamber vent control valve or circuit
71, 73, 74	Ignition pulse system

1983 through 1985 fuel-injected models

Code	Probable cause
11	Ignition pulse
12	Starter switch off
13	Starter switch on
14	Air flow meter or circuit
21	Seized air flow meter flap
22	Pressure or vacuum switches – fixed value
23	Idle switch – fixed value
24	Wide open throttle switch fixed
32	Oxygen sensor or circuit
33	Coolant sensor or circuit
35	Air flow meter or EGR solenoid switch or circuit
31,41	Atmosphere air sensor or circuit
42	Fuel injector – fixed value

Subaru

1986 fuel-injected models

Code	Probable cause
11	Ignition pulse
12	Starter switch off
13	Starter switch on
14	Air flow meter or circuit
15	Pressure switch – fixed value
16	Crank angle sensor or circuit
17	Starter switch or circuit
21	Seized air flow meter flap
22	Pressure or vacuum switches – fixed value
23	Idle switch – fixed value
24	Wide open throttle switch – fixed value
25	Throttle sensor idle switch or circuit
31	Speed sensor or circuit
32	Oxygen sensor or circuit
33	Coolant sensor or circuit
35	Air flow meter or EGR solenoid switch or circuit
41	Atmosphere pressure sensor or circuit
42	Fuel injector – fixed value
43, 55	KDLH control system
46	Neutral or parking switch or circuit
47	Fuel injector
53	Fuel pump or circuit
57	Canister control system
58	Air control system
62	EGR control system
88	TBI control unit

1987 fuel-injected models

Code	Probable cause
11	Ignition pulse
12	Starter switch or circuit
13	Crank angle sensor or circuit
14	Injectors 1 and 2
15	Injectors 3 and 4
21	Coolant temperature sensor or circuit
22	Knock sensor or circuit

1987 fuel-injected models (continued)

Code	Probable cause
23	Air flow meter or circuit
24	Air control
31	Throttle sensor or circuit
32	Oxygen sensor or circuit
33	Vehicle Speed Sensor (VSS) or circuit
35	Purge control solenoid or circuit
41	Lean fuel mixture indicated
42	Idle switch or circuit
45	Kick-down relay or circuit
51	Neutral switch or circuit
61	Parking switch or circuit

1988 and 1989 1200cc Justy model

Code	Probable cause
14	Duty solenoid valve or circuit
15	Coasting fuel cut system
21	Coolant temperature sensor or circuit
22	Vacuum line charging solenoid or circuit
23	Pressure sensor system
24	Idle-up solenoid or circuit
25	Fuel chamber vent solenoid or circuit
32	Oxygen sensor or circuit
33	Vehicle Speed Sensor (VSS) or circuit
35	Purge control solenoid or circuit
52	Clutch switch
62	Idle-up system
63	Idle-up system

1988 through 1990 1.7L four-cylinder, 1988 and 1989 six-cylinder engine

Code	Probable cause
11	Crank angle sensor or circuit
12	Starter switch or circuit
13	Crank angle sensor or circuit
14	Injectors 1 and 2
15	Injectors 3 and 4
21	Coolant temperature sensor or circuit

Subaru (continued)

1988 through 1990 1.7L four-cylinder, 1988 and 1989 six-cylinder engine (continued)

Code	Probable cause
22	Knock sensor or circuit
23	Air flow meter or circuit
24	Air control valve or circuit
31	Throttle sensor or circuit
32	Oxygen sensor or circuit
33	Vehicle Speed Sensor (VSS) or circuit
34	EGR solenoid or circuit
35	Purge control solenoid or circuit
41	Lean fuel mixture indicated
42	Idle switch or circuit
44	Duty solenoid valve (waste gate control) or circuit
45	Kick-down control relay or circuit
51	Neutral switch continuously in the on position
54	Neutral switch or circuit
55	EGR temperature sensor or circuit
61	Parking switch or circuit

1990 four-cylinder 2.1L engine

Code	Probable cause
11	Crank angle sensor or circuit
12	Starter switch or circuit
13	Cam position sensor or circuit
14	Injector #1
15	Injector #2
16	Injector #3
17	Injector #4
21	Coolant temperature sensor or circuit
22	Knock sensor or circuit
23	Knock sensor or circuit
24	Air control valve or circuit
31	Throttle position sensor or circuit
32	Oxygen sensor or circuit
33	Vehicle Speed Sensor (VSS) or circuit
35	Canister purge solenoid or circuit
41	Air/fuel adaptive control
42	Idle switch or circuit
45	Atmospheric pressure sensor or circuit
51	Neutral switch (MT), inhibitor switch (AT)
52	Parking switch

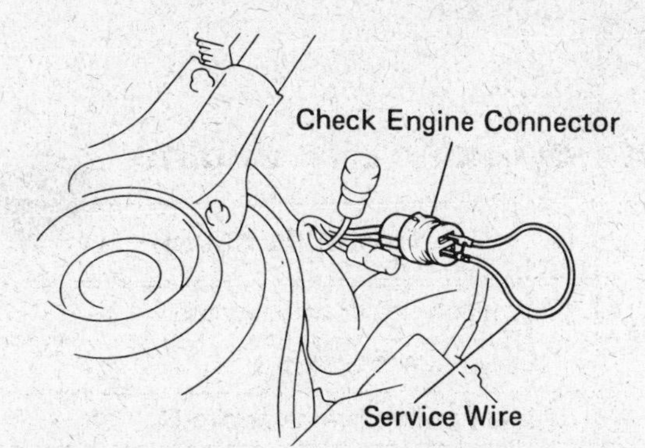

3.12a On 1984 Camrys, 1987 Corollas and 1986 and earlier Pick-ups, bridge the terminals of the round Check Engine connector with a jumper wire to obtain the diagnostic codes (Corolla model shown, others similar)

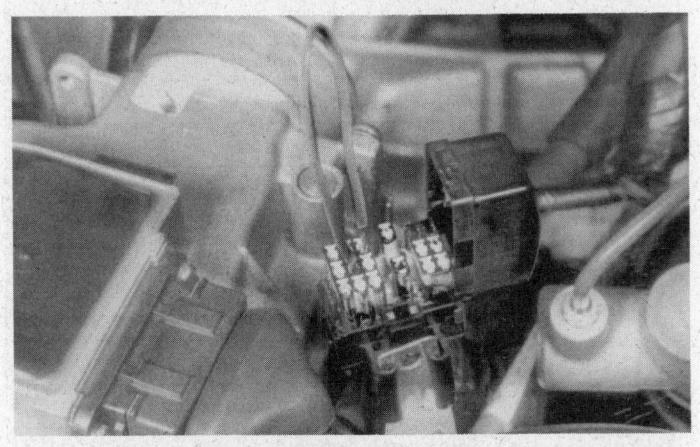

3.12b On other models, bridge terminals T and E1 of the service connector to output the codes

Note: *Because engine management systems may differ from year-to-year, certain trouble codes may indicate different problems from one year to the next. Since this is the case, it would be a good idea to consult your dealer or other qualified repair shop before replacing any electrical component, as they are usually expensive and can't be returned once they are purchased.*

1983 and later Camry, 1984 and later Pick-ups and 4-Runner (fuel-injected models only), 1987 and later Corolla (fuel-injected models only), 1985 through 1987 MR2

With the engine at normal operating temperature, turn the ignition switch On (engine Off). Use a jumper wire to bridge the two terminals of the round service connector (1984 Camry, 1987 Corolla and 1986 and earlier Pick-up) or terminals E1 and T (all other models) **(see illustrations)**. Read the diagnosis code as indicated on the "CHECK ENGINE" light on the dash. Normal system operation is indicated by code number 1 blinking on and off at a steady rate. If there is a malfunction, the light will blink the requisite number of times. After the codes have been displayed, there will be a delay, then the sequence will repeat. After making repairs, the memory must be cleared. With the ignition switch Off, remove the ECM or ECU fuse.

Toyota

Camry (1983 through 1986 models), Corolla (1987 models), Pick-ups and 4-Runner (1984 through 1987 models)

Code	Probable cause
1	Normal
2	Air Flow Meter Signal
3	Air Flow Meter Signal (1984 Trucks, 1983 through 1985 Camry)
3	No ignition signal from igniter
4	Water temperature sensor or circuit
5	Oxygen sensor or circuit
6	No ignition signal (1984 trucks, 1983 through 1985 Camry)
6	RPM signal (no signal to ECU)
7	Throttle Position Sensor (TPS) or circuit
8	Intake Air Temperature sensor or circuit
9	Vehicle Speed Sensor (VSS) or circuit (1986 Camry only)
10	Starter signal
11	Switch signal – air conditioning on during diagnosis check
12	Knock sensor or circuit
13	Knock sensor/CPU (ECU) faulty
14	Turbocharger pressure (22R-TE/Turbo 22R models) – over- boost (abnormalities in air flow meter may also be detected)

1987 and later Camry, 1988 and later models (all others)

Code	Probable cause
11	ECU (TB) momentary interruption in power supply to ECU
12	RPM signal/no NE or G Signal to ECU within several seconds after engine is cranked
13	RPM Signal/no signal to ECU when engine speed is above 1500 RPM
14	No ignition signal to ECU
21	Oxygen sensor circuit or oxygen sensor heater circuit failure

Toyota

1987 and later Camry, 1988 and later models (all others) (continued)

Code	Probable cause
22	Water temperature sensor circuit
23/24	Intake air temperature circuit
25	Air fuel ratio – lean condition indicated
26	Air fuel ratio – rich condition indicated
27	Oxygen sensor circuit (open or shorted)
31	Air flow meter or circuit
31	1989 through 1991 Corolla – vacuum sensor signal
32	Air flow meter or circuit
41	Throttle position sensor or circuit
42	Vehicle Speed Sensor (VSS) or circuit
43	Starter signal/no start signal to ECU
51	Switch signal/Neutral start switch off or air conditioning on during diagnostic check
51	Switch signal – no IDL signal, NSW or air conditioning signal to ECU (1988 through 1990 Corolla, 1988-1/2 through 1990 Camry models)
52	Knock sensor circuit
53	Knock sensor signal/faulty ECU
71	Exhaust Gas Recirculation (EGR) system malfunction

Volkswagen

Note: *Because engine management systems may differ from year-to-year, certain trouble codes may indicate different problems from one year to the next. Since this is the case, it would be a good idea to consult your dealer or other qualified repair shop before replacing any electrical component, as they are usually expensive and can't be returned once they are purchased.*

1988 through 1990 Golf, GTI and Jetta with 1.8L engine

Note: *Models with the 2.0L engine (1990 on) require a special tool to output trouble codes.*

Turn ignition switch On (engine Off), then depress and hold the "CHECK ENGINE" light rocker switch on the dash for at least four seconds. The codes will be displayed by the "CHECK ENGINE" light as a series of four-digit codes with a two or three second pause between digits. Hold the switch down an additional four seconds to retrieve any additional codes. When all of the codes have been accessed, the light will flash code 0000. The light will go On for 2.5 seconds then Off for 2.5 seconds.

To clear the memory after making repairs, turn the ignition switch Off. Hold the rocker switch down and turn the ignition switch On while continuing to hold the rocker switch for at least five seconds. Turn the ignition switch Off and release the rocker switch.

Code	Circuit	Probable cause
2142	Knock sensor	Defective sensor or circuit
2232	Air flow sensor potentiometer	Defective potentiometer or circuit.
2312	Coolant temperature sensor	Defective sensor or circuit
2322	Intake air temperature sensor	Defective sensor or circuit
2342	Oxygen sensor	Defective sensor or circuit
4444	System OK	No stored codes
0000	End of sequence	All fault codes have been displayed

3 System descriptions and servicing

1 Exhaust Gas Recirculation (EGR) system

What it does

The EGR system controls nitrogen oxide (NOx) emissions. The system works by allowing a specific amount of exhaust gas to pass from the exhaust manifold into the intake manifold to dilute the air/fuel mixture going to the cylinders.

EGR valves are designed specifically to recirculate the exhaust gas with the air/fuel mixture, thereby diluting the air/fuel mixture enough to keep the NOx compounds within breathable limits. It was discovered that short peak combustion temperatures create NOx. By blending the exhaust gas with the air/fuel mixture, scientists discovered that the rate of combustion slowed down, the high temperatures were reduced and the NOx compounds were kept within limits. Modern engines are equipped with oxidation/reduction catalysts and feedback carburetion or fuel-injection systems that keep the NOx compounds to a minimum. Even with these newer, more efficient systems, the EGR system is still necessary to reduce the excess emissions.

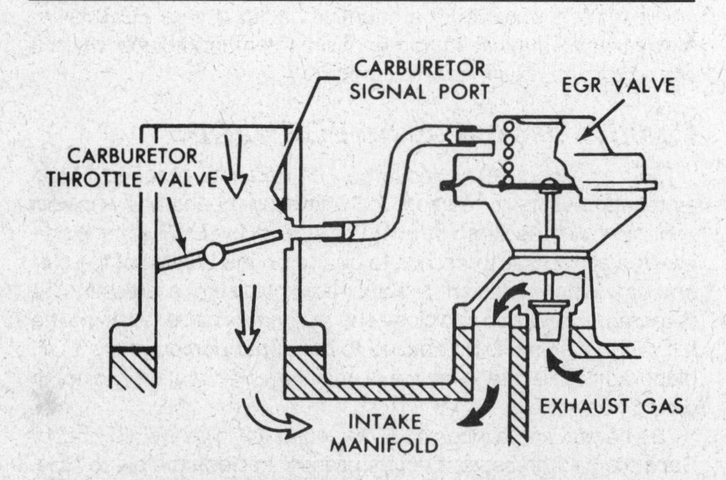

1.1 A cross-sectional diagram of an early EGR system in operation (thermostatic vacuum switch not shown for simplicity)

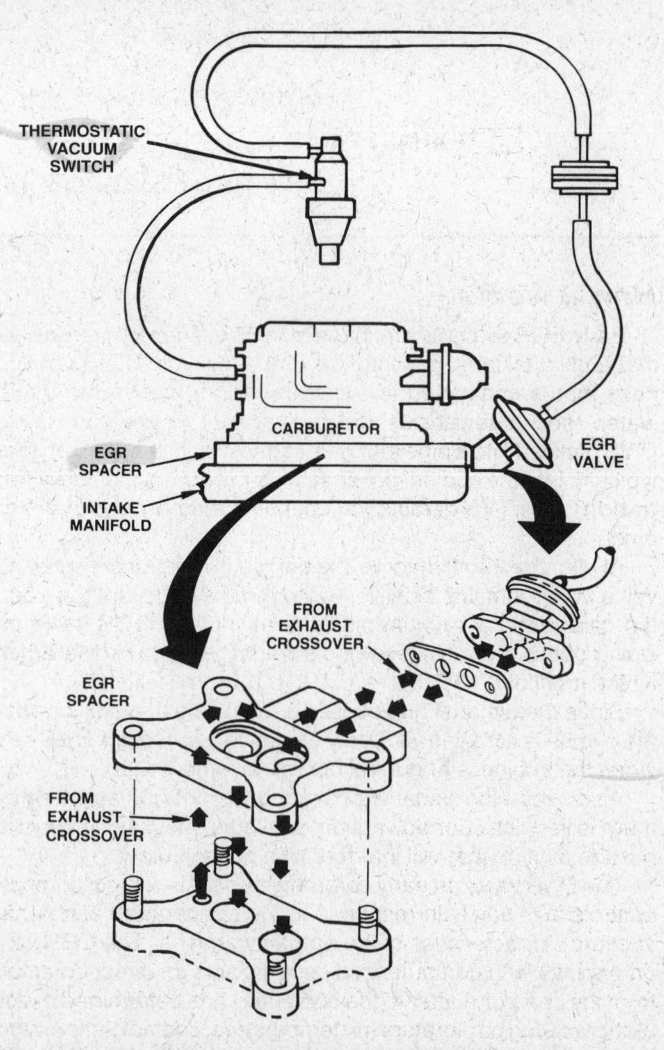

1.2 This diagram shows the thermostatic vacuum switch and exhaust flow details

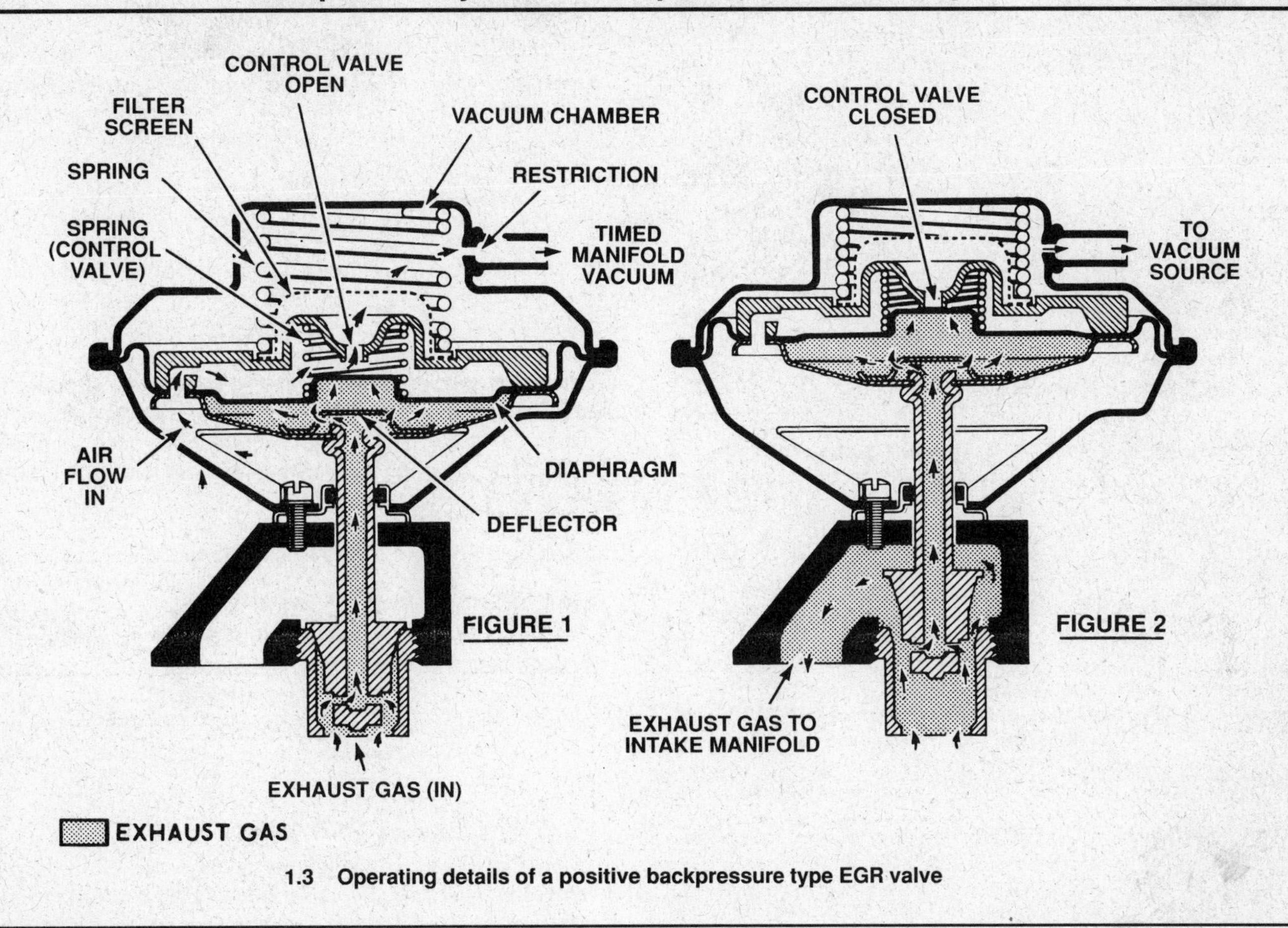

1.3 Operating details of a positive backpressure type EGR valve

How it works

Early EGR systems are made up of a vacuum-operated valve that admits exhaust gas into the intake manifold (EGR valve), a hose that is connected to a carburetor port above the throttle plates **(see illustration)** and a Thermostatic Vacuum Switch (TVS) spliced into a pipe that is threaded into the radiator or, more typically, into the coolant passage near the thermostat **(see illustration)**. The TVS detects the operating temperature of the engine.

At idle, the throttle blocks the port so no vacuum reaches the valve and it remains closed. As the throttle uncovers the port in the carburetor, a vacuum signal goes into the EGR valve and slowly opens the valve, allowing exhaust gases to circulate in the intake manifold.

Since the exhaust gas causes a rough idle and stalling when the engine is cold, the TVS only allows vacuum to the EGR valve when the engine is at normal operating temperature.

Also, when the pedal is pushed to the floor on acceleration, there is very little ported vacuum available, resulting in very little mixture dilution that will interfere with power output.

The EGR valve on early carbureted engines without computer controls acts solely in response to the temperature and venturi vacuum characteristics of the operating engine. The EGR valve on engines with computerized controls acts on direct command from the computer after it (the computer) has determined exactly all the working parameters (air temperature, coolant temperature, EGR valve position, fuel/air mixture etc.) of the engine. EGR valves on computerized vehicles normally have a computer-controlled solenoid in line between the valve and vacuum source. They also often have a position sensor on the EGR valve that informs the computer what position the EGR valve is in.

There are two common types of EGR valves; ported vacuum EGR valves and backpressure EGR valves. Besides the common ported type EGR valve described in the previous paragraphs, there are basically two types of backpressure EGR valves; The most common type is the positive backpressure valve and the other is the negative backpressure valve. It is important to know the difference between positive and negative backpressure valves because they work differently and they are tested differently. Never substitute a positive for a negative backpressure EGR valve. Always install original equipment from the manufacturer when it comes time to repair the EGR valve.

Positive backpressure EGR valve

This type of valve is used largely on domestic models. It uses exhaust pressure to regulate EGR flow by means of a vacuum control valve **(see illustration)**. The stem of the EGR valve is hollow and allows backpressure to bear onto the bottom of the diaphragm. When sufficient exhaust backpressure is present, the diaphragm moves up and closes off the control valve, allowing the full vacuum signal to be applied to the upper portion of the EGR diaphragm. This opens the valve and allows recirculation to occur during heavy loads.

Be careful not to incorrectly diagnose this type of EGR valve. Because backpressure must be present to close the bleed hole, it is not possible to operate the EGR valve with a vacuum pump at idle or when the engine is stopped. The valve is acting correctly when it refuses to move when vacuum is applied or it refuses to

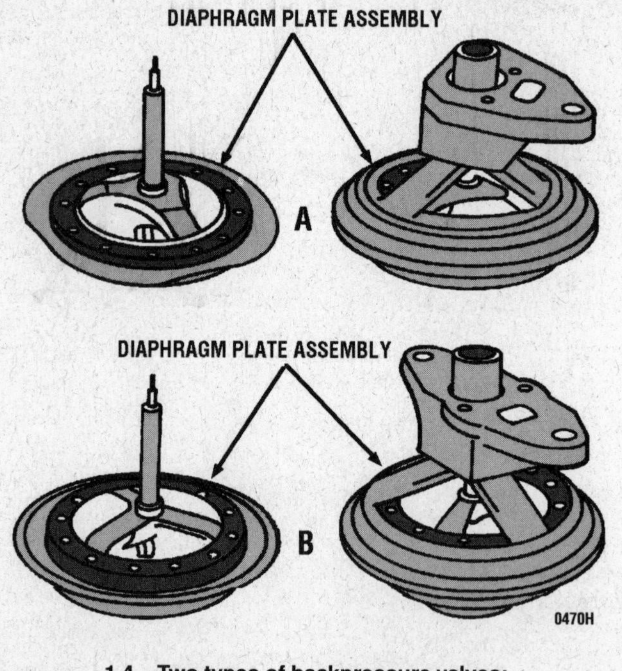

1.4 Two types of backpressure valves: (A) negative and (B) positive

hold vacuum. Remember that anything that changes the pressure in the exhaust stream will mess up the calibration of the backpressure system. This includes glass-pack mufflers, headers or even a clogged catalytic converter.

To distinguish this valve, turn the valve upside down and note the pattern of the diaphragm plate **(see illustration)**. Positive backpressure valves have a slightly raised X-shaped rib. Negative backpressure EGR valves are raised considerably higher. On some GM EGR valves, the only way to distinguish each type is by a letter next to the date code and part number. N means negative while P means positive.

Negative backpressure EGR valve

In this system, the bleed hole is normally closed. When exhaust backpressure drops (reduced load), the bleed valve opens and reduces the vacuum above the diaphragm, cutting the vacuum to the EGR valve. The negative backpressure EGR valve is similar to the positive backpressure EGR valve but operates in the OPPOSITE way. This type of valve is typically used on engines that have less than natural backpressure such as high-performance vehicles with free-flowing mufflers and large-diameter exhaust tubing.

Other types of EGR valves

Dual-diaphragm EGR valve

This EGR valve receives ported vacuum to the upper portion of the vacuum diaphragm while the lower portion receives manifold vacuum. The simultaneous response characteristics control both throttle position and engine load. The dual-diaphragm system is easily recognized by the two vacuum lines attached to the EGR valve.

Ford air pressure EGR valve

Most commonly installed on 1978 and 1979 Ford EEC-I systems, this type of EGR valve is operated by the thermactor air pump pressure instead of vacuum. Pump output is routed to the underside of the diaphragm. Some models are equipped with an EGR position sensor also.

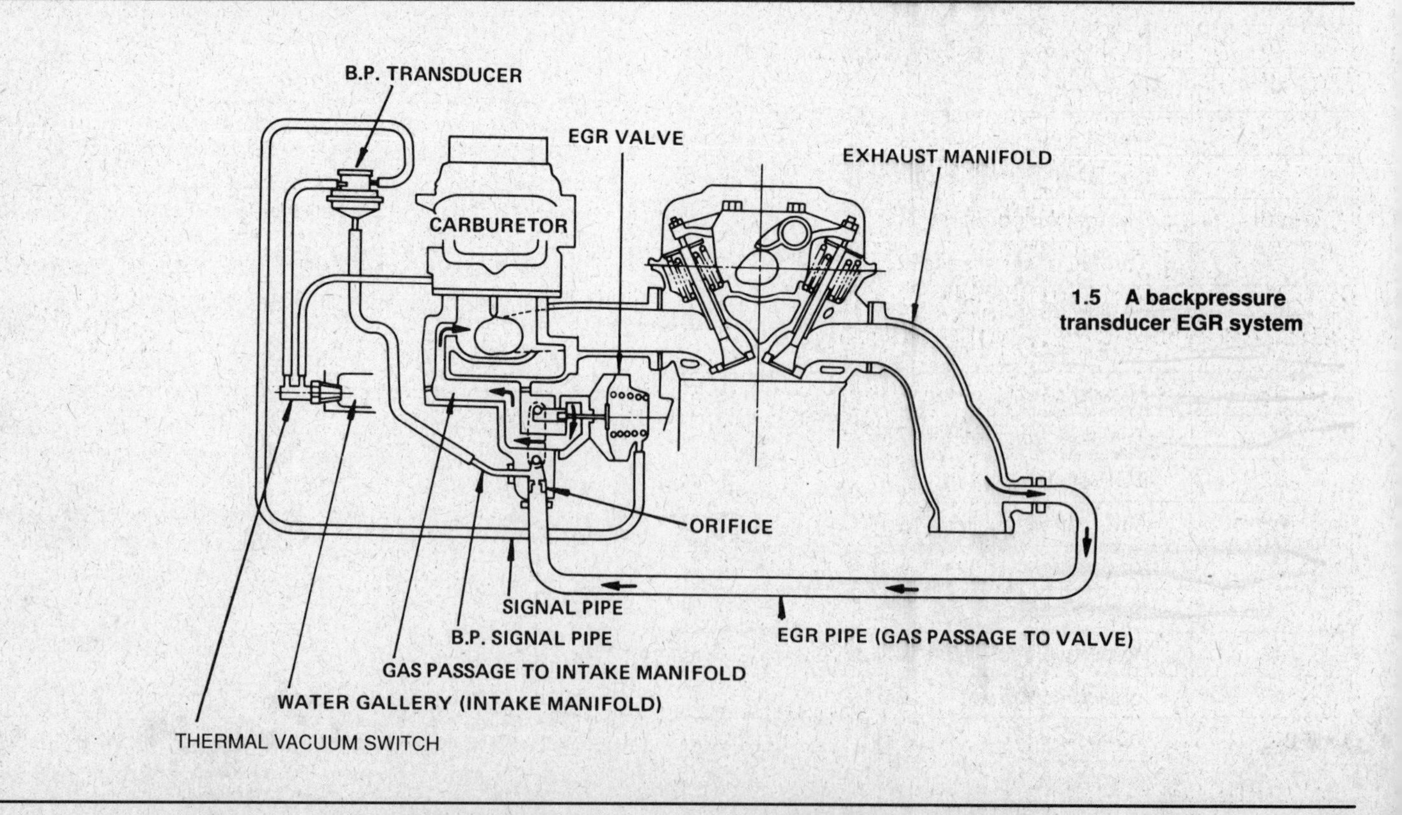

1.5 A backpressure transducer EGR system

Ford electronic control EGR valve

This EGR valve resembles the air pressure type but it is dependent on the computer and the EGR position sensors to detect the correct conditions and regulate the EGR valve angle.

Chrysler/Mitsubishi dual EGR valve

Most commonly equipped on the 2.6L silent shaft engine, this type of EGR valve uses both a primary and secondary EGR valve mounted at right angles to each other. This system allows for accurate measurement of the exhaust gases.

Computer controls

On the newer type computerized EGR systems, the EGR valve is regulated by the use of different sensors, transducers or vacuum solenoids directly linked to the EGR valve. Here is a list with a brief explanation for each type:

Remote backpressure transducer

This device is not mounted inside the EGR valve, but instead it is found in the vacuum line leading to the EGR valve **(see illustrations)**. At idle or light loads, the transducer bleeds off the signal to prevent recirculation to the EGR valve.

Electronic pressure sensor

This capacitive sensor converts exhaust system backpressure into an analog voltage signal that is sent directly to the computer for analysis. This type of pressure sensor is commonly found on newer EEC-IV Ford systems.

Venturi vacuum amplifier

Venturi vacuum from the carburetor indicates engine load and air consumption, but it is inherently too weak to transfer as information to the EGR system. By amplifying the venturi vacuum, the EGR valve is regulated by strong manifold vacuum. These systems also store vacuum in a reservoir for an extra supply when the engine is idling.

Wide open throttle valve

This device is located in-line between the EGR valve and the vacuum source. Controlled by a signal from the carburetor venturi, the wide open throttle valve bleeds off the signal to the EGR valve at wide open throttle to eliminate any mixture dilution and any power loss.

Air cleaner temperature sensor

This sensor cuts off vacuum to the EGR valve until a certain temperature is reached. Instead of reading coolant temperature, the sensor detects air temperature. This is commonly used on carburetor spark port vacuum systems.

Solenoid vacuum valve

This valve works directly with the computer to control the vacuum signal. It is found most commonly on the GM systems and it is referred to as "pulse width modulation."

Electronic vacuum regulator

Instead of the on/off function of a solenoid vacuum valve, the electronic vacuum regulator adjusts vacuum to the EGR valve by way of the pressure sensor and the computer. This device is most commonly found on Ford EEC-III and EEC-IV systems.

Delay timer

This valve interrupts the vacuum to the EGR valve to prevent stalling when the engine is cold. The actual delay time can be anywhere from 30 to 90 seconds after the engine is started. The delay timer works in conjunction with a solenoid vacuum valve.

Charge temperature switch

This switch senses the temperature of the intake system also, but it acts strictly as an ON/OFF switch to prevent current from reaching the delay timer when the temperature is below 60-degrees F. This prevents any EGR mixture and consequently rough idle or stalling when cold. This system is commonly found on Chrysler emissions systems.

1.6 This Isuzu P'up backpressure transducer is typical of many models – it's on the intake manifold, near the distributor

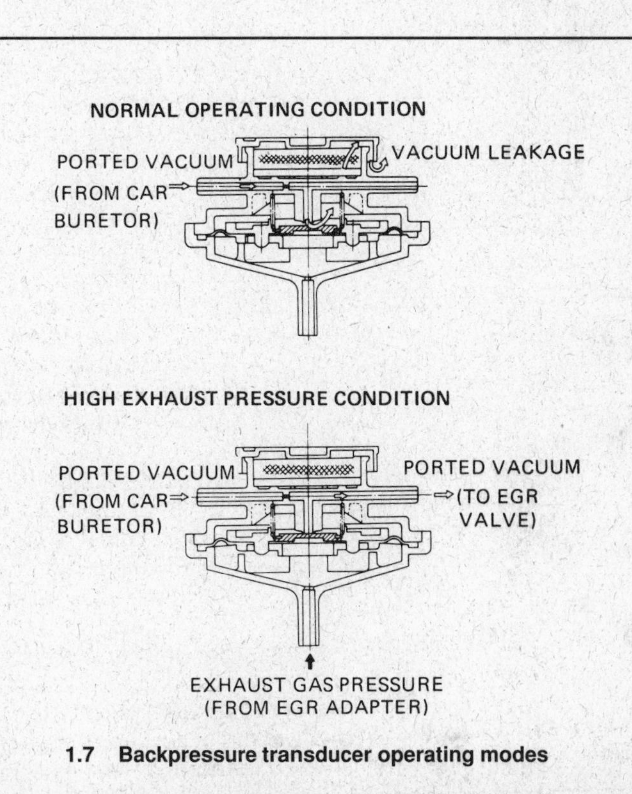

1.7 Backpressure transducer operating modes

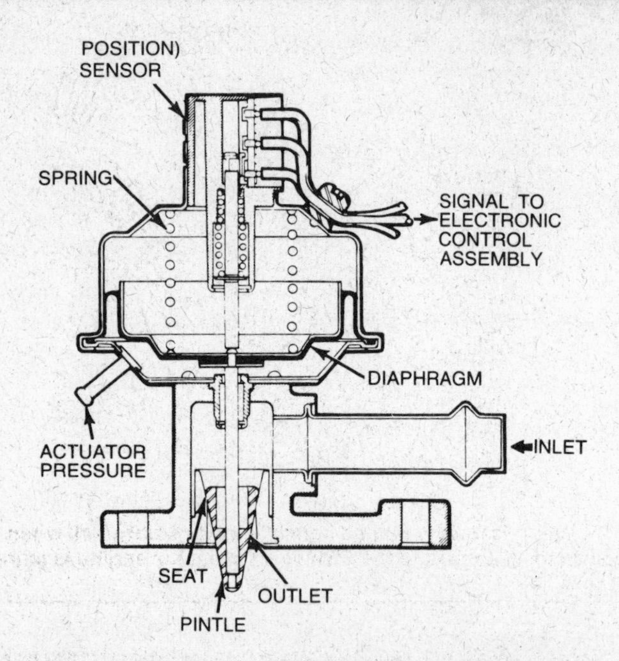

1.8 A cross-sectional view of an EGR valve equipped with an EGR valve position sensor mounted on the top – the mechanical motion of the pintle is converted to a voltage value and relayed to the computer

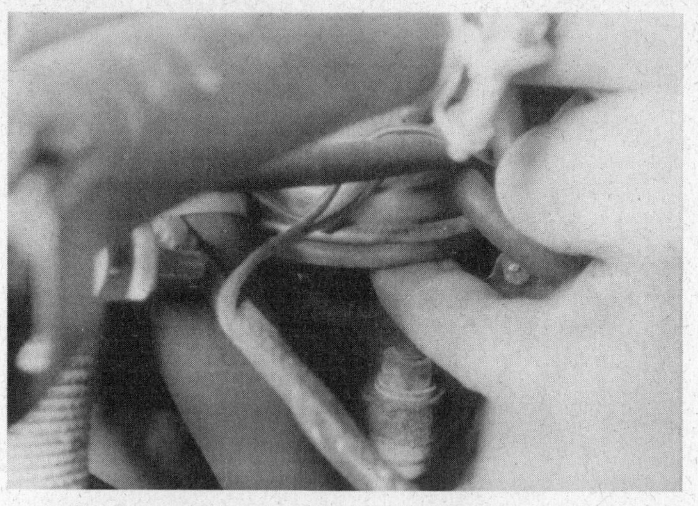

1.9 Use your finger to check for free movement of the diaphragm within the EGR valve

EGR valve position sensors

These sensors detect the exact position of the EGR valve and send the information to the computer **(see illustration)**. These sensors are discussed in detail in Section 8.

Electronic EGR valve

Some of the most recent designs (primarily from GM) employ an EGR valve that is not operated by vacuum at all – an electronic solenoid in the valve is operated electrically by the computer. Diagnosing systems with these valves is beyond the scope of the home mechanic.

Diagnosis and checking

Note: *The following procedures do not apply to vehicles equipped with electronic EGR valves. These vehicles must be taken to a dealer service department or other repair shop for diagnosis.*

Symptoms of bad EGR valves

EGR system malfunctions are recognized by common symptoms and running conditions that can be singled out with some basic testing and knowledge. The most common symptom is a rough idle or stalling when the engine is cold (and often when the engine is warmed up, too). Most problems in the EGR system are reduced to exhaust gases recirculating when they should not or not recirculating when they should. In the former case, the addition of exhaust gas to the intake air/fuel mixture when the engine is cold will make the engine run rough or even cause it to stall. On the other hand, without the exhaust gases recirculating after the engine is warm, the engine will react with detonation, increased combustion temperatures and a high output of NOx compounds.

One advantage with having your vehicle tested for emissions is that most professional mechanics will analyze the high NOx compound readings and make a quick check on the EGR valve. The test will vary the emissions levels if in fact the EGR valve and/or system components are faulty. Then it is just a matter of changing the faulty parts.

On the other hand, it is sometimes difficult to distinguish a fuel system problem from an EGR system problem. One typical example of this was a fellow who owned a Ford van with a 302 engine. After he had accelerated and passed a car on the interstate highway and slow downed to pull off the exit, the engine would start to mis-fire and sometimes even stall. Well, at this point he would pull over to the breakdown lane and turn the engine OFF and start it up again only to have the engine running smooth once more. After a multitude of tune-ups, carburetor overhauls and ignition system checks, a mechanic suggested he pull over into the breakdown lane, but instead of turning the engine OFF, simply pull the vacuum line off the EGR valve and see if the engine smooths back out. To make a long story short, the EGR valve (backpressure type) was sticking in the open position causing the exhaust gases to flow back into the intake system at low speed which in turn caused the rough engine conditions. When the fellow shut the engine OFF, the pressure causing the EGR valve to stick was removed and it closed itself. He replaced the EGR valve and the engine performed smoothly at all speeds.

Checking EGR systems

There are several basic system EGR checks that you can perform on your vehicle to pinpoint any problems. If the EGR valve stem is accessible, push it up or down (against spring pressure) to see if it can move and operate freely **(see illustration)**. If it is stuck, remove the valve and clean it thoroughly. If it still does not budge, replace it with a new unit.

If the EGR valve stem moves smoothly and the EGR system continues to malfunction, check for a pinhole vacuum leak in the diaphragm of the EGR valve. Obtain a can of spray carburetor cleaner and attach the flexible "straw" to the tip. Aim carefully into the diaphragm areas of the EGR valve and spray around the actuator shaft while the engine is running. Listen carefully for any changes in engine rpm. If there is a leak, the engine rpm will increase and surge temporarily. Then it will smooth back out to a constant idle. The only way to properly repair this problem is to replace the EGR valve with a new unit.

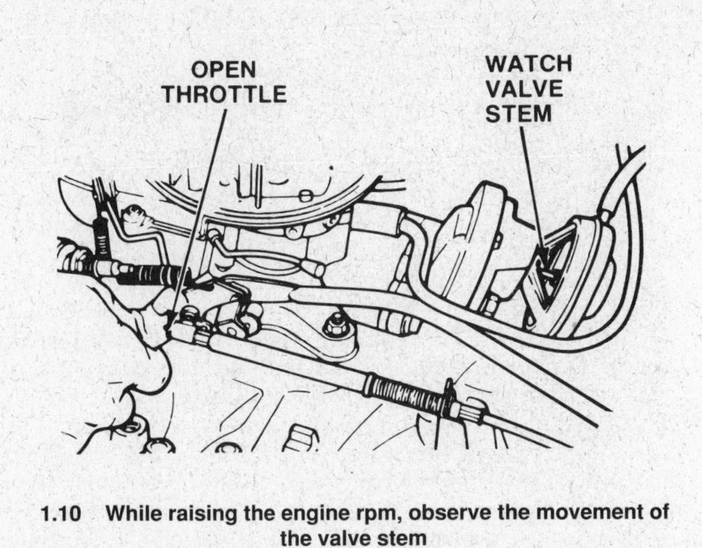

1.10 While raising the engine rpm, observe the movement of the valve stem

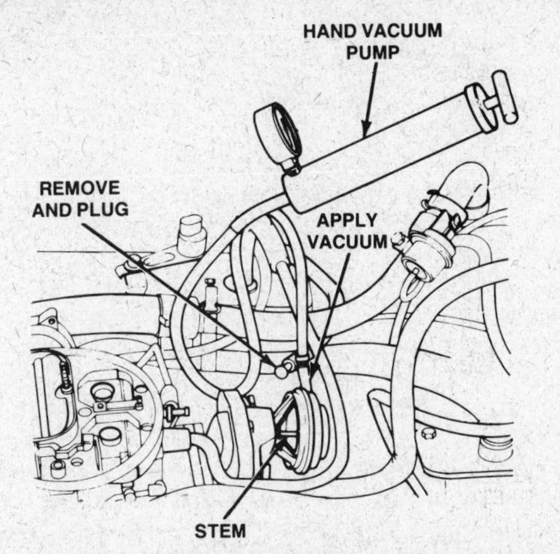

1.11 The engine should stumble and possibly stall when vacuum is applied to the EGR valve when the engine is idling

Another method is a visual check. After the engine has been warmed up to normal operating temperature, open the throttle to approximately 2,500 rpm and observe the EGR valve stem as it moves with the rise in engine rpm **(see illustration)**. Use a mirror if necessary. If it doesn't move, remove the vacuum hose and check for vacuum with a gauge or just feel with the tip of your finger. **Note:** *On the Ford pressure-activated EGR valves, you should feel pressure instead. If there is no vacuum or pressure, check the EGR system controls.*

Test the EGR valve with the engine running and at normal operating temperature. This test will tell you if the gas flow passages are open and if the gas flow is proper. Remove the vacuum line from the EGR valve and plug it with a golf tee or other suitable device. Attach a hand vacuum pump to the EGR valve **(see illustration)**. With the engine idling, slowly apply 15 in-Hg of vacuum to the valve and watch the valve stem for movement. If the gas flow is good, the engine will begin to idle rough or it may even stall. If the stem moves but the idle does not change, there is a restriction in the valve (see *Cleaning the EGR valve*, below), spacer plate or passages in the intake manifold. If the valve stem does not move or the EGR valve diaphragm does not hold vacuum, replace the EGR valve with a new part.

Another common check is for the thermostatic vacuum switch (TVS). This switch is usually regulated by a bimetal core that expands or contracts according to the temperature. The valve remains closed and does not operate as long as the coolant temperature is below 115 to 129-degrees F. As the coolant temperature rises, the valve will open and the EGR system will operate. Remove the switch and place it in a pan of cool water and check the valve for vacuum **(see illustration)** – vacuum should

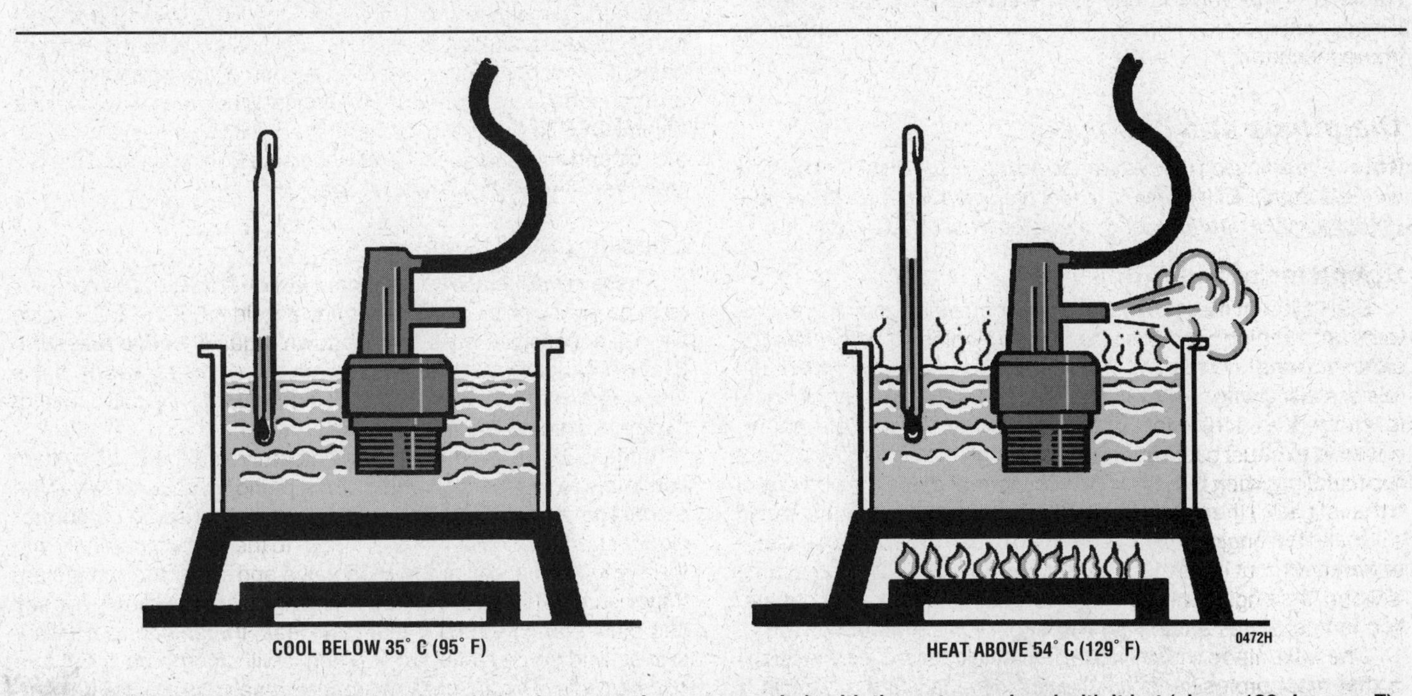

1.12 Check to see if vacuum flows through the thermostatic vacuum switch with the water cool and with it hot (above 129-degrees F)

not pass through the valve. Heat the water to the the specified temperature (over 129-degrees F) and make sure the valve opens and allows the vacuum to pass. If the switch fails the test, replace it with a new part.

A vacuum gauge can be used to check for excessive exhaust backpressure by observing any vacuum variation. Disconnect a vacuum line connected to an intake manifold port. Install a vacuum gauge between the disconnected vacuum line and the intake manifold port. Block the wheels and set the parking brake. Start the engine and gradually increase speed to 2,000 rpm with the transmission in Neutral. The reading from the vacuum gauge should be above 16 in-Hg. If not, there could be excessive backpressure in the exhaust system. To verify an excessive backpressure problem or a vacuum leak, perform the following:

1 Turn the ignition key OFF.
2 Disconnect the exhaust system at the exhaust manifold.
3 Start the engine (despite the loud exhaust roar) and gradually increase the engine speed to 2,000 rpm.
4 The reading from the exhaust manifold vacuum gauge should be above 16 in-Hg.
5 If 16 in-Hg. is not attained, the exhaust manifold may be restricted (or the valve timing or ignition timing may be late, or there could be a vacuum leak).
6 If 16 in-Hg. is attained, the blockage is most likely in the muffler, exhaust pipes, or catalytic converter. Also, if the catalytic converter debris has entered the muffler, have it replaced also.

Remember, don't condemn a backpressure-type valve until you're sure the exhaust system is stock, has no leaks and it is not clogged or restricted. Also, don't try to get the positive backpressure-type to hold vacuum with the engine off or idling.

Cleaning the EGR valve

The bottoms of EGR valves often get covered with carbon deposits, causing them to restrict exhaust flow or leak exhaust. The valve must be removed so the bottom can be cleaned. There are important points that must be observed when cleaning EGR valves: Never use solvent to dissolve deposits on EGR valves unless you are extremely careful not to get any on the diaphragm. Clean the pintle and valve seat with a dull scraper and wire brush and knock out loose carbon by tapping on the assembly. Some EGR valves can be disassembled for cleaning, but be sure the parts are in alignment before assembly.

2 Evaporative emissions control (EVAP, EEC or ECS) system

What it does

Evaporating fuel accounts for up to 20% of a vehicle's potential pollution, so since 1971 Federal law has required Evaporative emissions control systems on most vehicles. The system traps fuel vapors that would normally escape into the atmosphere and re-routes them back into the engine where they are burned.

The system consists of a fuel tank with an air space for heat expansion that allows the vapors to collect and flow to the charcoal canister, the tank cap and associated hoses and tubes. The cap contains a check valve to provide pressure and vacuum relief to the system. On carbureted models, the float bowl has a vent which connects to the canister by a tube.

How it works

With the engine off, the vapors flow from the tank (and carburetor float bowl, on models so equipped) to the canister where they are absorbed in the charcoal until the engine is started. When the engine is running, the vapors are then purged from the canister and routed to the intake manifold or air cleaner and into the combustion chambers where they are burned.

The system operates using a purge control valve which allows engine vacuum to suck the vapors from the canister at the appropriate time while outside air enters the canister by way of a tube or filter **(see illustration)**. This purge valve is usually mounted on the canister body, but can also be located remotely or in a hose. Some earlier models have an air intake and filter at the bottom of the canister.

The operation of the purge valve is controlled on some models by solenoids and/or delay valves that make sure the vapors will be purged when the engine can burn them most efficiently. On later models, the system is controlled by the computer and operates in slightly different ways, depending on manufacturer.

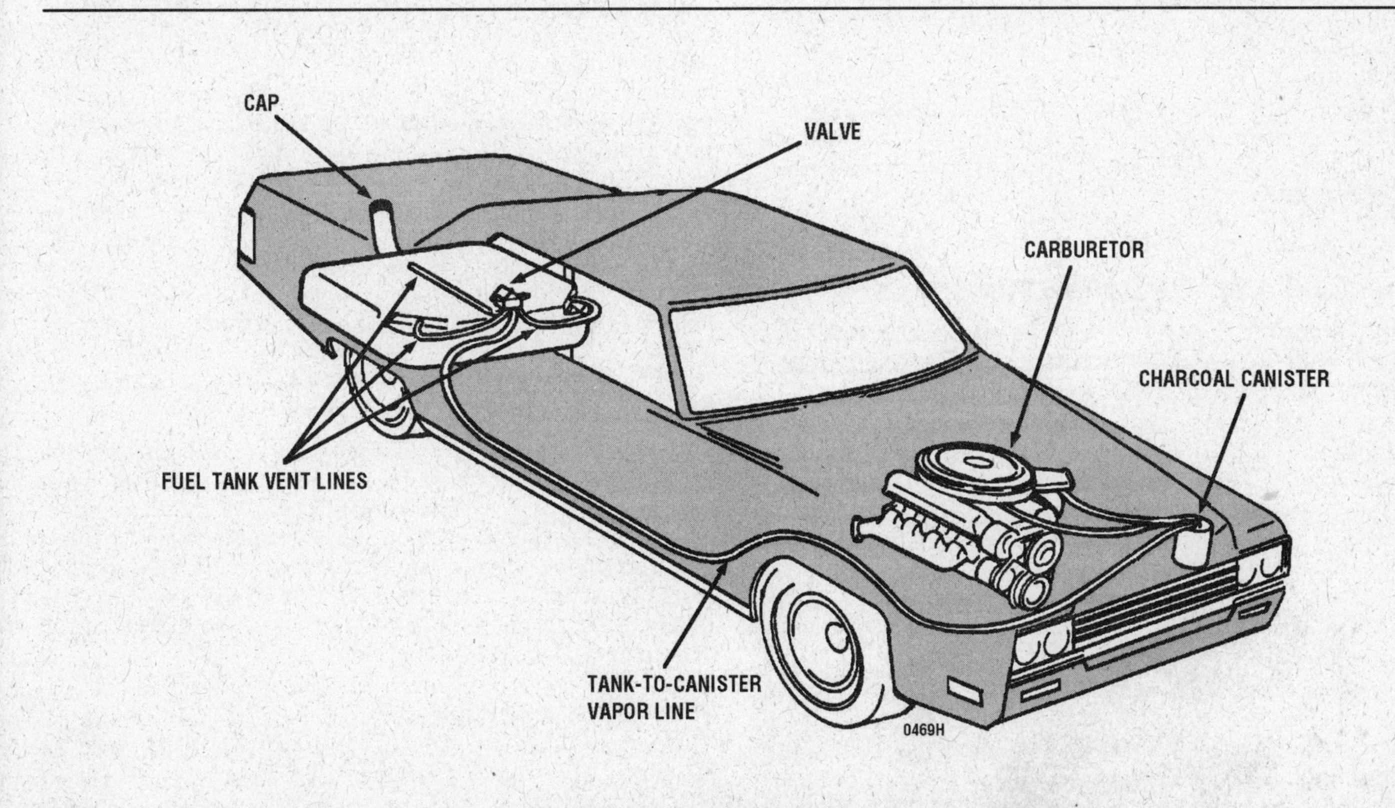

2.1 Evaporative emissions control system details (typical)

2.2 Apply vacuum to the signal hose port on the purge control valve and check that air flows through the purge hose port (the larger port below the signal hose port which is shown with the hose still connected)

On Chrysler models with the Single Module Engine Controller (SMEC) system, the controller grounds a solenoid when the engine is below operating temperature so no vacuum can reach the purge valve. When operating temperature is reached, this solenoid is de-energized so vacuum can then purge the fuel vapors through the fuel injection system or carburetor.

The computer on later model GM vehicles also uses a solenoid valve to operate the purge valve when the engine is hot, after it has been running for a specified period, and at certain speeds and throttle positions. The purging increases until the computer receives a rich fuel condition signal from the oxygen sensor (the vapors are burned), then is regulated until the signal decreases.

Operation of the Ford EEC IV system is similar to the GM system. It purges whenever the engine is at normal operating temperature and off idle.

Checking and component replacement

Note: *Symptoms of problems with the evaporative emissions control system include poor idle, stalling, generally poor driveability and a strong gasoline smell.*

Canister

The canister is usually found in the engine compartment, but may also be located under the vehicle. Some models have more than one canister.

Look the canister over for cracks and damage. If the canister is cracked or the inside is soaked with gas, it will have to be replaced. Reach underneath to see if there is a filter in the bottom (later model canisters are sealed). If the fiberglass filter is dirty, it will have to be replaced. Most auto parts stores carry inexpensive emissions replacement parts such as filters. You'll probably have to go to a dealer for a replacement canister, but some well-equipped auto parts stores may stock them. To replace the canister, mark the hoses with tape before disconnecting them. Remove the mounting bolts or disconnect the clip and detach the canister.

Purge valve

There are many variations in the operations of the purge valve, but a simple check to determine if it's working can be made using a hand-operated vacuum pump. With the engine at normal operating temperature, remove the vacuum signal hose (the hose that's usually near the top of the valve and connected to engine vacuum) and the larger purge hose from the valve. Apply suction to the purge hose port on the valve – the valve should hold suction. Apply about 16 in-Hg of vacuum to the signal hose port on the valve and again apply suction to the purge hose port on the valve – the valve should flow air (not hold suction) **(see illustration)**. If the valve does not perform as described, it is probably faulty. Never apply vacuum to the carburetor bowl vent hose outlet of the valve.

Replacing a purge valve located in a hose is a simple matter of detaching the valve (note which direction it faces) from the hoses. A faulty canister-mounted valve will probably require replacement of the canister.

Purge solenoid

Follow the hoses from the top of the canister until you locate the purge solenoid (sometimes the solenoid is located on top of the canister). These solenoids are usually controlled by the computer or other devices so about the only check that can be made is to determine that the connector is securely plugged in and the vacuum hose is not damaged or leaking. Further testing will have to be done by a mechanic or a dealer service department. To replace, unplug the hose and electrical connector and detach the solenoid.

Carburetor float bowl vent valve

Inspect the hose between the carburetor and the canister for cracks or damage. If you suspect that a solenoid-type bowl vent valve is malfunctioning, you'll have to remove it from the carburetor, then apply about ten volts to jumper wires inserted in the connector. The valve should close, preventing air from passing through. If it doesn't, replace the solenoid with a new one.

Vacuum-actuated vent valves are mounted remotely from the carburetor and the charcoal canister. Although these valves operate in different ways, depending on design, they are generally kept closed by engine vacuum. Disconnect the vacuum source hose, connect a hand vacuum pump and make sure the valve holds vacuum. If it doesn't, mark and detach the hoses and install a new valve.

Filler cap and relief valve

Remove the cap. On some models it's possible to detach the valve by unscrewing it from the cap. Look for a damaged or deformed gasket and make sure the relief valve is not stuck open. If either is not in good condition, replace the filler cap with a new one. Replacement caps are generally available at auto parts stores. Make sure you get the right cap because the wrong one may not let the system vent properly, causing fuel starvation and could even collapse of the fuel tank.

Vacuum delay valve

If the engine is hard to start when hot, the delay valve could be faulty. Disconnect the vacuum hose, connect a vacuum pump and see if the valve will hold vacuum. If it doesn't, replace the canister with a new one.

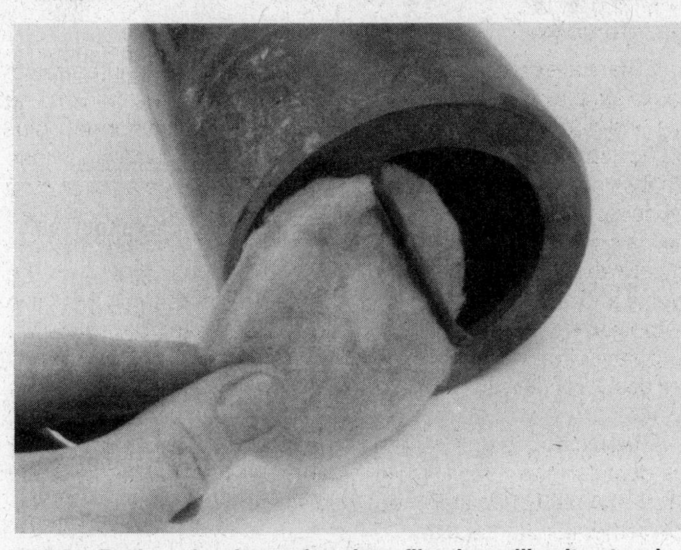

2.3 Replace the charcoal canister filter by pulling it out and inserting a new one

Canister filter

Some models are equipped with a filter at the bottom of the charcoal canister which should be replaced when it gets dirty **(see illustration)**.

Hoses

Trace the hoses leading to and from the canister to make sure they aren't disconnected or split. Your sense of smell can be a valuable diagnostic tool here because gasoline and it's residue have a strong smell which can tip off the location of even a small crack in a hose.

Always mark the hoses with tape before disconnecting them because even one misrouted hose can cause major problems. Make sure to use only hoses designed for fuel system use.

3 Positive Crankcase Ventilation (PCV) system

What it does

When the engine is running, a certain amount of the fuel/air mixture escapes from the combustion chamber past the piston rings into the crankcase as blow-by gases. The Positive Crankcase Ventilation (PCV) system is designed to reduce the resulting hydrocarbon emissions (HC) by routing them from the crankcase to the intake manifold and combustion chambers, where they are burned during engine operation.

How it works

The PCV system is basically a check valve with hoses for directing crankcase blowby back into the combustion chambers in the engine. It consists of a hose which directs fresh air from the air cleaner into the crankcase, the PCV valve (basically a one-way valve that allows the blow-by gases to pass back into the engine) and associated hoses **(see illustration)**. On some models a separate filter for the PCV system is located in the air cleaner housing. Some models have a fixed orifice (usually in a hose) instead of a PCV valve that must be kept clear or rough idling and stalling can result.

FRESH AIR HOSE

PCV VALVE

3.1 A typical PCV system component layout

Checking

Note: *Symptoms of problems with the PCV system include rough idling or high idle speed and stalling. A clogged PCV system can cause oil leaks around the PCV valve, oil filler and dipstick.*

PCV valve

The PCV valve is usually located in the valve cover, in the intake manifold, in the oil filler cap or on the side of the engine block. Remove the valve and shake it; it should rattle freely **(see illustration)**. If it is stuck or dirty, replace it with a new one. To check the valve operation with the engine running at idle, pull the PCV valve out of the mount and place your finger over the valve inlet. A strong vacuum will be felt and a hissing noise will be heard if the valve is operating properly **(see illustration)**. Replace the valve with a new one if it is not functioning as described. Do not attempt to clean the old valve. PCV valves for most models are available inexpensively at auto parts stores.

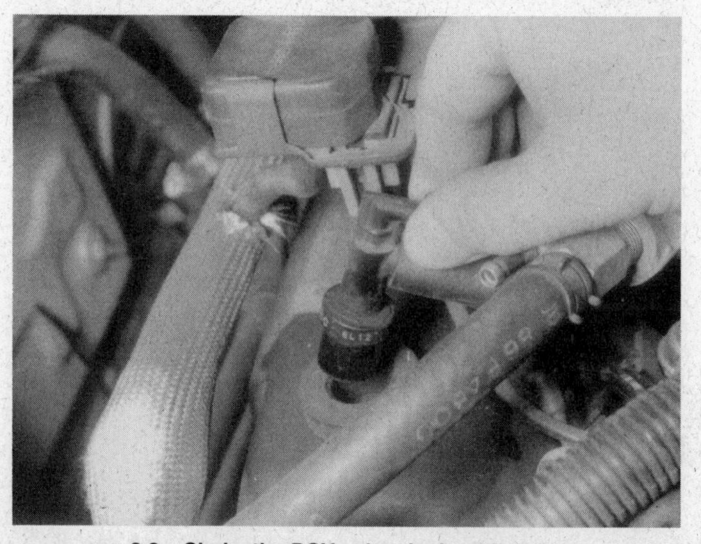

3.2 Shake the PCV valve; it should rattle

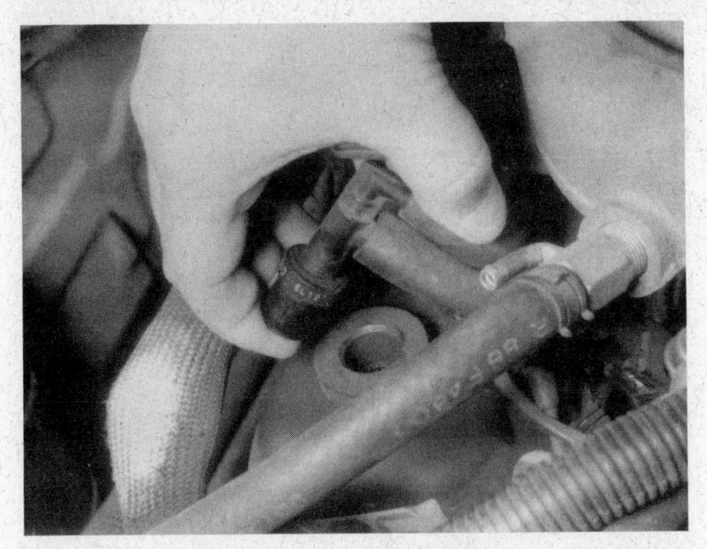

3.3 With the engine running, put your finger over the end of the PCV valve; you should feel vacuum

3.4 On some models, you can pull the PCV filter out with your fingers

Hoses

Trace the hoses from the valve cover to the other connections and check for cracks and leaks. Mark the hose connections with tape so you know where they go and disconnect them. Check the hoses to make sure they aren't clogged (a major PCV system problem). Clean a clogged hose by blowing compressed air through it or by using a long wire and solvent. If any hoses are soft, collapsed, cracked or deteriorated, replace them with new ones. Be sure to use oil-resistant hose of the same type. This hose is available in bulk at auto parts stores, although you may have to get replacement molded hoses at a dealer.

PCV filter

Whenever the air cleaner element is replaced, check the PCV filter **(see illustrations)**. Removable filters can be washed in solvent, squeezed out and reinstalled. If the filter is clogged, replace it with a new one. They are available at most auto parts stores. Filters made of wire mesh can simply be cleaned, re-oiled and reinstalled.

3.5 On this type of PCV filter, the housing is held in place by a clip; it should be replaced as a unit

4 Air injection systems

What it does and how it works

This Section deals with the air injection systems that are present on earlier carbureted engines as well as some updated computerized engines. The air injection system on most vehicles is simply a specialized series of components (e.g., an air pump, pulley, drivebelt, injection tubes and several different types of air management valves) attached to the engine for the purpose of injecting air into the exhaust downstream from the exhaust ports to help complete the combustion of any unburned gases after they leave the combustion chamber **(see illustrations)**.

The formal names of these systems include the "AIR" (Air Injection Reaction) by GM, "Thermactor" by Ford and "Air Injection" by Chrysler. Mechanics commonly refer to this system as the "smog pump."

Regardless of the names, their functions are the same. The air

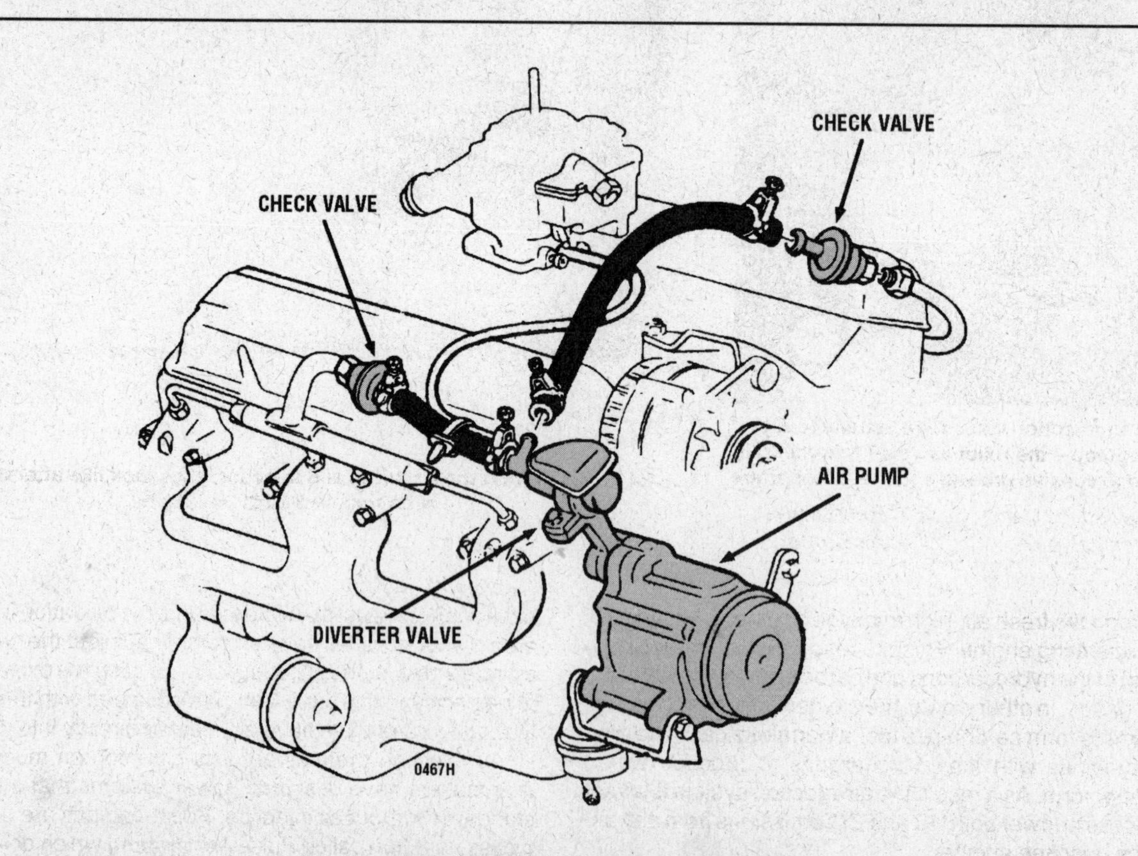

4.1 A typical air injection system on a GM vehicle

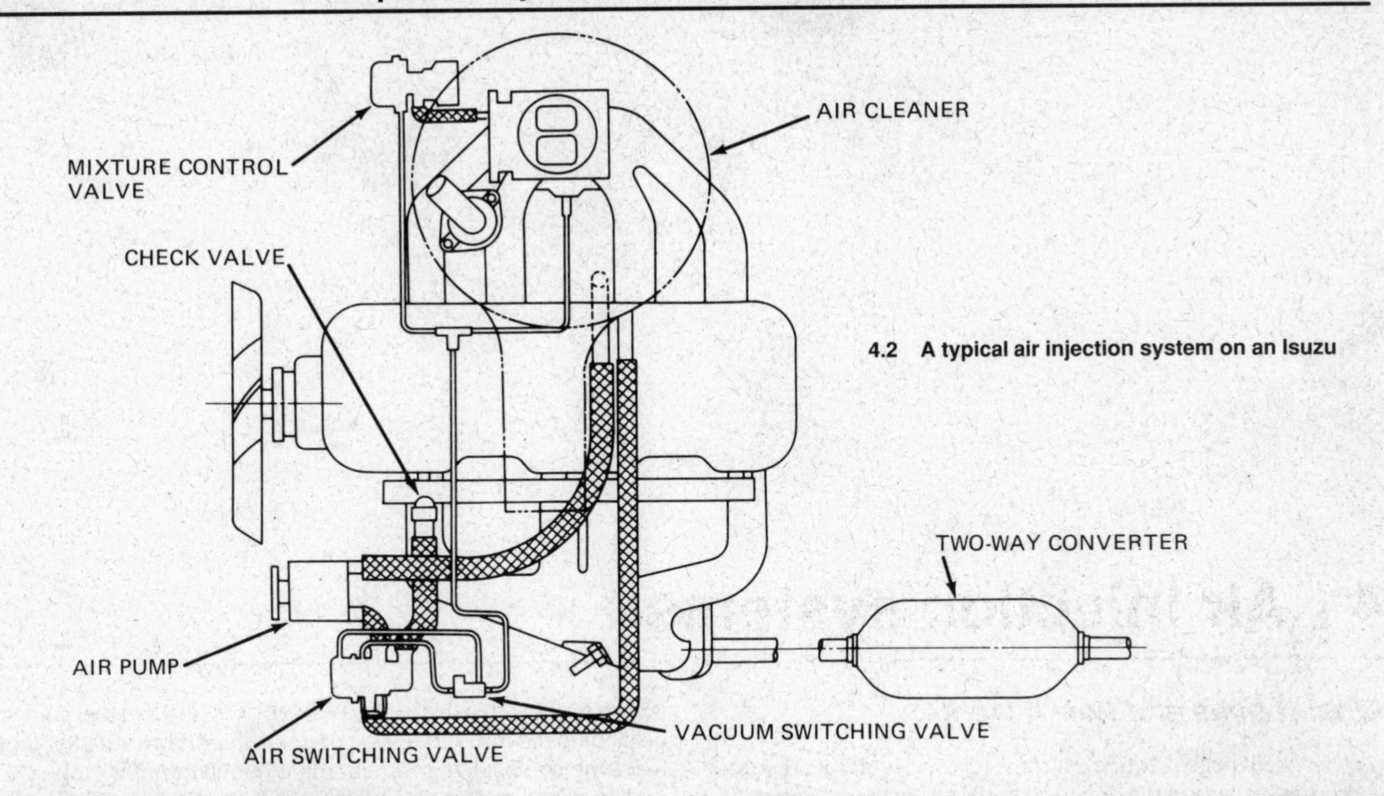

4.2 A typical air injection system on an Isuzu

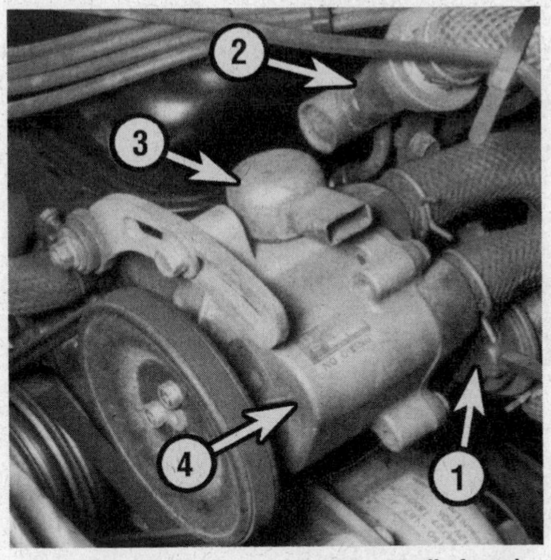

4.3 The air injection valves are usually found near the air pump – the relief valve on this pump is used to vent excessive pressure to the atmosphere

1 Air switching valve
2 Check valve
3 Relief valve
4 Air pump

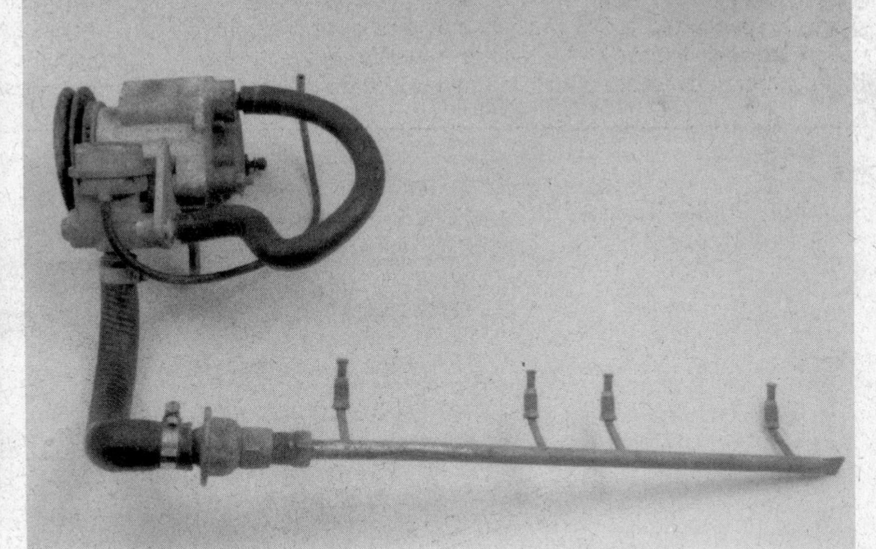

4.4 This is what the air pump and injection lines look like after they have been removed from the engine

injection introduces fresh air, high in oxygen content, into the exhaust of an operating engine. This process causes further oxidation (burning) of the hydrocarbons and carbon monoxide left in the hot exhaust gases. In other words, the oxygen unites with the carbon monoxide to form carbon dioxide, a harmless gas. The oxygen also combines with the hydrocarbons to produce water, usually in vapor form. As a result, the air injection system is a very efficient process to lower both HC and CO emissions from any automotive type gasoline engine.

In some vehicles, the air injection system directs the air into the base of the exhaust manifold to assist the oxidation process in this area. Other vehicles have systems that inject the air through the cylinder head, at the exhaust ports, causing the oxidation process to begin within this area. Vehicles equipped with three-way catalytic converters often have air injected directly into the converter.

Air injection systems are less common on modern engines. Automakers have designed newer systems that meet emission standards without air injection. Some engines are equipped with a passive (often called Pulse Air) system, which does not use an air pump. In this system, positive and negative exhaust pressures

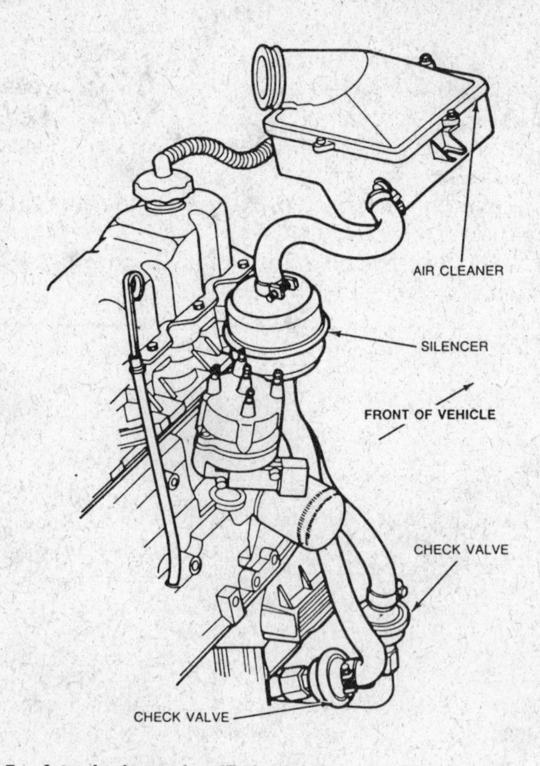

4.5 A typical passive (Pulse air) system (this one's from a Ford) – this system does not use an air pump

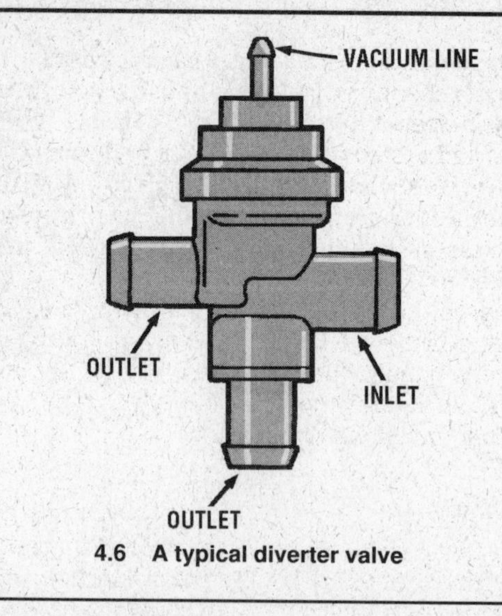

4.6 A typical diverter valve

pull air into the exhaust system by way of special reed or check valves **(see illustration)**.

All things considered, the different systems serve one common purpose: to inject air into the exhaust system and help burn any fuel that did not ignite while in the combustion chamber.

By today's standards, the "smog pump" is considered a performance robber or gas mileage reducer. For many years it was quite common to find many of the air injection systems completely removed from the engine and the exhaust ports capped with brass plugs! Eventually, the law required all air injection systems to be intact and ready to perform as originally intended. It became necessary to dig into the garage or even hunt wrecking yards for the pump, hoses and valves that were originally installed on the vehicle.

Checking the air injection system

When the air injection system is working correctly, you should hear a slight whirring sound that rises in pitch as the engine rpm's increase. Common problems associated with an air injection system are excessively noisy pumps, screeching or whining drivebelts, backfiring (diverter valve), or failing the emissions test.

Another common problem is an exhaust leak in or around the air injection components. First check all the fittings in the exhaust manifold that attach the injection lines. Stripped threads or broken tubes (usually at the bends) cause the exhaust gases to enter the engine compartment and eventually the passenger compartment. Next check the exhaust manifold itself for cracks, warpage or burned-out gaskets. Replace any necessary component to seal the exhaust system.

Checking air injection systems usually involves checking the valves (check valve, diverter valve, bypass valve and switching valve) for proper functions and response. Each different air injection system has a different combination of these valves, so it is best to familiarize yourself with their differences.

Diverter and switching valves

There are instances where the injected air can cause problems. When the engine is decelerating, the fuel mixture in the combustion chamber is rich and air injected into the exhaust can cause backfiring.

To prevent this problem, the system requires a diverter valve. This valve reroutes the injection system air away from the exhaust system during deceleration. The diverter valve is located downstream of the air pump. On some vehicles the diverter valve is mounted on the fenderwell or on the firewall and in combination with other types of management valves.

On some GM engines, the switching valve is located downstream of the diverter valve. On computer-controlled models, this device switches air to the catalytic converter when the engine is in closed loop. In open loop, the switching valve sends air to the exhaust manifold. The main purpose is to get the engine management system into closed loop (normal) operation as soon as possible by heating the oxygen sensor.

Typically, the air must pass through the diverter valve before it goes anywhere else in the system. Most of the time the air is directed one way, except when the vehicle is decelerating. In this situation, the air gets diverted into the air cleaner or into the atmosphere.

There are different methods of controlling the diverter valve, depending on the age of the system. On older carbureted engines, a vacuum line running from the carburetor signals the valve to switch airflow while on computerized engines, the computer decides when the best time is depending upon the information it receives from the throttle position sensor, temperature sensor etc.

Diverter valves are commonly called "anti-backfire" valves. When the valve is not working properly, the engine will sputter and pop as the vehicle is decelerating. Also, the valve might get stuck in the divert position and cause the CO and HC levels to increase abnormally high. In either case, check carefully to make sure the diverter valve is not stuck in any one position!

Another quick check is to find out if there is an ample amount of fresh air coming from the air pump. Squeeze the main hose that sends fresh air to the diverter valve and feel for a steady pulsation that increases when the engine rpm's are increased. Also, check the outgoing hoses after the diverter valve for a steady pulsation of air as it gets channeled into the exhaust system (acceleration) or into the air cleaner (deceleration).

On older carbureted engines, remove the vacuum line **(see illustration)** to the diverter valve and check to make sure that the pressurized air does not get diverted but instead continues to flow into the exhaust system. Install the vacuum line and check that the air is diverted on deceleration.

On newer computerized engines, check the electrical connections on the switching solenoids to verify complete contact and proper voltage signals. These systems might suffer from electronic problems, so we recommend they be diagnosed by a dealer service department or other repair shop.

Check valve

The check valve is a simple device that allows air to flow one direction (to the exhaust manifold) but not the other direction (exhaust backpressure toward the air pump) **(see illustration)**. If a check valve fails in an air pump system, exhaust gas will escape and contaminate the pump and hoses. This can also cause backpressure loss and offset the EGR valve operation.

Check valve failure is obvious when exhaust soot is detected in and around the air pump. To test a check valve, remove it and blow through it toward the manifold end, then attempt to suck back through the valve. Normal air flow should be in one direction only.

Bypass valve

Bypass valves and combination bypass/diverter valves can cause a loss of air to the exhaust if they fail in the bypass mode. If they fail open, they can put air into the exhaust system when it is rich. This often causes backfire and can melt the catalytic converter substrate.

Check a normally closed bypass valve **(see illustration)** with the engine at a fast idle. Remove the small vacuum hose from the diaphragm upper portion of the bypass valve and feel for air to vent out the bottom (to the atmosphere).

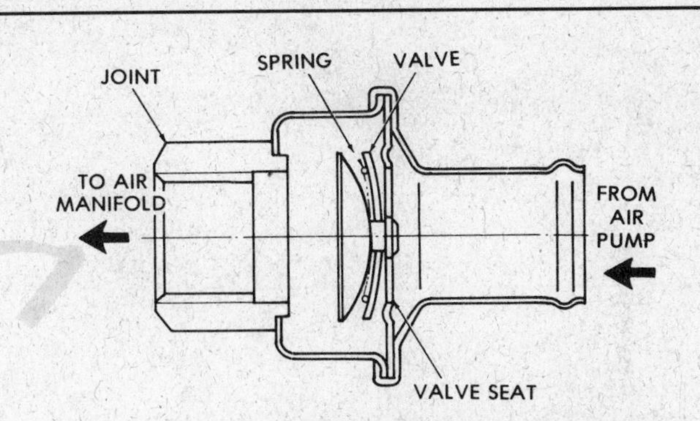

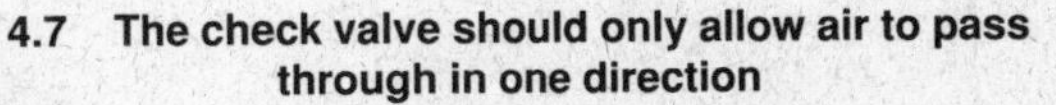
4.7 The check valve should only allow air to pass through in one direction

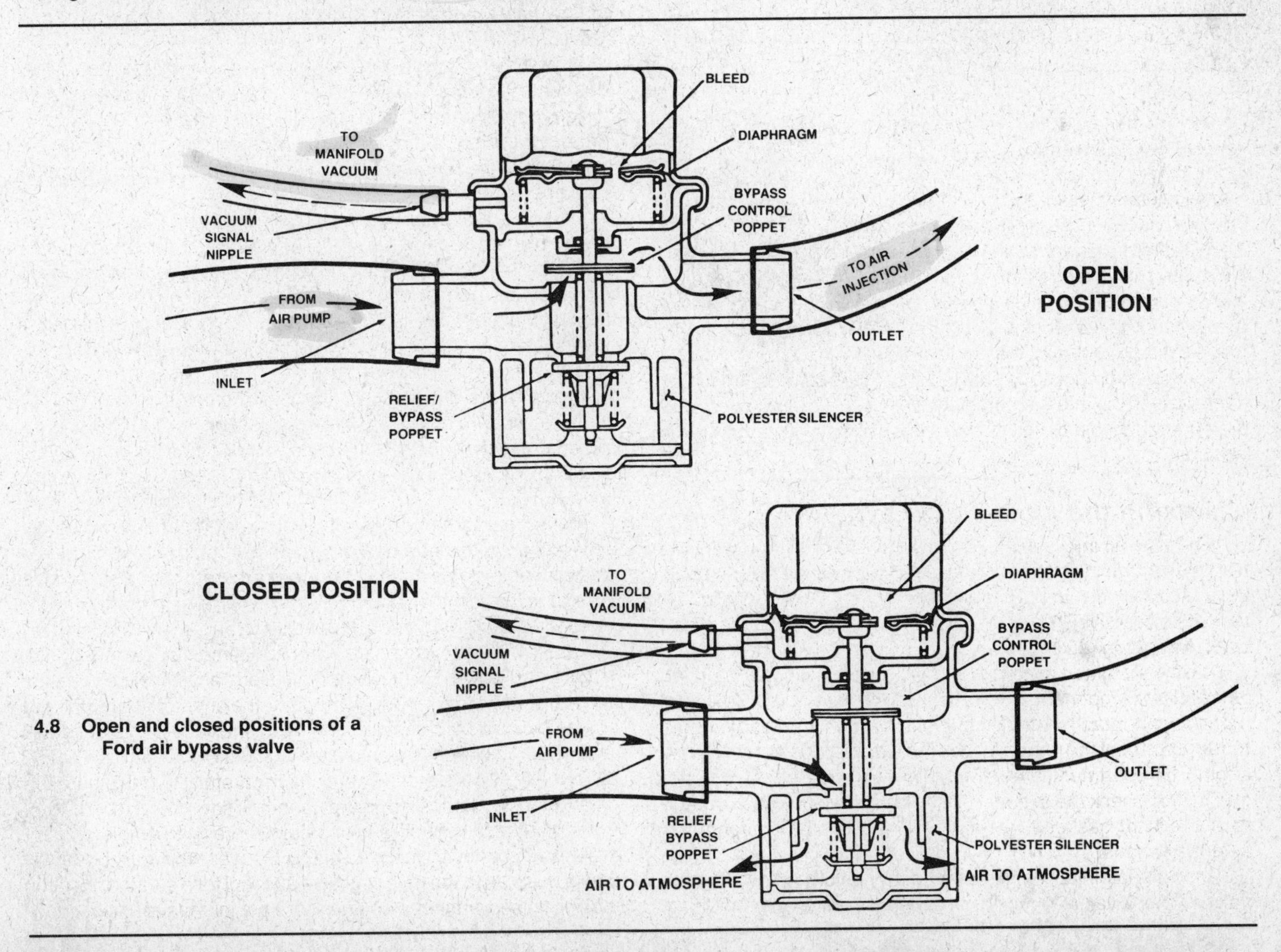

4.8 Open and closed positions of a Ford air bypass valve

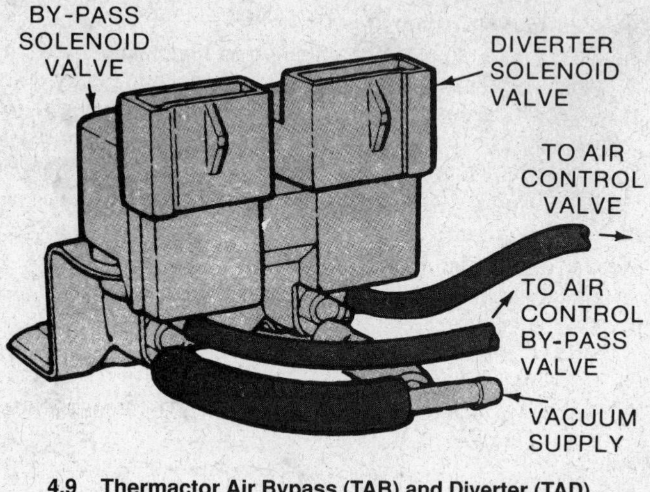

4.9 Thermactor Air Bypass (TAB) and Diverter (TAD) solenoid valves

Check a normally open bypass valve with the vacuum line attached to the bypass valve. The air should flow toward the outlet.

If the bypass valve fails the test, replace it with a new unit. If the vacuum source from the engine has been cut off, check the vacuum control system.

Air control solenoids on Fords

Many Ford vehicles are equipped with a Thermactor air control system **(see illustration)** that uses an air control valve to channel air from the air pump to either the exhaust manifold, the catalytic converter or back into the atmosphere, depending on the operating conditions of the engine. When the engine is cold, the air control valve directs air to the exhaust manifold to reduce HC and CO emissions **(see illustration)**. When the engine is warm, the air control valve directs air to the three-way catalytic converter to control NOx compounds as well as HC and CO. If the engine idles for an unusually long time, the air control valve directs the air back into the atmosphere to avoid overheating the catalytic converter.

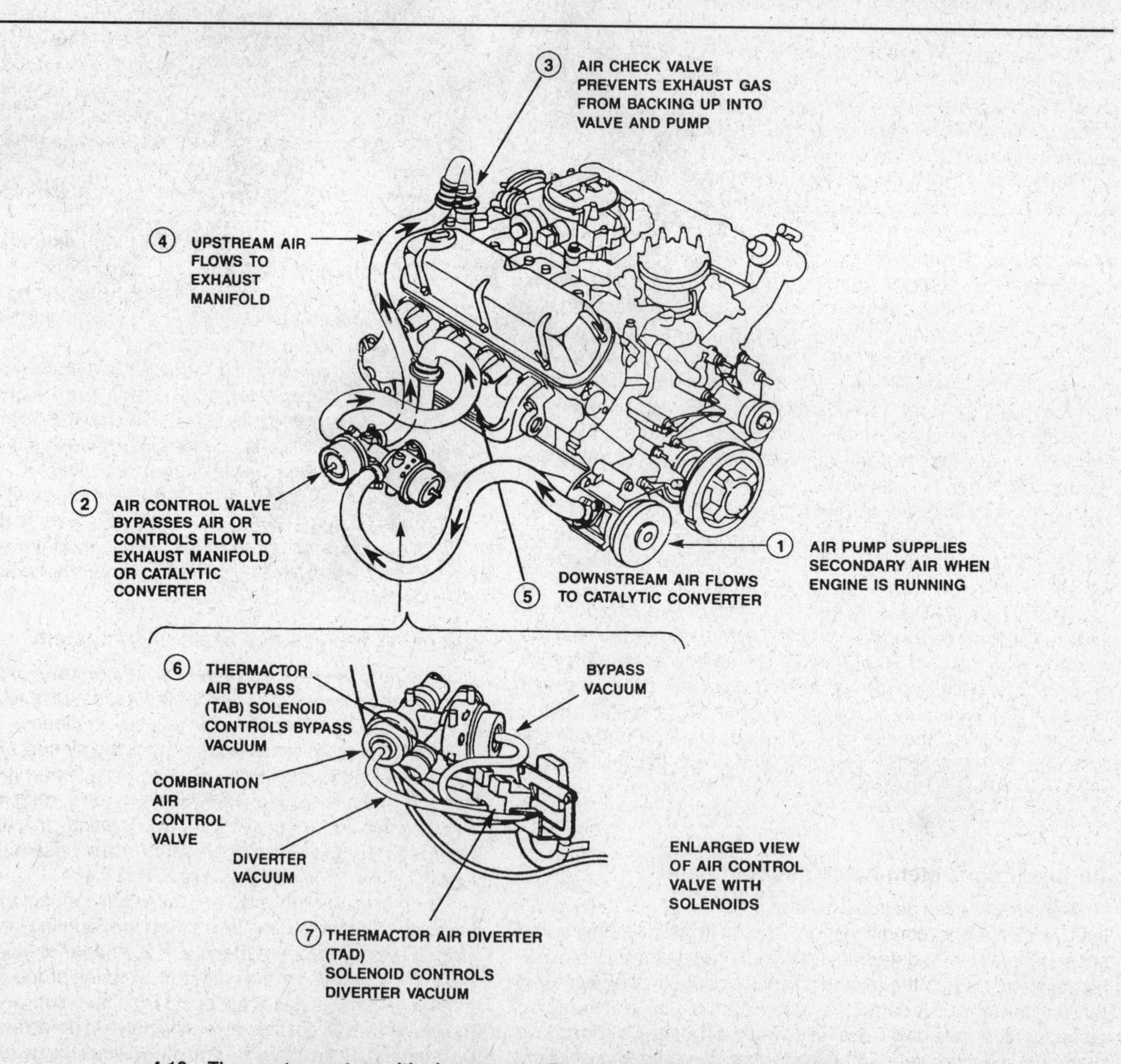

4.10 Thermactor system with air pump and Electronic Engine Controls (EEC)

The operation of the air control valve is regulated by the computer by way of two vacuum control solenoid valves; The Thermactor Air Bypass (TAB) and the Thermactor Air Diverter (TAD). The TAB and TAD valves are directly connected to engine vacuum and the valves are in complete control of the air control valve functions by their ability to switch the vacuum supply to certain portions of the valve. The TAB and TAD valves are normally CLOSED, and vent vacuum only when they are de-energized.

The computer controls manifold vacuum to the air control valve bypass circuit by activating the TAB valve. This allows air from the air pump to vent into the atmosphere by way of the bypass circuit. When the computer activates the TAD valve, manifold vacuum passes from the diverter valve, through the air control valve (one unit) and from there the air goes to the catalytic converter instead of the exhaust manifold.

The TAB and TAD valves can fail and prevent the catalytic converter from receiving extra oxygen. This off-balance situation can increase the emission levels of HC, CO and NOx compounds.

There are several checks you can perform if a system with these valves is suspect:

Turn the ignition key OFF and disconnect the electrical connectors from the TAB and TAD valves. Use an ohmmeter and check the resistance of the valves. It should range from 50 to 110 ohms. More or less resistance indicates that the solenoid is faulty and must be replaced with a new part.

Turn the key ON (engine not running) – the TAB and TAD valves should both be receiving battery voltage. If the reading is less, check the harness for continuity. This is a difficult procedure and requires a thorough wiring diagram.

Check the mechanical operation of the TAB and TAD valves with a vacuum pump. Connect the pump to the inlet port on the valve (manifold) and plug the outlet port (air control valve). Apply vacuum. With the key OFF, the valve should lose vacuum (bleed down). With the key ON, the TAB and TAD valves should open and hold vacuum (only if the outlet port is securely plugged). If necessary, use a jumper wire to apply battery voltage to the TAB and TAD valves. This test will indicate if the valve is directing the vacuum properly.

With the engine running, check the vacuum supply at the TAB and TAD valve. Each valve should receive approximately 10 in-Hg. of manifold vacuum at idle. If the reading is low or zero, check the vacuum lines for leaks and correct hose routing.

The TAB and TAD valves are emissions control components and they are covered under the vehicle manufacturer's emissions warranty of 5 years or 50,000 miles. Be sure to check with your dealership service department before replacing either of these valves. If you absolutely have to change the TAB and TAD valves, on some vehicles, the two valves are separate parts (as in the 2.3L four-cylinder engines) while other models, the valves are a single unit (V6 and V8 engines).

Air injection system noise check

The pump-driven air injection system is not completely noiseless. Under normal conditions, noise rises in pitch as the engine speed increases. To determine if noise is the fault of the air injection system, detach the injection pump drivebelt (after verifying that the belt tension is correct) and operate the engine. If the noise disappears, proceed with the following diagnosis. **Caution:** *The pump must accumulate 500 miles before the following check is valid.*

If the belt noise is excessive:

a) Check for a loose belt and tighten as necessary.
b) Check for a seized pump and replace it if necessary.
c) Check for a loose pulley. Tighten the mounting bolts as required.
d) Check for loose, broken or missing mounting brackets or bolts. Tighten or replace as necessary.

If there is excessive mechanical noise:

a) Check for an overtightened mounting bolt.
b) Check for an overtightened drivebelt.
c) Check for excessive flash on the air pump adjusting arm boss and remove as necessary.
d) Check for a distorted adjusting arm and, if necessary, replace the arm.

If there is excessive noise (whirring or hissing sounds):

a) Check for a leak in the hoses (use a soap and water solution to find the leaks) and replace the hose(s) as necessary.
b) Check for a loose, pinched or kinked hose and reassemble, straighten or replace the hose and/or clamps as required.
c) Check for a hose touching other engine parts and adjust or reroute the hose to prevent further contact.
d) Check for an inoperative bypass valve and replace if necessary.
e) Check for an inoperative check valve and replace if necessary.
f) Check for loose pump or pulley mounting fasteners and tighten as necessary.
g) Check for a restricted or bent pump outlet fitting. Inspect the fitting and remove any casting flash blocking the air passageway. Replace bent fittings.
h) Check for air dumping through the bypass valve (only at idle). On many vehicles, the system has been designed to dump air at idle to prevent overheating the catalytic converter. This condition is normal. Determine that the noise persists at higher speeds before proceeding.
i) Check for air dumping through the bypass valve (the decel and idle dump). On many vehicles, the air is dumped into the air cleaner or the remote silencer. Make sure that the hoses are connected properly and are not cracked.

Checking the passive (Pulse Air) system

If the reed or check valve fails, you'll normally hear excessive exhaust system noise from under the hood and notice hardening of the rubber hose from the valve to the air cleaner.

If exhaust noise is excessive, check the air supply tube-to-exhaust manifold joint and the valve and air cleaner hose connections for leaks. If the manifold joint is leaking, retighten the tube fitting. If the hose connections are leaking, install new hose clamps (if the hose hasn't hardened). If the hose has hardened, replace it with a new one as well.

To determine if the valve has failed, disconnect the hose from the inlet. With the engine idling (transmission in Neutral), hold a strip of paper in front of the inlet – the paper should be sucked against the opening of the valve if it's working properly. If a steady stream of exhaust gas is escaping from the inlet (which will blow the paper away from the valve), the valve is defective and should be replaced with a new one. **Warning:** *Don't use your hand to feel for the exhaust pulses – the exhaust gas is very hot!*

Component replacement

To replace the air bypass valve, air supply control valve, check valve, combination air bypass/air control valve or the silencer, label and disconnect the hoses leading to them, replace the faulty component and reattach the hoses to the proper ports. Make sure the hoses are in good condition. If not, replace them with new ones.

To replace the air supply pump, first loosen the appropriate engine drivebelts, then remove the faulty pump from the mounting bracket. Label all wires and hoses as they're removed to facilitate installation of the new unit.

If you're replacing either of the check valves on a Pulse Air System (Thermactor II), be sure to use a back-up wrench.

After the new pump is installed, adjust the drivebelts to the specified tension.

Servicing air pumps

Many GM systems include air pumps that have a filter element that must be replaced periodically. Check your owner's manual for the mileage intervals. First remove the air pump drivebelt and the pulley. Use a pair of needle-nose pliers to remove the filter element **(see illustration)**. Use a factory filter replacement to avoid any problems with size and exact fit.

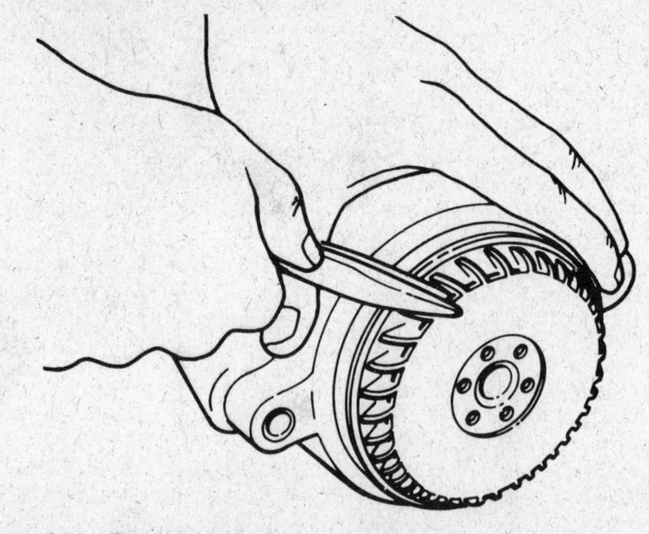

4.11 On many GM vehicles, the filter can be pulled out of the air pump with needle-nose pliers

5 Heated air intake (Thermac and EFE) systems

What it does

Although coming under different names and using different techniques, these systems produce the same result: improving engine efficiency and reducing hydrocarbons during the initial warm-up period. Two different methods are used to achieve this goal: Thermostatic air intake (Thermac) and Early Fuel Evaporation (EFE). Thermac warms the air as it enters the air cleaner while EFE heats the air/fuel mixture in the intake manifold. Virtually all later models use some form of Thermac and/or EFE system.

How it works

Thermac

The Thermostatic air intake system improves driveability, reduces emissions and prevents carburetor icing in cold weather by directing hot air from around the exhaust manifold to the air cleaner intake **(see illustration)**. The Thermac system is made up of the air cleaner housing, a temperature sensor, a vacuum-operated damper door mechanism in the air cleaner snorkel, a flexible tube connected to the exhaust manifold and associated vacuum hoses **(see illustration)**.

When the engine is cold, the temperature sensor in the air cleaner is closed and full vacuum reaches the vacuum motor which holds the damper door shut so only air heated by the exhaust manifold can enter the snorkel. As the engine warms up, the temperature sensor opens, bleeding off the vacuum motor vacuum and allowing its internal spring to push the door down. This closes off the heated air and allows only cold outside air to enter the snorkel. The vacuum motor spring and vacuum balance one another so the air entering the air cleaner is always at optimum temperature for the best fuel vaporization.

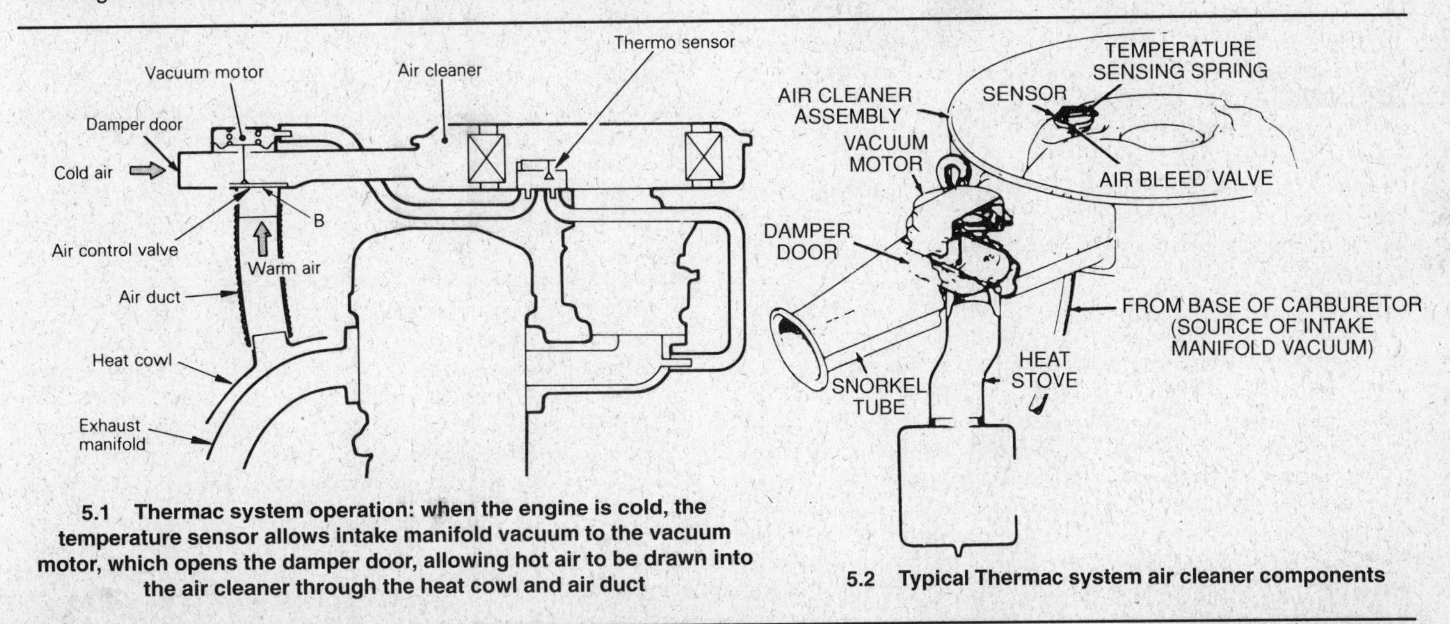

5.1 Thermac system operation: when the engine is cold, the temperature sensor allows intake manifold vacuum to the vacuum motor, which opens the damper door, allowing hot air to be drawn into the air cleaner through the heat cowl and air duct

5.2 Typical Thermac system air cleaner components

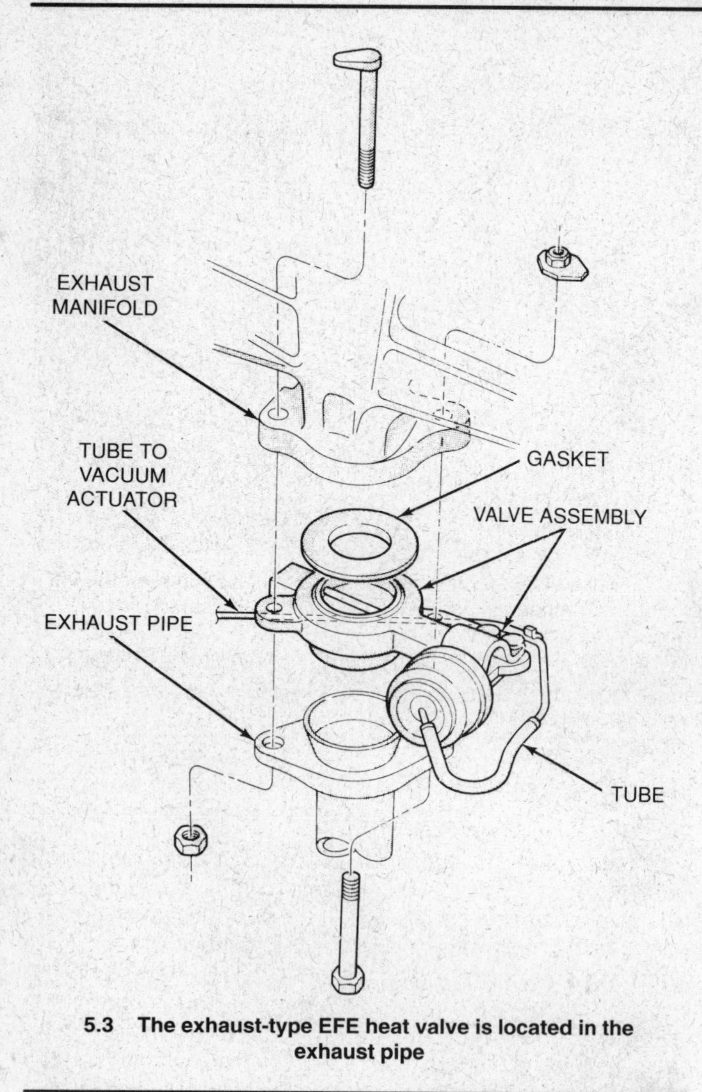

5.3 The exhaust-type EFE heat valve is located in the exhaust pipe

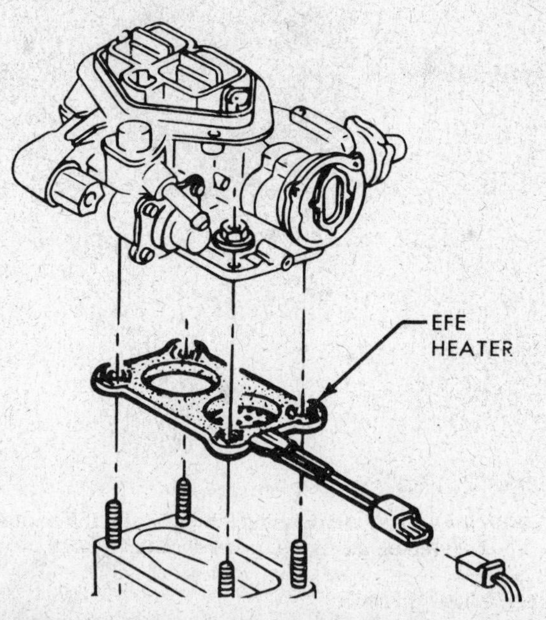

5.4 The electrical EFE heats the air and fuel mixture as it enters the intake manifold

EFE

Two types of EFE are used to heat the vaporized fuel in the intake manifold for improved driveability and emissions during the warm-up period after the engine is first started. One type routes exhaust heat from the exhaust manifold to warm the intake manifold, while the other electrically heats the fuel/air mixture as it enters the manifold.

The exhaust-type EFE uses a valve in the exhaust manifold to recirculate hot exhaust gases which are then used to pre-heat the carburetor and choke for better driveability and emissions. When the engine is cold, the valve is shut, forcing hot exhaust gases to heat the intake manifold until the engine warms up and the valve opens.

The exhaust-type EFE system consists of a heat riser valve in the exhaust manifold, a thermostatic actuator and heat shroud or duct which directs heat to the intake manifold and carburetor. On some models the actuator is simply a counterweighted heat riser with a thermostatic coil spring that contracts when cold, closing the valve and relaxes and opens it when hot. On others the actuator is operated by engine vacuum **(see illustration)**. On this type, when the engine is cold, the vacuum actuator on the valve is held closed by vacuum from a thermostatic switch in a coolant passage. As the coolant heats up, the switch opens, cutting off the vacuum and the actuator opens the valve.

The electrical-type EFE system is quite simple. It consists of an electrical heating element between the carburetor or fuel injection throttle body which heats and vaporizes the air/fuel mixture as it is drawn into the intake manifold **(see illustration)**. The system is made up of the electrical grid which is activated by a thermostatic switch screwed into a coolant passage or by the computer.

Checking and component replacement

Note: *Symptoms of problems with the heated air intake system(s) include:*

When cold: Poor idle, uneven acceleration, hesitation on acceleration, generally poor performance.

When hot: Overheating, hard starting, poor idle, stalling and detonation

Thermac

With the engine cold, make sure the damper door in the snorkel is in the up (heat on) position. Then, start the engine and allow it to warm up. The door should slowly open, indicating that it is operating properly **(see illustration)**. If there is a problem, check the system components, as described below.

Damper door and vacuum motor

Check the door for binding and make sure it moves freely on the hinge pin. Use spray cleaner to remove any foreign matter and penetrating oil to lubricate it. Disconnect the vacuum hose and connect a vacuum pump to the motor **(see illustration)**. The door should be in the up (heat on) position with the vacuum applied and must hold vacuum.

Replacing the motor is a simple matter of removing the screws or rivets and detaching it from the air cleaner housing **(see illustration)**. On some later Japanese vehicles the motor "screws" into place and can be rotated counterclockwise and lifted out. Replacement motors are available at auto parts stores and dealers.

5.5 With the engine at operating temperature, the damper door in the air cleaner snorkel should be open

5.6 Use a vacuum pump to check that the vacuum motor will close the damper door and hold it closed

Temperature sensor

Detach the hoses and use a vacuum pump to make sure the sensor holds vacuum when it's cold and passes vacuum when the engine is hot. Most sensors are held in the air cleaner by clips, so, when replacing, all you have to do is pry up on the tabs of the clip with a screwdriver to detach the valve from the air cleaner. Dealers and auto parts stores carry these valves.

Hot air duct

Check the duct to make sure it is properly connected and free of tears **(see illustration)**. Replacement ducting hose is available at auto parts stores. If your duct is missing, you can make a new one with the proper length ducting, using large-diameter radiator hose clamps to fasten it in place. Missing plastic or metal ducts must be replaced with the same type (available from a dealer), not ducting hose.

Vacuum hoses

Trace the hoses from the temperature sensor to the manifold and the vacuum motors to make sure they are secure and undamaged. Auto parts stores usually carry vacuum hose in various sizes, so bring the old hose to the store and make sure replacement has the same inside diameter and length.

EFE

Exhaust type

Thermostatic spring-actuated

The heat riser is located in the exhaust pipe and is open to the elements, so corrosion can not only keep it from operating freely but can even freeze it in position. If the riser pivot bushings are worn the riser will make a knocking noise when the engine is started. With the engine cold, try moving the counterweight. The valve should move easily with no binding. Start the engine and make sure the valve moves to the closed position and then slowly opens as the engine warms up. A stuck or binding heat riser can often be loosened by soaking the valve in solvent or penetrating oil. Tapping lightly with a hammer can help loosen a stuck valve.

If the heat riser is frozen or the bushings are worn, you'll have to unbolt the exhaust pipe and install a new unit (available from a dealer or auto parts store).

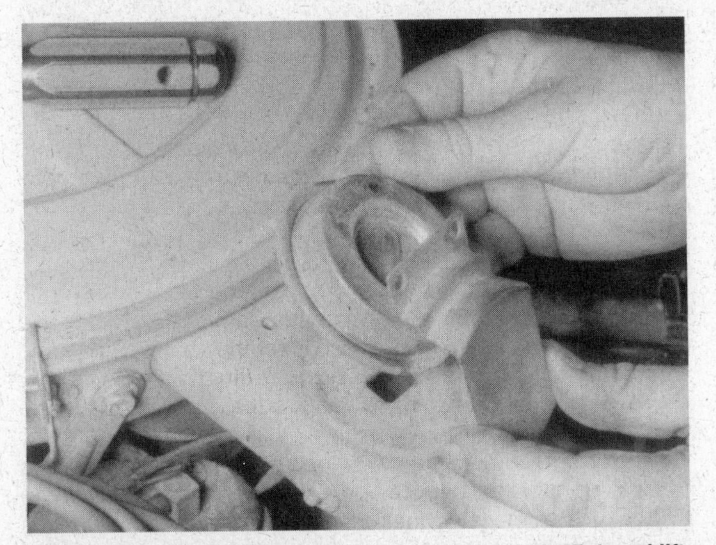

5.7 After removing the screws or rivets, detach the link and lift the vacuum motor out of the air cleaner housing

5.8 Inspect the hot air duct hose for damage and leaks

Vacuum-actuated

Unplug the vacuum hose from the actuator and with the engine cold, start the engine. Put you finger over the end of the hose and determine that vacuum is felt. If there is no vacuum, the hose is plugged or the thermostatic switch is no good. If the hose is in good shape, unscrew the switch and replace it with a new one (available at a dealer or auto parts store). It may be necessary to drain the coolant to a level below the switch to avoid a mess as it runs out during the replacement.

To check the vacuum motor itself, connect a vacuum pump and apply about ten inches of vacuum, which should operate the valve. If the motor holds vacuum but doesn't operate the valve, the valve could be frozen. The motor can be unbolted and detached from its mount. To replace a frozen heat valve, you'll have to unbolt the exhaust pipe and detach the valve. Both the valve and actuator are available at a dealer or auto parts store.

Electrical type

Element

With the engine cold, unplug the EFE heating element electrical connector and check it with an ohmmeter. Check the connector on the unit to make sure there is around three ohms resistance. Also check for voltage at the wiring-harness-side of the connector with the ignition on and the engine cold. There should be voltage. Replacing the heating element will require removal of the carburetor or fuel-injection throttle-body unit.

Temperature switch

With the engine cold, unplug the EFE temperature switch and connect a test light to the terminals of the connector. With the ignition switch on and the engine off, the test light should glow. Replace the switch by draining the coolant to below the switch level and unscrewing it with a wrench **(see illustration)**. Before installing the new switch, apply gasket sealant to the threads (but not to the end of the sensor).

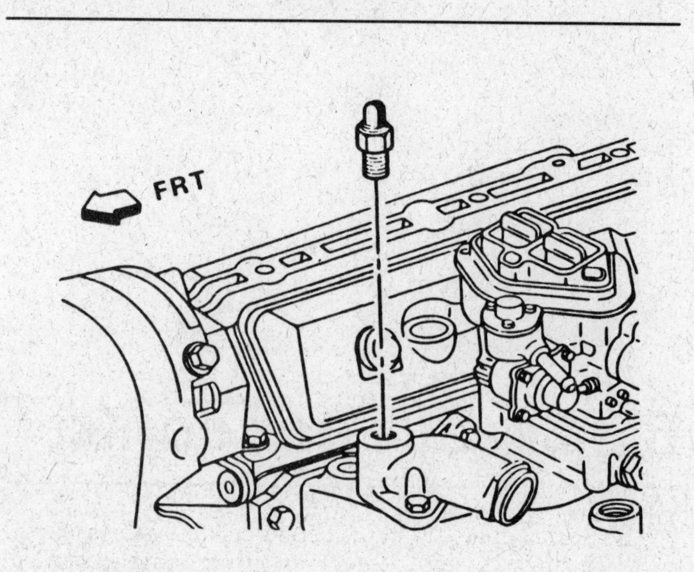

5.9 The temperature switch is often located in the thermostat housing or intake manifold

6 Carburetor control systems

This Section deals with the carburetor control systems that are installed onto carbureted engines to help control deceleration, acceleration, backfiring and emissions requirements. The carburetor controls installed on various carburetors will vary, but the purpose for each system is essentially similar. Here are some common carburetor control systems along with a brief explanation and some easy checks and adjustments.

Dashpot

The dashpot system installed on early carbureted engines slows the closing of the throttle on deceleration. This allows the carburetor to switch from the main fuel jets to the idle system, thereby preventing stalling due to an excessively rich air/fuel mixture. Also, the amount of HC (hydrocarbons) emissions is reduced. The air/fuel mixture richens when the intake manifold vacuum suddenly rises when the throttle is closed. The high vacuum will draw fuel into the carburetor from the float bowl without any dilution of air from the air horn (venturi).

ADJUSTMENT LOCKNUT
PLUNGER
THROTTLE LEVER
0465H
DASHPOT
THROTTLE LINKAGE

6.1 A typical carburetor dashpot

The dashpot is made up of a small chamber with a spring-loaded diaphragm and a plunger. The dashpot plunger is in contact with the throttle lever during the last stages of deceleration **(see illustration)**. When the lever contacts the plunger on deceleration, the lever exerts force on the plunger and air or hydraulic fluid (depending on the type of dashpot) slowly leaks out of the diaphragm through a small hole. This allows the throttle plate to close slowly.

Some dashpot components are adjustable while others are not **(see illustration)**. Consult a factory service manual or other automotive repair manual for the exact procedure. Keep in mind that dashpots are often combined with other throttle devices into the same component **(see illustration 6.3)**.

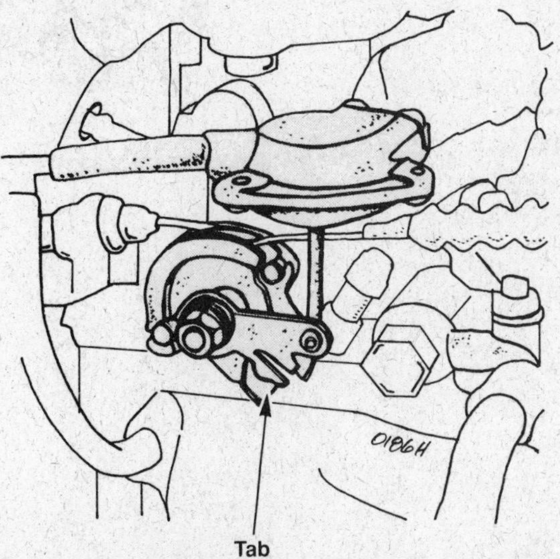

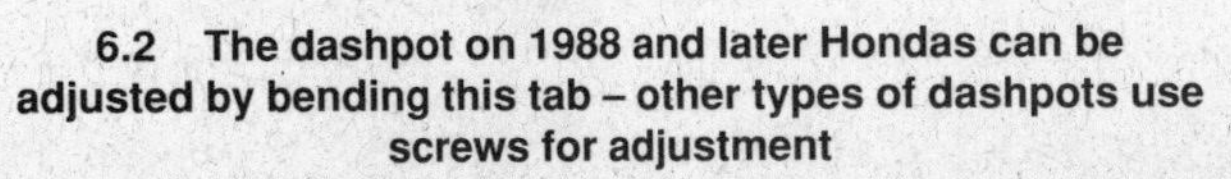
6.2 The dashpot on 1988 and later Hondas can be adjusted by bending this tab – other types of dashpots use screws for adjustment

Throttle positioner

What it is and how it works

Throttle positioners are used to control the engine's idle speed under various conditions. Some designs are vacuum actuated and others are electric solenoids. On vacuum-actuated positioners (which usually look the same as dashpots, except there is one or more vacuum hose(s) connected to it), some use vacuum to turn the positioner on while others use vacuum to turn it off. Be aware of this when checking vacuum conditions.

One basic type throttle positioner functions to prevent dieseling (engine run-on). This type is called a throttle stop solenoid or an idle stop solenoid. When the engine is started, the solenoid is energized and the plunger extends out, pushing against the throttle linkage. This forces the throttle plate to open slightly to the curb idle position. When the ignition switch is turned off, the throttle position solenoid is de-energized and the plunger returns to the normal position. The throttle plate closes completely and the air/fuel supply is cut off, effectively preventing dieseling.

Some throttle positioners are used to increase the curb idle speed to compensate for extra loads on the engine. In this situation, the throttle positioner is referred to as an idle speed-up solenoid or a throttle kicker **(see illustration)**. This type is often used on air-conditioned vehicles. When the air conditioner is switched on, the relay energizes the solenoid, which extends its plunger farther onto the throttle plate, thereby raising the idle speed. This keeps the engine running at a higher rpm to control emissions levels.

Throttle positioners are also used to control idle speed when the automatic transmission is engaged. A relay in the Park/Neutral switch signals the solenoid to raise the idle speed when the transmission is in gear. This opens the throttle slightly to compensate for the increased load on the engine.

Another type of system is sometimes used on vehicles equipped with power steering. When the steering wheel is turned while the vehicle is stationary and idling, the positioner solenoid raises the idle speed, compensating for the additional load the power steering pump is placing on the engine. This type of system has a switch located on the steering gear, power steering pump or power steering pressure hose. The switch completes the circuit to the solenoid when there's power steering fluid pressure at the switch.

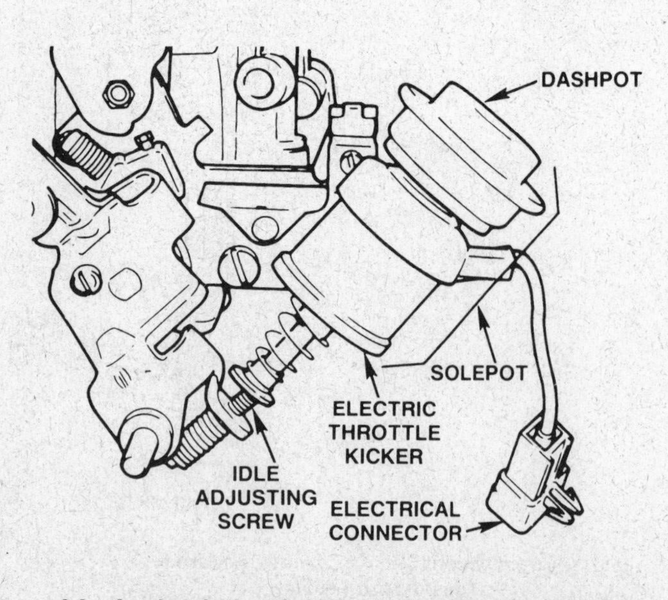

6.3 An electric throttle kicker (solenoid) with a dashpot mounted on the top

Checking

Note: *A single throttle positioner may be used to operate the throttle under more than one condition (for example, when the air conditioning is switched on and when the automatic transmission is placed in gear). Be sure to check the positioner's function under all the conditions it is designed to operate.*

To check for proper positioner response, first locate the positioner on the carburetor, then start the engine and induce the appropriate load onto the engine and carefully watch the positioner as it is forced to activate. Depending on the type of system, this can be done by either switching the air conditioning system ON, turning the engine OFF (anti-dieseling solenoid), placing the transmission in Drive (wheels blocked, parking brake set and an assistant holding down the brake pedal) or turning the steering wheel from side to side. If the positioner does not actuate to increase the engine speed, first check the vacuum or electrical connections. Make sure the wires or hoses are in good shape. Then check the positioner plunger to make sure it moves freely and is not "frozen." If it is, replace the positioner. Next, apply battery voltage (electrical type) or vacuum to the positioner and see if it operates. If it does not, replace it. If it does operate, the problem is in the circuit or switches.

ISC (Idle Speed Control) motor

What it is and how it works

The ISC motor is a more advanced version of a throttle positioner (see above). The motor is under direct control of the computer, which has the desired idle speed programmed into its memory. The computer compares the actual idle speed from the engine (taken from the distributor or crankshaft position sensor ignition impulses) to the desired rpm reference in memory. When the two do not match, the ISC plunger is moved in or out. This automatically adjusts the throttle to hold an idle speed independent of engine loads.

Many ISC motors have a throttle contact switch at the end of the plunger **(see illustration)**. The position of the switch deter-

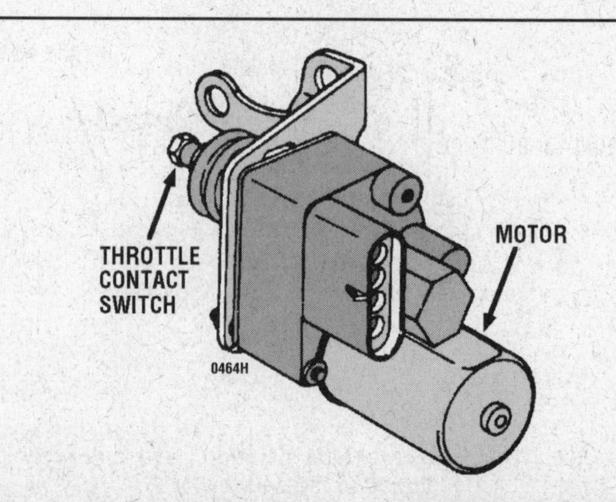

6.4 A typical ISC motor (1981 GM model shown)

mines whether or not the ISC should control idle speed. When the throttle lever is resting against the ISC plunger, the switch contacts are closed, at which time the computer moves the ISC motor to the programmed idle speed. When the throttle lever is not contacting the ISC plunger, the switch contacts are open and the ECM stops sending idle speed commands and the driver controls engine speed.

Checking

With the engine warmed to normal operating temperature, remove the air cleaner assembly and any other components that obscure your view of the ISC motor. Hook up a tachometer in accordance with the manufacturer's instructions and check the VECI label under the hood to determine what the correct idle rpm should be.

Have an assistant start the engine. Check that the engine rpm is correct. Have your assistant turn on the air conditioning (if equipped), headlights and any other electrical accessories. If the vehicle is equipped with power steering, have your assistant turn the steering wheel from side to side. Note the reading on the tachometer. The engine speed should remain stable at the correct idle speed. If the vehicle is equipped with an automatic transmission, block the wheels and have your assistant set the parking brake, place his/her foot firmly on the brake pedal and place the transmission in Drive. Again, the engine rpm should remain stable at the correct speed. **Warning:** *Do not stand directly in front of the vehicle during this test.*

If the ISC motor is not functioning as it should, first check the condition of the wiring and electrical connector(s). Make sure the connector is securely attached and there is no corrosion at the terminals. For further diagnosis of this system, refer to the factory service manual for your particular vehicle or take the vehicle to a dealer service department or other qualified shop.

Fuel deceleration valve

The fuel deceleration valve is designed to prevent backfire during deceleration. This device opens a separate air/fuel mixture passage in the carburetor to dilute the fuel charge with additional air. When the intake manifold vacuum rises, the valve moves up to allow a mixture of air and fuel from the carburetor to flow into the intake manifold **(see illustration)**. The valve provides enough mixture to maintain proper combustion and prevent unburned fuel from being released out the tailpipe.

Some deceleration valves are not attached directly to the carburetor but accomplish the same results **(see illustration)**. This type of valve has a diaphragm housing on one end. A control manifold vacuum line is attached to a port on the lower portion of the valve. Other ports on the valve are connected to the intake manifold and air cleaner. When deceleration causes an increase in manifold vacuum, the diaphragm opens the deceleration valve and allows air to pass from the air cleaner to the intake manifold, leaning out the fuel mixture and preventing exhaust system backfire.

Automatic choke

What it is and how it works

Automatic choke systems use a bi-metal, heat-sensitive element to control choke valve position, and most modern choke systems also have an electric heater to speed warming up the bi-metal element (this causes the choke to disengage more quick-

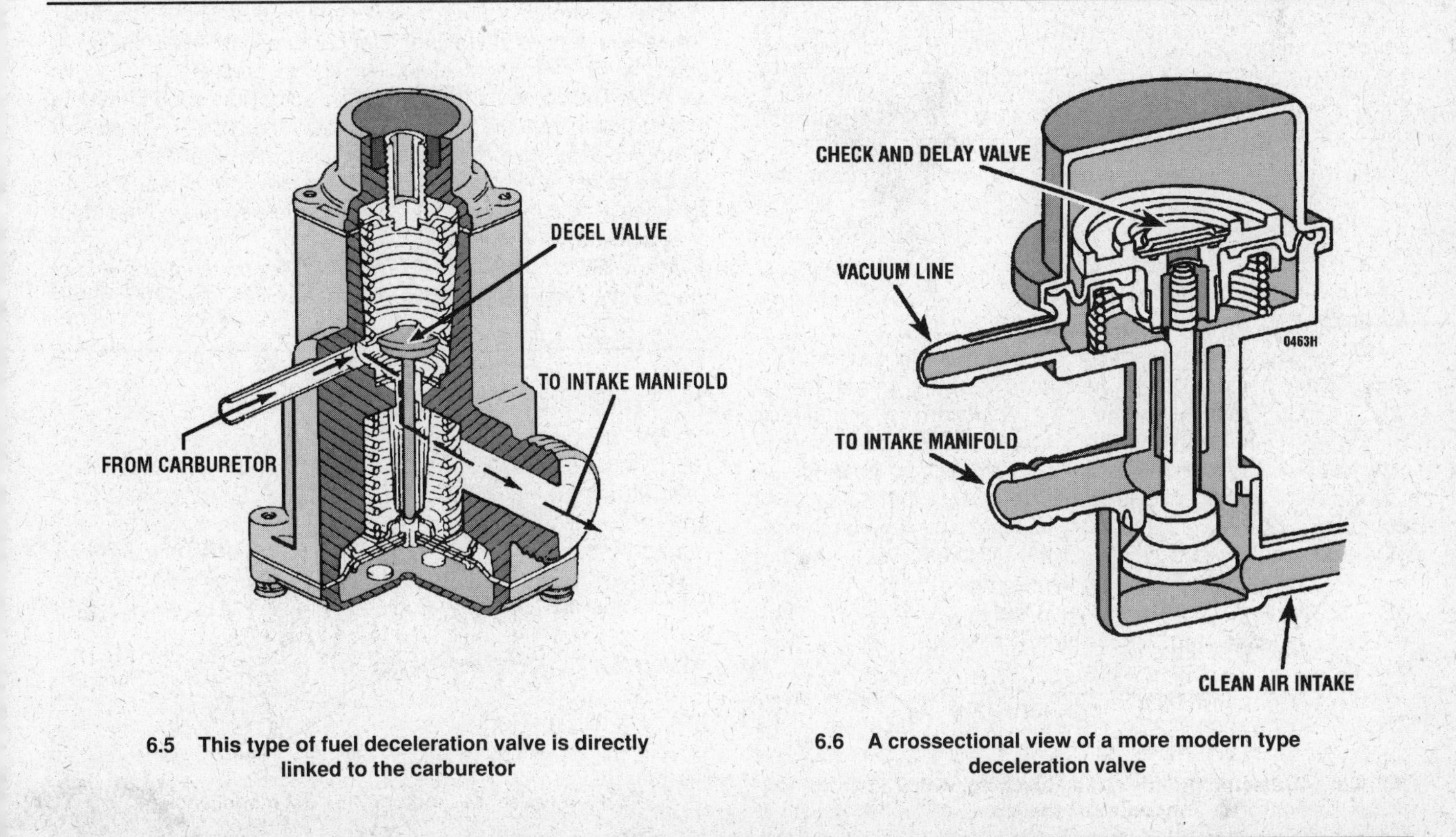

6.5 This type of fuel deceleration valve is directly linked to the carburetor

6.6 A crossectional view of a more modern type deceleration valve

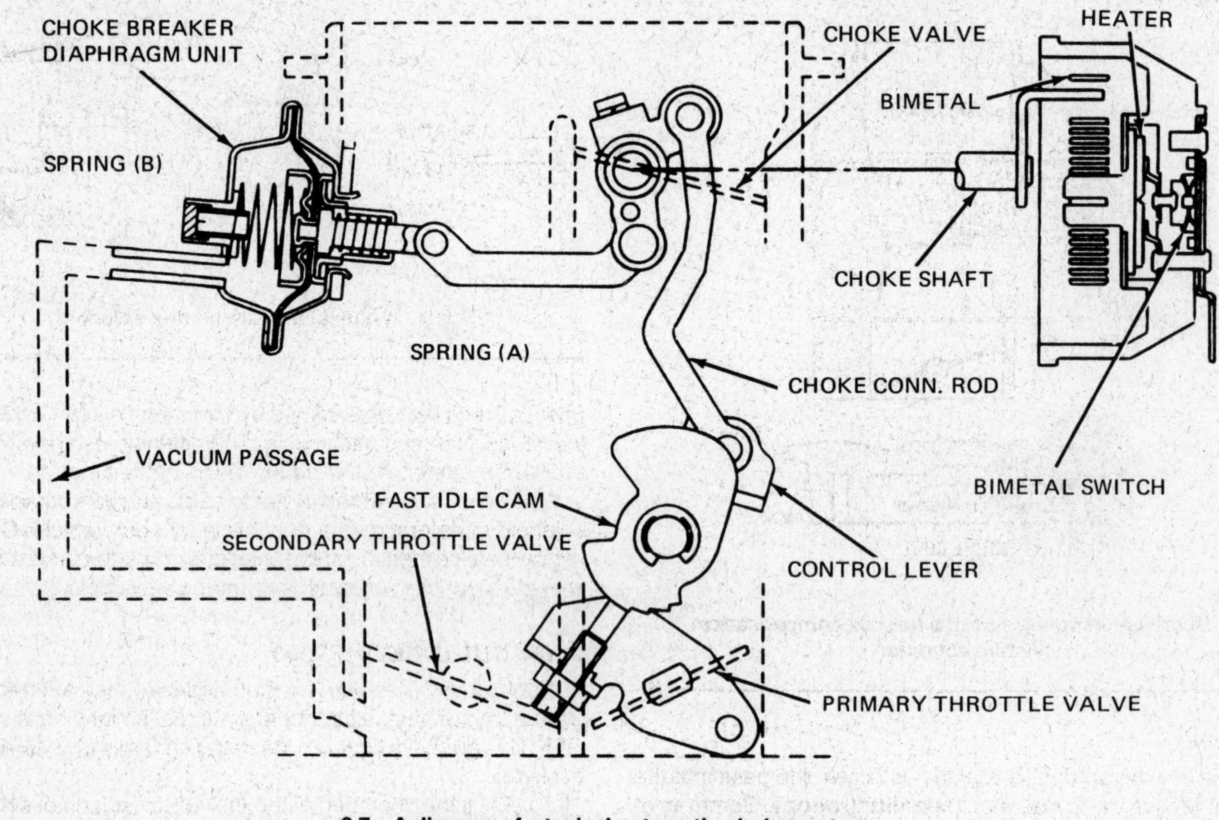

6.7 A diagram of a typical automatic choke system

ly, helping reduce emissions) **(see illustration)**. The bi-metal element operates a choke valve which closes the the carburetor air horn and is synchronized with the throttle plate(s). When the engine is cold, the choke valve closes and the throttle plate opens (operated by the fast idle cam) sufficiently to provide a rich mixture and an increased idle speed for easy starting. Many automatic choke systems are equipped with a choke breaker diaphragm that opens the choke valve when the engine is accelerating and the engine is cold. This prevents the engine from bogging from insufficient airflow.

Checking

The choke only operates when the engine is cold, so this check should be performed before the engine has been started for the day. Open the hood and remove the air cleaner cover and filter from the top of the carburetor. Locate the choke plate (the flat plate attached by small screws to a pivot shaft) in the carburetor throat.

Operate the throttle linkage and make sure the plate closes completely. Start the engine and watch the plate – when the engine starts, the choke plate should open slightly. Allow the engine to continue running at idle speed. As the engine warms up to operating temperature, the plate should slowly open. After several minutes, the choke plate should be fully open to the vertical position.

Note that the engine speed corresponds to the plate opening angle. With the plate closed, the engine should run at a fast idle speed. As the plate opens, the engine speed will decrease. The fast idle speed is controlled by the fast idle cam, and, even though the choke plate is open completely, the idle speed will remain high until the throttle plate is opened, releasing the fast idle cam. Check the drop in idle speed as the choke plate opens by occasionally tapping the accelerator.

If the choke doesn't work as described, shut off the engine and check the shaft and linkage for deposits which could cause binding. Use a spray-on choke cleaning solvent to remove the deposits as you operate the linkage. This should loosen up the linkage and the shaft and allow the choke to work properly. If the choke still fails to function correctly, the choke bimetal assembly is malfunctioning or in need of adjustment.

Some chokes use warm coolant to change the position of the butterfly valve. In this case, check for the proper water circulation to the choke heating element.

HIC (Hot Idle Compensator) valve

On some vehicles, when the engine is excessively hot, a hot idle compensator opens an air passage to lean the fuel/air mixture **(see illustration)**. This increases the idle speed, which in turn cools the engine and prevents excess fuel vaporization and consequently the release of unburned hydrocarbons. The compensator is controlled by a bimetallic strip which bends when it senses high temperatures, thereby opening the air passage.

Mixture control solenoid

In the late 1970's, the feedback carburetor was introduced to reduce emissions on carbureted vehicles. This system incorporates a computer which controls certain solenoids and valves on the carburetor. The main solenoid controlled by computer is the mixture control solenoid. It is an electronically controlled metering

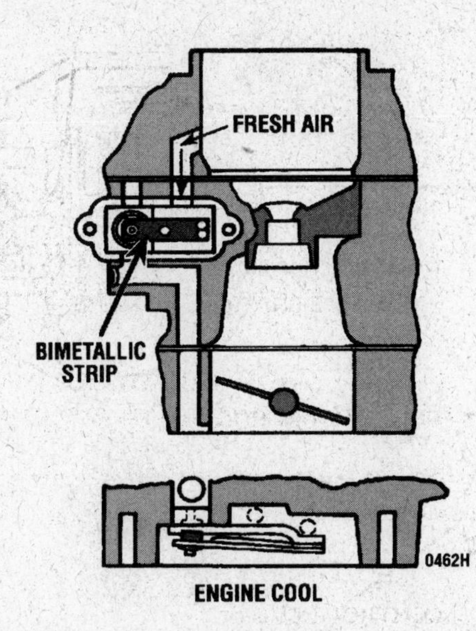

6.8 A cross-sectional view of a hot idle compensation system in operation

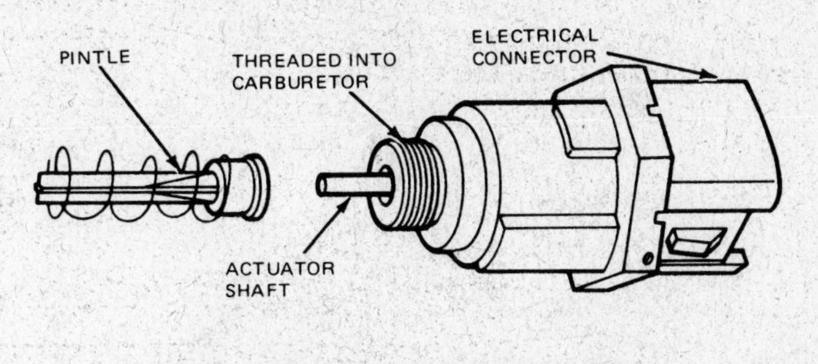

6.9 A typical mixture control solenoid

rod that varies the amount of fuel that is allowed to pass into the main fuel jets of the carburetor **(see illustration)**. Some solenoids are mounted vertically and others are mounted horizontally. The computer is programmed to turn the solenoid ON and OFF (cycle) ten times per second. These solenoids are generally referred to as duty-cycle solenoids. Each cycle lasts about 100 milliseconds. The amount of fuel metered into the main fuel jet or passage is directly determined by how many milliseconds the solenoid is ON during each cycle. The solenoid can be ON almost 100% of the time or OFF nearly 100% of the time.

Use only a specialized meter to calculate the frequency of the solenoid to determine the duty-cycle for your vehicle. Consult a factory service manual for the checking procedure and correct duty-cycle measurements for your particular vehicle.

Fuel cut-off solenoid

Fuel cut-off solenoids are mounted onto the carburetor to instantly shut off the fuel to the main jet as the ignition is switched OFF. This prevents engine run-on and unnecessary vibration and backfire.

To check the operation of the fuel cut-off solenoid, simply turn the ignition key to ON and check for battery voltage at the solenoid connector. If the voltage is present, the solenoid should make a slight "clicking" noise as the plunger is energized. When the ignition is turned OFF the solenoid should click again, as the plunger extends into the carburetor.

7 Catalytic converter

What it does

The catalytic converter is a unique device because it promotes a reaction which changes the exhaust gases flowing through it without being affected itself. This catalytic reaction reduces the level of three major pollutants: Hydrocarbon (HC), Carbon Monoxide (CO) and Oxides of Nitrogen (NOx). By removing these major pollutants, the catalytic converter system allows the other fuel and emissions systems to be fine-tuned for optimum operation and driveability. These are controlled on later models by the computer and a network of engine sensors. This is called a "feedback" or "closed loop" system.

Catalytic converters are mounted in the exhaust system between the exhaust manifold and the muffler. Because they generate a lot of heat, they are surrounded by heat shields.

The catalytic elements in the converter are palladium, platinum and rhodium. By coating ceramic pellets in the bed of the converter or a ceramic honeycomb, a large surface area is provided for the gases to react on as they pass through the converter.

There are two basic types of converters: oxidation and reduction. On later models, they are combined into one unit called a three-way converter. An oxidation converter uses platinum and palladium to oxidize (add oxygen to) hydrocarbons and carbon monoxide, converting them to water vapor. Since oxidization converters have little effect on NOx, a reduction converter containing rhodium and platinum is used to convert (reduce) the oxygen in the NOx into nitrogen and carbon dioxide.

One stage of a typical three-way converter contains a reduction-oxidation catalyst using rhodium and platinum which controls NOx, HC and CO emissions. The second stage has only a platinum catalyst for controlling the remaining HC and CO emissions. On some models air is pumped directly into a chamber between the two stages **(see illustration)**.

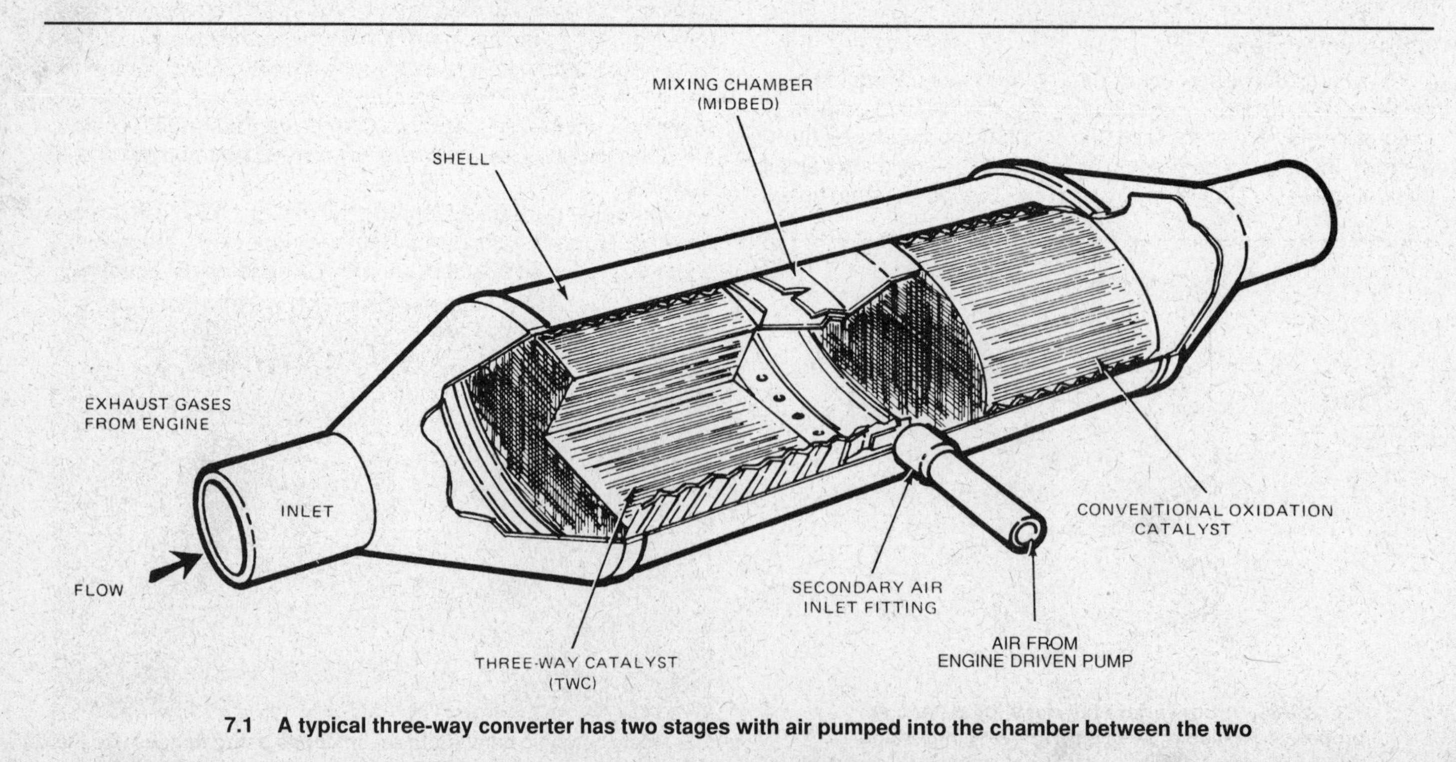

7.1 A typical three-way converter has two stages with air pumped into the chamber between the two

How it works

As the gases flow through the converter, they start to burn rapidly at temperatures reaching 1600-degrees F. The extra oxygen needed to support such high temperatures is provided by the air injection system which pumps air into the exhaust system or the converter itself, or by a lean air/fuel ratio. Three-way catalysts use air switching valves to direct air to the manifold during the high-emissions warm-up mode to help burn the HC and CO. It then shifts the air injection to the chamber in the middle of the converter when NOx production begins (normal operating temperature is reached).

Checking

Note: *About the only common noticeable catalytic converter problem is a plugged converter. This will cause a noticeable drop in power and, if severe enough, stalling or hard starting because of the extreme exhaust back-pressure. Melting of the catalyst pellets or disintegration of the ceramic honeycomb matrix is due to overheating caused by a rich mixture or a misfiring cylinder.*

Catalytic converter

Testing the catalytic system requires special equipment, but there are a couple of simple ways to check for a clogged converter. Tap the converter with a rubber mallet and listen for rattling which indicates that the catalysts have flaked off or the ceramic matrix is broken, meaning that replacement is in order (does not apply to ceramic bead-type converters).

Use a vacuum gauge to check for vacuum drop caused by a plugged converter. Connect the gauge to a direct intake manifold vacuum source and run the engine up to a speed around 3000 rpm and hold it there. If the exhaust system is blocked, the gauge will drop considerably at this speed. For example, if your gauge reads 18 inches of vacuum and you rev the engine up, the reading should be about 15 inches. If the gauge reading drops, there is a blockage.

Catalytic system

Every time the vehicle is raised, you can easily check the converter and components for damage. Check the heat shields to make sure they are all in place and securely mounted **(see illustration)**. Inspect the converter connections to the exhaust pipes for damage or missing bolts. Trace the air injection system hoses and pipes from the converter, making sure they aren't disconnected or leaking.

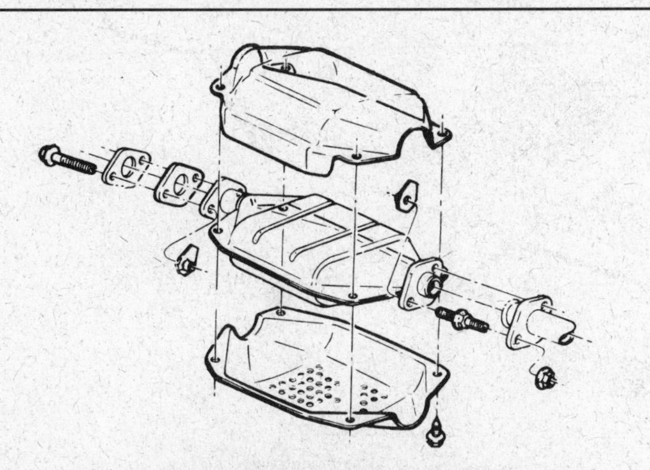

7.2 Catalytic converters put out a lot of heat, so properly installed heat shields are very important

About replacement catalytic converters

The Environmental Protection Agency (EPA) closely regulates replacement catalytic converters, so be sure to familiarize yourself with their requirements before starting work. Not following the EPA guidelines is considered "tampering" with the emissions system and is punishable by a hefty fine. The EPA does realize that factory replacement catalytic converters are expensive so they have approved the installation of more reasonably priced aftermarket units under certain conditions.

The EPA says basically that you must use only a factory replacement converter on any vehicle which is still under the federally mandated emissions warranty (in which case the manufacturer would generally replace it at no charge anyway). You can use an approved aftermarket converter on any vehicles which are out of warranty and have damaged or non-functioning units.

Aftermarket converters and installation kits are available at auto parts stores. Also, most auto exhaust shops now install these aftermarket converters. Originally there were only a few "universal" units, but now the full range of aftermarket catalytic converters covers virtually every make and model. Be sure to check with the installer or auto parts store to make sure you are getting the right converter or kit for your car and that everything is in compliance with EPA regulations to avoid trouble later. The auto parts store should have all the necessary information along with advice on the hardware and pipes you'll need to do the job. Remember also that these replacement units come with a mandated lifetime guarantee.

Replacement

Raise the vehicle and support it securely on jackstands. The vehicle must be cold before starting work (let it set for several hours because converters can hold their heat for some time). Before beginning, check to make sure all the parts required for installation, as well as replacement clamps, bolts and nuts are with the converter. Removal of the existing hardware usually destroys it, so plan ahead. If your converter has air hoses, it's also a good idea to get new ones while you're at it. On the vehicle, squirt penetrating oil on the bolts or nuts which will have to be removed and let it soak in.

Most converters are attached to the exhaust pipes with clamps and flanges **(see illustration)**. Remove these clamps (not always a simple matter, so a nut splitter may be necessary). Some con-

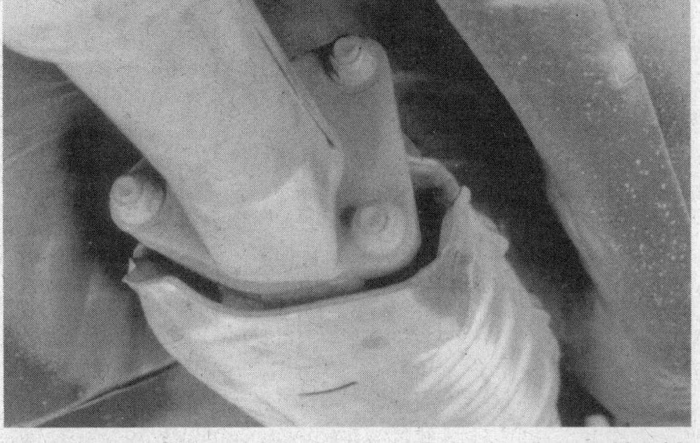

7.3 Most catalytic converters are installed using flanges like this

verters are welded in place. On these you'll have to cut the pipes next to the welds with a pipe cutter or hacksaw. Detach the air hose and lower the converter.

Install the new one in reverse order of the previous steps, clamping the new air hoses in place. Assemble everything loosely, making sure that all the hardware and gaskets are in place. All slip fittings must insert at least two inches over each other with the clamp over the center of the overlap. It's a good idea to apply anti-seize compound to each joint, to prepare for the day when you might have to disassemble the exhaust system. Make sure everything is lined up properly before tightening the bolts, clamps and hangers.

After installation, be sure to dispose of the old converter properly. Most wrecking yards will probably pay you for it because there is an ongoing program to reclaim the rhodium, platinum and palladium. Lastly, be sure to fix whatever condition caused the converter to go bad in the first place, so you won't have to repeat the job later.

8 Engine management systems

Note: *Some of the procedures in this Section require you to operate the vehicle after disconnecting a portion of the engine management system (such as a sensor or a vacuum line). This may set trouble codes in the computer. Be sure to clear any trouble codes (see Chapter 2) before returning the vehicle to normal service.*

This Section deals with the engine management systems used on modern, computer-controlled vehicles to meet new low-emission regulations. The system's computer, information sensors and output actuators interact with each other to collect, store and send data. Basically, the information sensors collect data (such as the intake air mass and/or temperature, coolant temperature, throttle position, exhaust gas oxygen content, etc.) and transmit this data, in the form of varying electrical signals, to the computer. The computer compares this data with its "map," which tells what these data should be under the engine's current operating conditions. If the data does not match the map, the computer sends signals to output actuators (fuel injectors or carburetor mixture control solenoid, Electronic Air Control Valve (EACV), Idle Speed Control (ISC) motor, etc.) which correct the engine's operation to match the map **(see illustration)**.

When the engine is warming up (and sensor input is not pre-

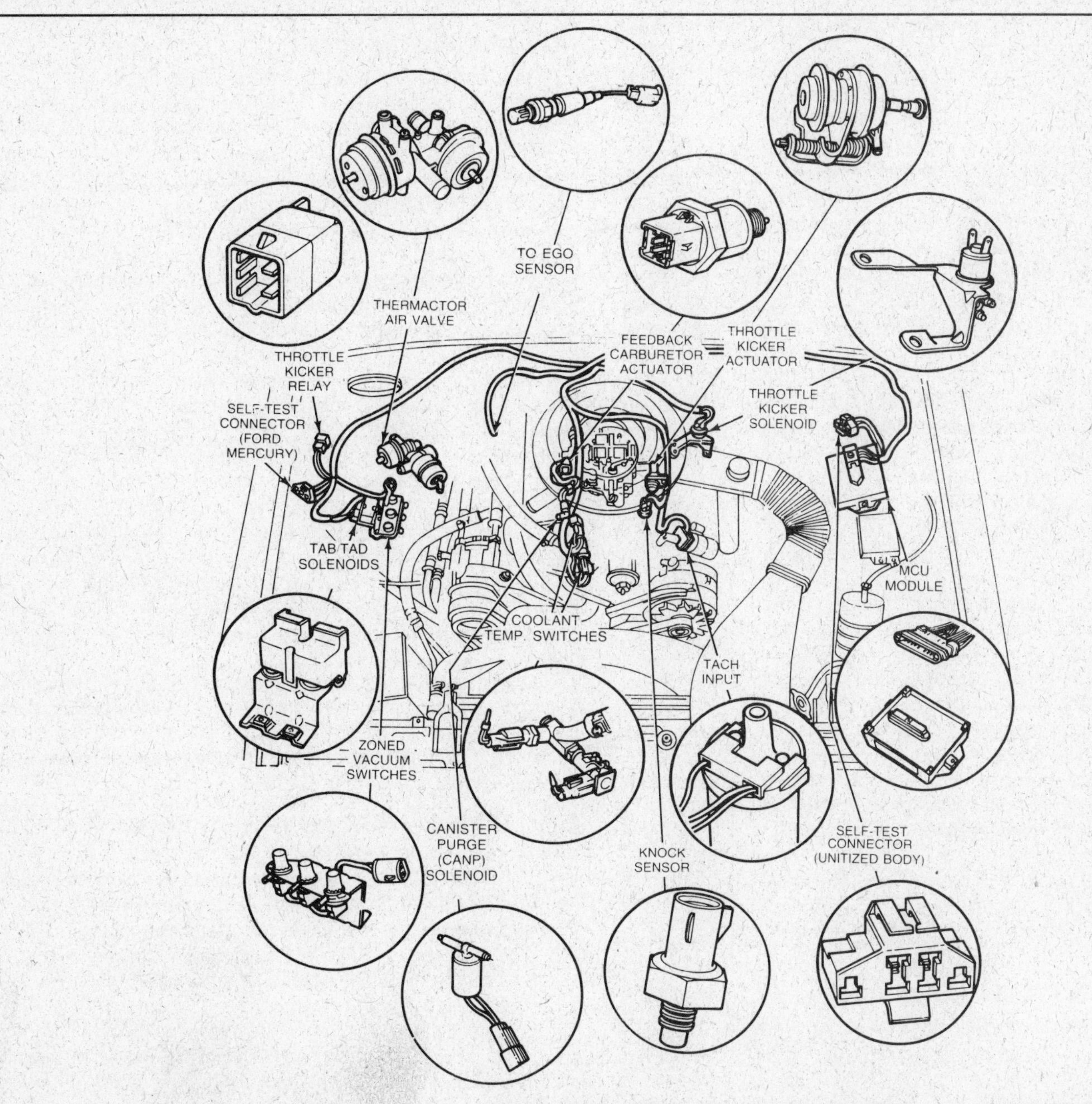

8.1 Overall view of an engine management system, including the computer, information sensors and output actuators (Ford EEC system shown)

cise) or there is a malfunction in the system, the system operates in an "open loop" mode. In this mode, the computer does not rely on the sensors for input and sets the fuel/air mixture rich so the engine can continue operation until the engine warms up or repairs are made. **Note:** *The engine's thermostat rating and proper operation are critical to the operation of a computer-controlled vehicle. If the thermostat is rated at too low a temperature, is removed or stuck open, the computer may stay in "open loop" operation and emissions and fuel economy will suffer.*

The automotive computer

Automotive computers come in all sizes and shapes and are generally located under the dashboard, around the fenderwells or under the front seat. The Environmental Protection Agency (EPA) and the Federal government require all automobile manufacturers to warranty their emissions systems for 5 years or 50,000 miles. This broad emissions warranty coverage will allow most computer malfunctions to be repaired by the dealership at their cost. Keep this in mind when diagnosing and/or repairing any emissions systems problems.

Computers have delicate internal circuitry which is easily damaged when subjected to excessive voltage, static electricity or magnetism. When diagnosing any electrical problems in a circuit connected to the computer, remember that most computers operate at a relatively low voltage (about 5 volts).

Observe the following precautions whenever working on or around the computer and engine management system circuits:

1) Do not damage the wiring or any electrical connectors in such a way as to cause it to ground or touch another source of voltage.
2) Do not use any electrical testing equipment (such as an ohmmeter) that is powered by a six-or-more-volt battery. The excessive voltage might cause an electrical component in the computer to burn or short. Use only a ten megaohm impedence multimeter when working on engine management circuits.
3) Do not remove or troubleshoot the computer without the proper tools and information, because any mistakes can void your warranty and/or damage components.
4) All spark plug wires should be at least one inch away from any sensor circuit or control wires. An unexpected problem in computer circuits is magnetic fields that send false signals to the computer, frequently resulting in hard-to-identify performance problems. Although there have been cases of high-power lines or transformers interfering with the computer, the most common cause of this problem in the sensor circuits is the position of the spark plug wires (too close to the computer wiring).
5) Use special care when handling or working near the computer. Remember that static electricity can cause computer damage by creating a very large surge in voltage (see *Static electricity and electronic components* below).

Static electricity and electronic components

Caution: *Static electricity can damage or destroy the computer and other electronic components. Read the following information carefully.*

Static electricity can cause two types of damage. The first and most obvious is complete failure of the device. The other type of damage is much more subtle and harder to detect as an electrical component failure. In this situation the integrated circuit is degraded and can become weakened over a period of time. It may perform erratically or appear as another component's intermittent failure.

The best way to prevent static electricity damage is to drain the charge from your body by grounding your body to the frame or body of the vehicle and then working strictly on a static-free area. A static-control wrist strap properly worn and grounded to the frame or body of the vehicle will drain the charges from your body, thereby preventing them from discharging into the electronic components. Consult your dealer parts department for a list of the static protection kits available.

Remember, it is often not possible to feel a static discharge until the charge level reaches 3,000 volts! It is very possible to be damaging the electrical components without even knowing it!

Information sensors

The information sensors are a series of highly specialized switches and temperature-sensitive electrical devices that transform physical properties of the engine such as temperature (air, coolant and fuel), air mass (air volume and density), air pressure and engine speed into electrical signals that can be translated into workable parameters for the computer.

Each sensor is designed specifically to detect data from one particular area of the engine; for example, the Mass Airflow Sensor is positioned inside the air intake system and it measures the volume and density of the incoming air to help the computer calculate how much fuel is needed to maintain the correct air/fuel mixture.

Diagnosing problems with the information sensors can easily overlap other management systems because of the inter-relationships of the components. For instance, if a fuel-injected engine is experiencing a vacuum leak, the computer will often release a diagnostic code that refers to the oxygen sensor and/or its circuit. The first thought would be "Well, I'd better change my oxygen sensor." Actually, the intake leak is forcing more air into the combustion chamber than is required and the fuel/air mixture has become lean. The oxygen sensor relays the information to the computer which cannot compensate for the increased amount of oxygen and, as a result, the computer will store a fault code for the oxygen sensor.

The testing information in the following sections is generalized and applies to most fuel injection and feedback carburetor components. In order to solidify your diagnosis, it may be necessary to consult a factory service manual for the exact specification(s) for your vehicle.

MAP (Manifold Absolute Pressure) sensor

What it is and how it works

The MAP sensor reports engine load to the computer which uses the information to adjust spark advance and fuel enrichment **(see illustration)**. The MAP sensor measures intake manifold pressure and vacuum on the absolute scale (from zero instead of from sea-level atmospheric pressure [14.7 psi] as most gauges and sensors do). The MAP sensor reads vacuum and pressure through a hose connected to the intake manifold. A pressure-sensitive ceramic or silicon element and electronic circuit in the sensor generates a voltage signal that changes in direct proportion to pressure.

Under low-load, high-vacuum conditions, the computer leans the fuel/air mixture and advances the spark timing for better fuel

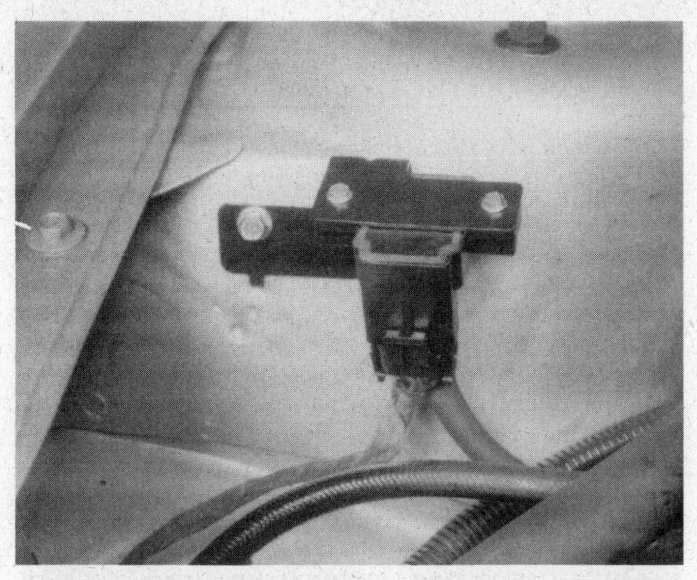

8.2 Here's a typical MAP sensor – this one, on a Plymouth Sundance, is located on the firewall, near the shock tower

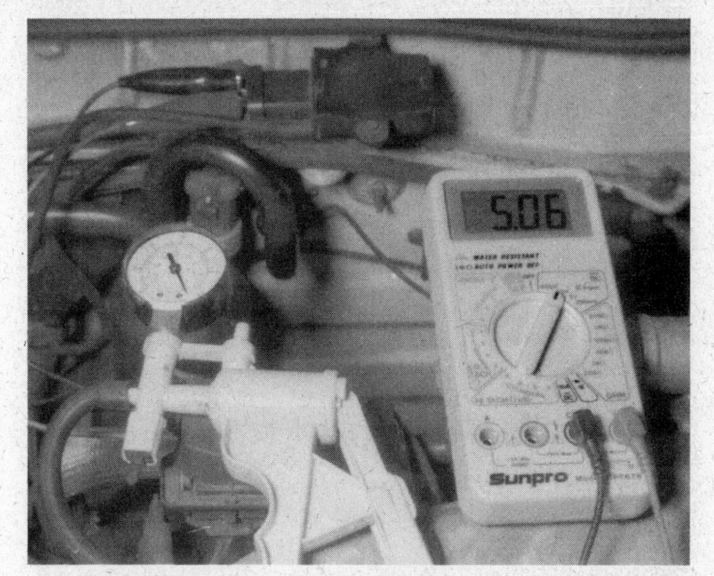

8.3 The MAP sensor voltage (measured at the signal wire) will decrease as vacuum is applied to the sensor

economy. Under high-load, low-vacuum conditions, the computer richens the fuel/air mixture and retards timing to prevent detonation. The MAP sensor serves as the electronic equivalent of both a vacuum advance on a distributor and a power valve in the carburetor.

Checking

Anything that hinders accurate sensor input can upset both the fuel mixture and ignition timing. This includes the MAP sensor itself as well as shorts or opens in the sensor wiring circuit and/or vacuum leaks in the intake manifold or vacuum hose. Some of the most typical driveability symptoms associated with problems in the MAP sensor circuit include:

1) Detonation and misfire due to increased spark advance and a lean fuel mixture.
2) Loss of power and/or fuel economy and sometimes even black smoke due to retarded ignition timing and a very rich fuel mixture.
3) Poor fuel economy.
4) Hard starts and/or stalling.

Note: *A vacuum leak in the hose to the MAP sensor causes the MAP sensor to indicate a higher than normal pressure (less vacuum) in the manifold, which makes the computer think the engine is under much more load than it really is. As a result, the ignition timing is retarded and the fuel mixture is richened.*

When the MAP sensor trouble code is detected, be sure to first check for vacuum leaks in the hoses or electrical connectors or wiring damage in the MAP sensor circuit. Kinks in the line, blockage or splits can occur and deter the sensor's ability to respond accurately to the changes in the manifold pressure. Check for anything that is obvious and easily repaired before actually replacing the sensor itself.

A MAP sensor will typically produce a voltage signal that will drop with decreasing manifold pressure (rising vacuum). Test specifications will vary according to the manufacturer and engine type. A typical MAP sensor will read 4.6 to 4.8 volts with 0 in-Hg vacuum applied to it **(see illustration)**. Raise it to 5 in-Hg vacuum and the reading should drop to about 3.75 volts. Raise it up again to 20 in-Hg and the reading should drop to about 1.1 volts.

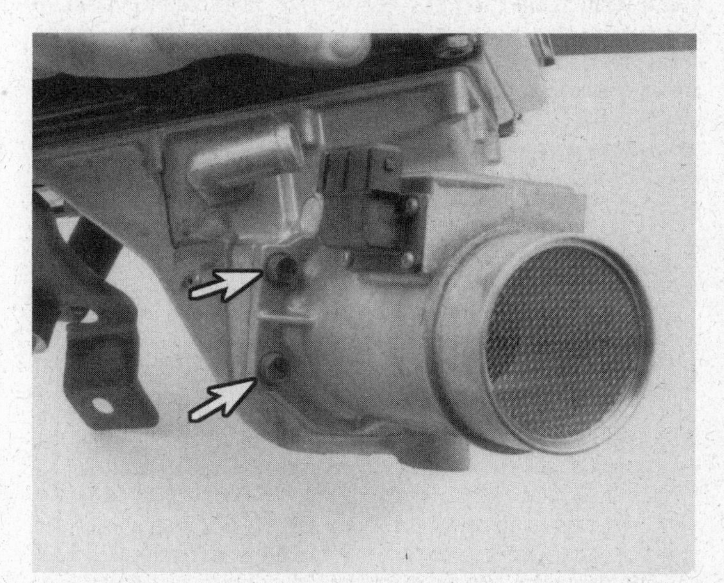

8.4 Here's a typical air flow sensor (this one's from a Nissan Maxima) – to remove it, remove the bolts (arrows)

MAF (Mass Air Flow) sensor

What it is and how it works

The MAF sensor is positioned in the fresh air intake **(see illustration)**, and it measures the amount of air entering the engine. Mass airflow sensors come in two basic varieties; hot wire and hot film. Both types work on the same principle, though they are designed differently. They measure the volume and density of the air entering the engine so the computer can calculate how much fuel is needed to maintain the correct fuel/air mixture.

MAF sensors have no moving parts. Contrary to the vane air flow sensors (see below) that use a spring-loaded flap, MAF sensors use an electrical current to measure airflow. There are two types of sensing elements; platinum wire (hot wire) or nickel foil grid (hot film). Each one is heated electrically to keep the temperature higher than the intake air temperature. With hot-film MAF sensors, the film is heated 170-degrees F warmer than the incom-

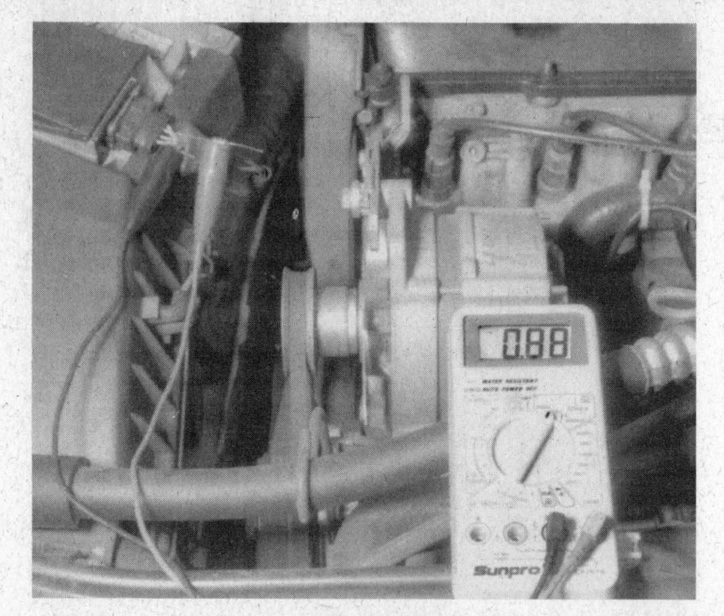

8.5 The signal voltage on a typical Bosch MAF sensor will read 0.60 to 0.80 volts at idle and . . .

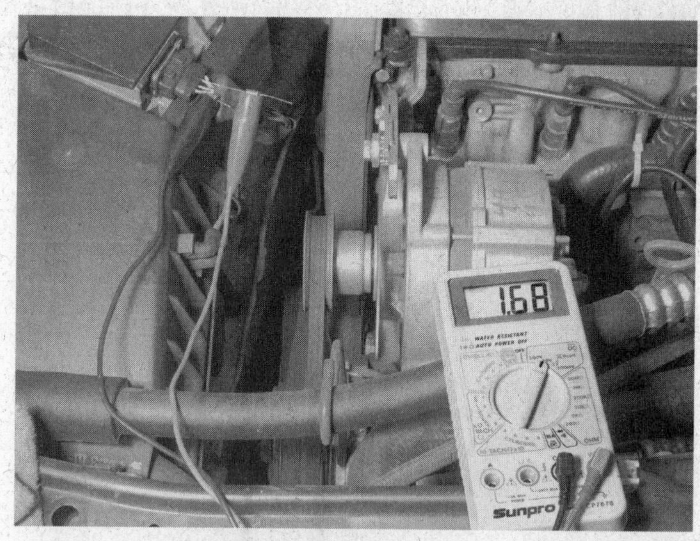

8.6 . . . when the engine's rpm are raised to 2,500 to 3,500, the voltage increases to approximately 1.50 to 2.20 volts

ing air temperature. On hot-wire MAF sensors, the wire is heated to 210-degrees F above the incoming air temperature. As air flows past the element it cools the element and thereby increases the amount of current necessary to heat it up again. Because the necessary current varies directly with the temperature and the density of the air entering the intake, the amount of current is directly proportional to the air mass entering the engine. This information is fed into the computer and the fuel mixture is directly controlled according to the conditions.

Checking

The most effective method for testing the MAF sensor is measuring the sensor's output or its effect on the injector pulse width. On Bosch or Ford hot-wire systems, the voltage output can be read directly with a voltmeter by probing the appropriate sensor terminals **(see illustrations)**. Refer to the your factory service manual for the correct terminal designations and specifications. If the voltage readings are not within range or the voltage fails to INCREASE when the throttle is OPENED with the engine running, the sensor is faulty and must be replaced with a new part. A dirty wire or a contaminated wire (a direct result of a faulty self-cleaning circuit) will deliver a slow response of the changes in airflow to the computer. Also, keep in mind that the self-cleaning circuit is controlled by relays. So, check the relays first if the MAF sensor appears to be sluggish or not responsive. Proper diagnosis of the MAF sensor is very important because this part is usually available only from a dealership parts department and the cost can be somewhat expensive. Be sure to check the diagnostic codes, if available, to make sure they indicate a problem with the MAF sensor rather than the MAF sensor circuit. If the wiring checks out and all other obvious areas are checked carefully, replace the sensor.

Another way to check MAF sensor output is to see what effect it has on injector pulse width (if this specification is available). Using a multimeter or oscilloscope that reads milliseconds, connect the positive probe directly to any injector signal wire and the negative probe to a ground terminal **(see illustration)**. Remember that one injector terminal is connected to the supply voltage (battery voltage) and the other is connected to the computer (signal wire) which varies the amount of time the injector is grounded. **Note:** *Typically, if by chance you connect to the wrong side of the injector connector, one wire will give you a steady reading (battery voltage) while the signal wire will fluctuate slightly.* Look at the pulse width at idle or while cranking the engine. The injector pulse width will vary with different conditions. If the MAF sensor is not producing a signal, the pulse width will typically be FOUR times longer than the correct width. This will indicate an excessively rich fuel/air mixture.

8.7 Checking an MAF sensor (this one's on a Ford) – this test requires a special multimeter that detects pulse width variations

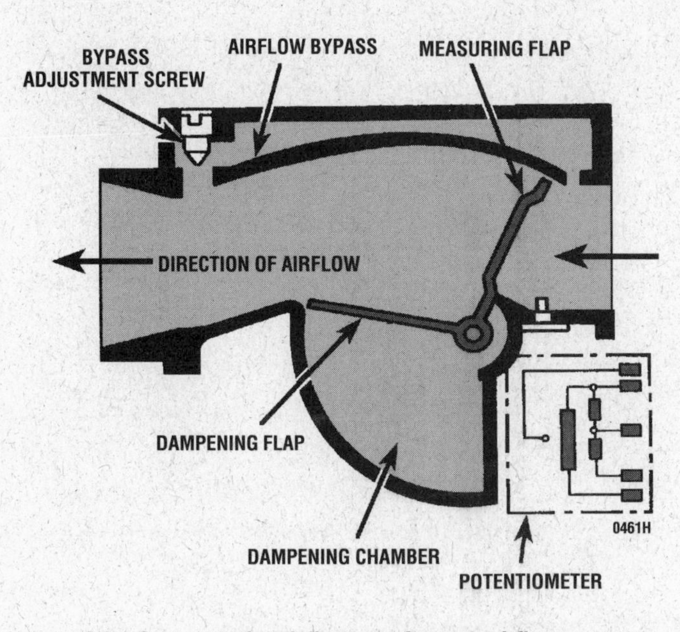

8.8 A crossectional diagram of a vane airflow sensor

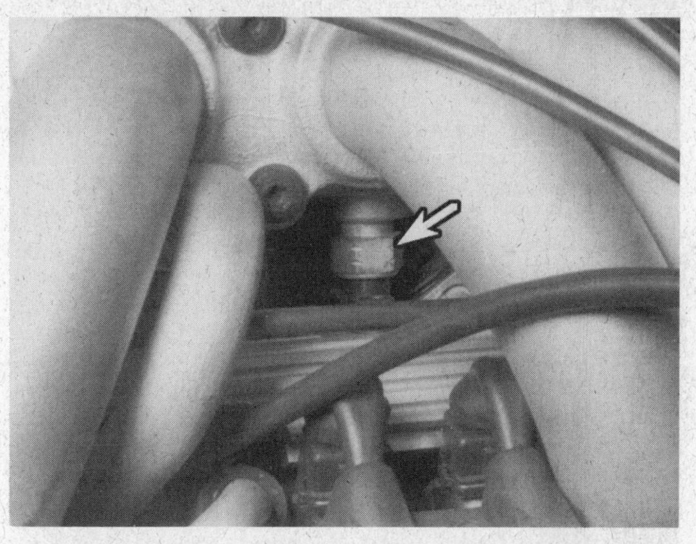

8.9 Here's a typical MAT sensor (this one's on a 1985 or later Corvette) – it's located in the underside of the air intake plenum

VAF (Vane Air Flow) sensor

What it is and how it works

VAF sensors are positioned in the air intake stream ahead of the throttle, and they monitor the volume of air entering the engine by means of a spring-loaded flap **(see illustration)**. The flap is pushed open by the air entering the system and a potentiometer (variable resistor) attached to the flap will vary the voltage signal to the computer according to the volume of air entering the engine (angle of the flap). The greater the airflow, the further the flap is forced open.

VAF sensors are used most commonly on Bosch L-Jetronic fuel injection systems, Nippondenso multi-port fuel injection systems and certain Ford multi-port fuel injection systems (Thunderbird, Mustang and Probe).

Checking

Diagnosing VAF sensors is quite different from diagnosing MAF or MAP sensors. Vane airflow sensors are vulnerable to dirt and grease. Unfiltered air that gets by a dirty or torn air filter will build up on the flap hinge or shaft, causing the flap to bind or hesitate as it swings. Remove the air intake boot and gently push open the flap with your finger; it should open and close smoothly. If necessary, spray a small amount of carburetor cleaner on the hinge and try to loosen the flap so it moves freely.

Disconnect the electrical connector to the VAF sensor. Install an ohmmeter to the electrical connector on the VAF sensor; the resistance should vary evenly as the flap opens and closes. If the resistance changes erratically or skips and jumps, you will have to replace the VAF with a new unit. **Note:** *Be sure to use an ANALOG ohmmeter for this check, since a digital meter will not usually register the rapid resistance changes that occur during this test.*

Another common problem to watch out for with VAF sensors is a bent or damaged flap caused by backfiring in the intake manifold. Some VAF sensors incorporate a "backfire" valve in the sensor body that prevents damage to the flap by venting any explosion. If the "backfire" valve leaks, the valve will cause the sensor to read low, consequently causing the engine to run on a rich fuel/air mixture.

The VAF sensor is manufactured as a sealed unit, preset at the factory with nothing that can be serviced except the idle mixture screw. Do not attempt to disassemble the unit if it is still under warranty, because tampering with the unit will void the warranty. Here again, be sure the diagnosis of the VAF sensor is complete and correct before buying a new unit because most often they are available only at a dealership parts department (and are expensive).

Air temperature sensor

What it is and how it works

The air temperature sensor is also known a a Manifold Air Temperature (MAT) sensor, an Air Charge Temperature (ACT) sensor, a Vane Air Temperature (VAT) sensor, a Charge Temperature Sensor (CTS), an Air Temperature Sensor (ATS) and a Manifold Charging Temperature (MCT) sensor. The sensor is located in the intake manifold or air intake plenum **(see illustration)** and detects the temperature of the incoming air. The sensor usually consists of a temperature sensitive thermistor which changes the value of its voltage signal as the temperature changes. The computer uses the sensor signal to richen or lean the fuel/air mixture, and, on some applications, to delay the EGR valve opening until the manifold temperature reaches normal operating range.

Checking

The easiest way to check an air temperature sensor is to remove it from the manifold, then hook up an ohmmeter to its terminals and check the resistance when the sensor is cold. Then warm up the tip of the sensor with a blow drier (never a propane torch!) and watch for a decrease in resistance. No change in resistance indicates the sensor is defective. When reinstalling the sensor, be sure to use sealant on the threads so you don't end up with a vacuum leak.

On most GM vehicles equipped with the MAT sensor, a Code 23 or 25 will indicate a fault in the sensor (see Chapter 2). Be aware that problems with the EGR system might be caused by a defective MAT sensor.

TPS (Throttle Position Sensor)

What it is and how it works

The TPS or Throttle Position Sensor is usually mounted externally on the throttle body or carburetor. Some are inside the throttle body or carburetor. The TPS is attached directly to the throttle shaft and varies simultaneously with the angle of the throttle. Its job is to inform the computer about the rate of throttle opening and relative throttle position. A separate Wide Open Throttle (WOT) switch may be used to signal the computer when the throttle is wide open. The TPS consists of a variable resistor that changes resistance as the throttle changes its opening. By signaling the computer when the throttle opens, the computer can richen the fuel mixture to maintain the proper air/fuel ratio. The initial setting of the TPS sensor is very important because the voltage signal the computer receives tells the computer the exact position of the throttle at idle.

Checking

Throttle position sensors typically have their own types of driveability symptoms that can be distinguished from other information sensors. The most common symptom of a faulty or misadjusted sensor is hesitation or stumble during acceleration. The same symptom of a bad accelerator pump in a carbureted engine.

There are basically two voltage checks you can perform to test the Throttle Position Sensor. **Note:** *It is best to have the correct wiring diagram for the vehicle when performing the following checks.*

The first test is for the presence of voltage at the TPS sensor supply wire with the ignition key ON. The throttle position sensor cannot deliver the correct signal without the proper supply voltage. You can determine the function of each individual wire (ground, supply, signal wire) by probing each one with a voltmeter and checking the different voltages. The voltage that remains constant when the throttle is opened and closed will be the supply voltage. If there's no voltage at any of the wires, there's probably an open or short in the wiring harness to the sensor.

The second check is for the proper voltage change that occurs as the throttle opens and closes. As the throttle goes from closed-to-wide open, the voltage at the signal wire should typically increase smoothly from 1 volt to 5 volts. **Note:** *An alternate method for checking the range is the resistance test. Hook up an ohmmeter to the supply and signal wires. With the ignition key OFF, slowly move the throttle through the complete range* **(see illustration)**. *Observe carefully for any unusual changes in the resistance (the change should be smooth) as it increases from low to high.*

Also, check your diagnostic codes for any differences in the circuit failures versus the actual sensor failure. Be sure you have checked all the obvious items before replacing the throttle position sensor.

Adjusting

TPS's seldom need adjustment. However, most TPS's must be adjusted when they are replaced. Since different makes and models of vehicles have different specifications and procedures for adjusting the TPS, we recommend you refer to a factory service manual for your specific vehicle to adjust the TPS. Also, dealer service departments or other qualified shops can usually adjust the TPS for you for a nominal fee. **Note:** *The adjustment information in the following paragraph may not be applicable to your vehicle. It is only intended to familiarize you with a typical procedure.*

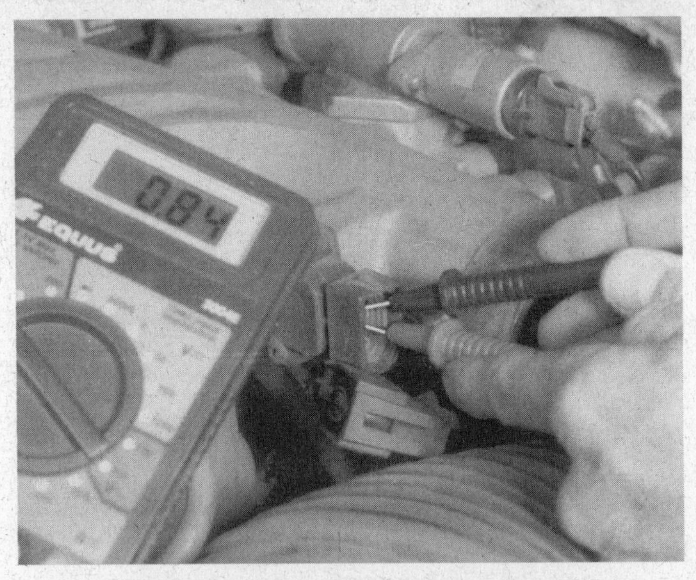

8.10 Slowly move the throttle and observe the resistance readings on the display – there should be a smooth transition as the resistance increases

Normally, you'll only need a voltmeter to adjust the TPS. Hook the meter up to the terminals specified in the manual and loosen the mounting screws. With the throttle in the specified position (usually against the throttle stop), rotate the sensor clockwise or counterclockwise until the specified voltage is obtained **(see illustration)**. Then retighten the mounting screws and check the voltage again.

Oxygen sensor

What it is and how it works

The oxygen sensor (also known as a Lambda or EGO sensor) is located in the exhaust manifold (or in the exhaust pipe, near the exhaust manifold) and produces a voltage signal proportional to the content of oxygen in the exhaust **(see illustration)**. A higher oxygen content across the sensor tip will vary the oxygen differential, thereby lowering the sensor's output voltage. On the other hand, lower oxygen content will raise the output voltage. Typically the voltage ranges from 0.10 volts (lean) to 0.90 volts (rich). The computer uses the sensor's input voltage to adjust the air/fuel mixture, leaning it out when the sensor detects a rich condition or enrichening it when it detects a lean condition. When the sensor reaches operating temperature (600-degrees F), it will produce a variable voltage signal based on the difference between the amount of oxygen in the exhaust (internal) and the amount of oxygen in the air directly surrounding the sensor (external). The ideal stochiometric fuel/air ratio (14.7:1) will produce about 0.45 volts.

There are basically two types of oxygen sensors on the market. The most popular type uses a zirconia element in its tip. The latest type of oxygen sensor uses a titania element. Instead of producing its own voltage, the titania element resistance will alter a voltage signal that is supplied by the computer itself. Although the titania element works differently than the zirconia element, the results are basically identical. The biggest difference is that the titania element responds faster and allows the computer to maintain more uniform control over a wide range of exhaust temperatures.

Contamination can directly affect the engine performance and life span of the oxygen sensor. There are basically three types of

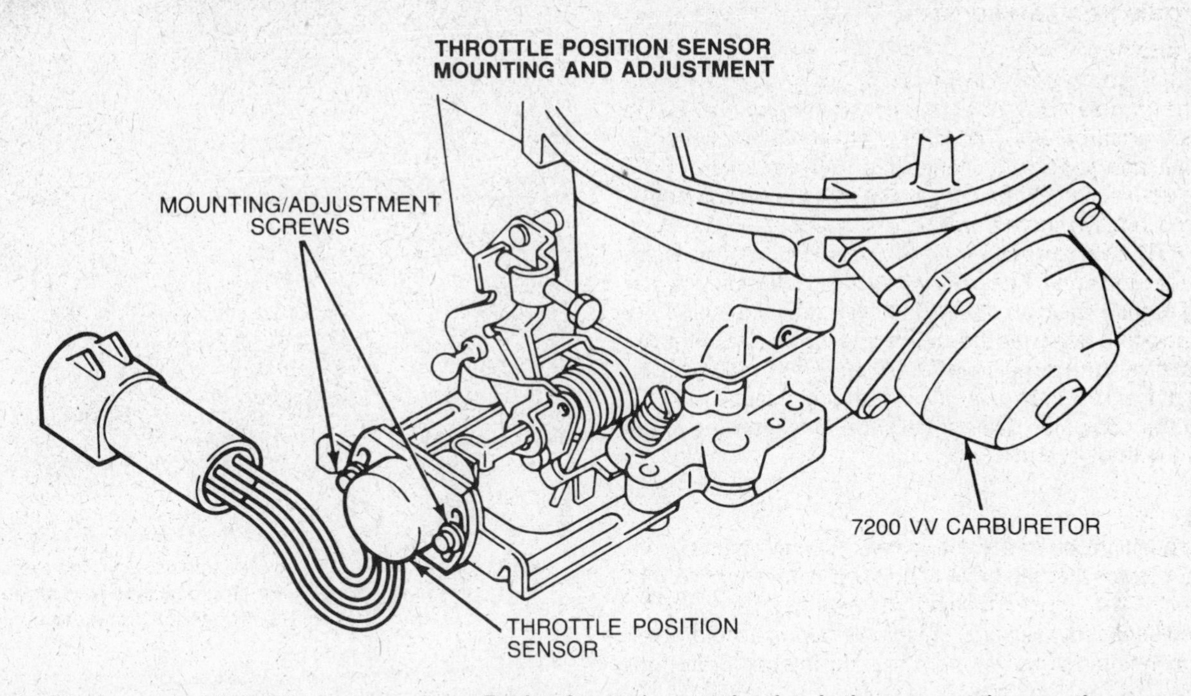

8.11 Here's a typical TPS (this one's used on a Ford carburetor) – note the slots in the sensor at the mounting screws; these allow for adjustment

contamination; carbon, lead and silicon. Carbon buildup due to a rich-running condition will cause inaccurate readings and increase the problem's symptoms. Diagnose the fuel injection system or carburetor feedback controls for correct fuel adjustments. Once the system is repaired, run the engine at high rpm without a load (parked in the driveway) to remove the carbon deposits. Avoid leaded gasoline as it causes contamination of the oxygen sensor. Also, avoid using old-style silicone gasket sealant (RTV) that releases volatile compounds into the crankcase which eventually wind up on the sensor tip. Always check to make sure the RTV sealant you are using is compatible with modern emission systems.

Before an oxygen sensor can function properly it must reach a minimum operating temperature of 600-degrees F. The warm-up period prior to this is called "open loop." In this mode, the computer detects a low coolant temperature (cold start) and wide open throttle (warm-up) condition. Until the engine reaches operating temperature, the computer ignores the oxygen sensor signals. During this time span, the emission controls are not precise! Once the engine is warm, the system is said to be in "closed loop" (using the oxygen sensor's input). Some manufacturers have designed an electric heating element to help the sensor reach operating temperature sooner. A typical heated sensor will consist of a ground wire, a sensor output wire (to the computer) and a third wire that supplies battery voltage to the resistance heater inside the oxygen sensor. Be careful when testing the oxygen sensor circuit! Clearly identify the function of each wire or you might confuse the data and draw the wrong conclusions.

Checking

Sometimes an apparent oxygen sensor problem is not the sensor's fault. An air leak in the exhaust manifold or a fouled spark plug or other problem in the ignition system causes the oxygen sensor to give a false lean-running condition. The sensor reacts only to the content of oxygen in the exhaust, and it has no way of knowing where the extra oxygen came from.

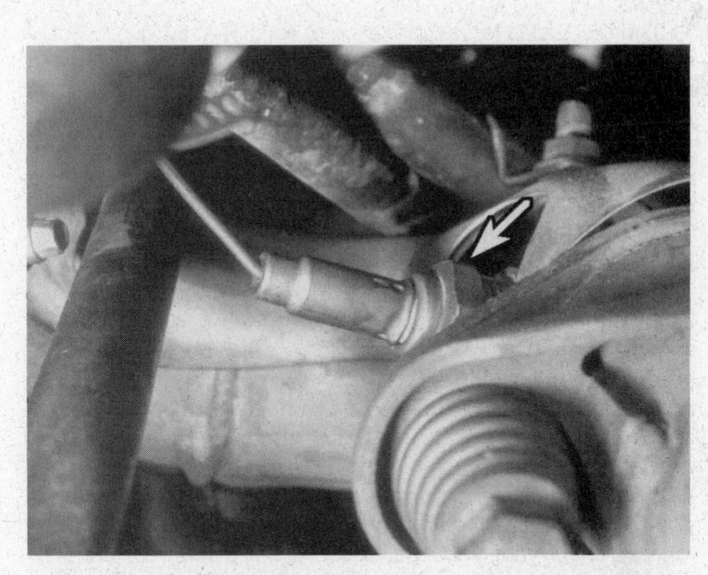

8.12 This oxygen sensor (arrow) is screwed into the exhaust manifold (GM 3.0L V6 engine shown)

When checking the oxygen sensor it is important to remember that a good sensor produces a fluctuating signal that responds quickly to the changes in the exhaust oxygen content. To check the sensor you will need a 10 megaohm digital voltmeter. Never use an ohmmeter to check the oxygen sensor and never jump or ground the terminals. This can damage the sensor.

Connect the meter to the oxygen sensor circuit. Select the mV (millivolt) scale. If the engine is equipped with a later style (heated

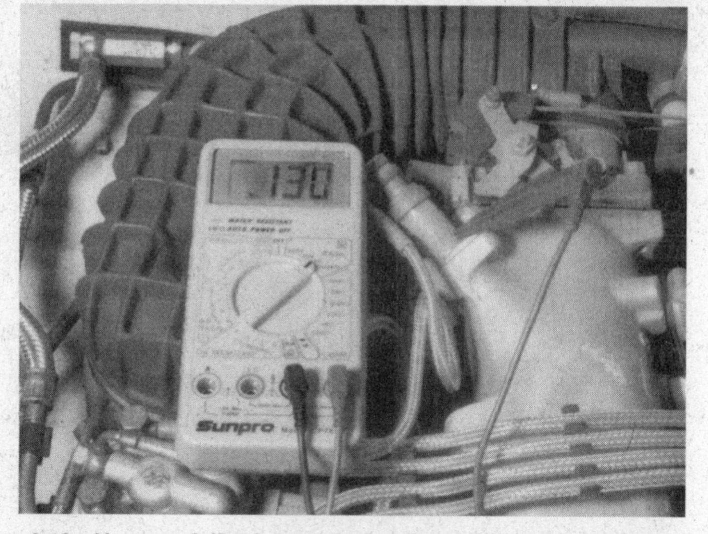

8.13 Very carefully observe the readings as the oxygen sensor cycles – note on paper the high and low values and try to come up with an average – also, if the VOM does not have a millivolt scale, just move the decimal point over; for example, 0.130 volt = 130 millivolts

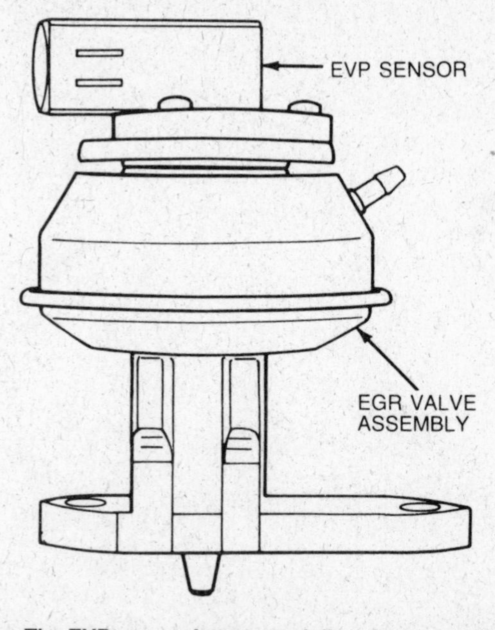

8.14 The EVP sensor is mounted directly on top of the EGR valve

oxygen sensor), be sure you are connected to the signal wire and not one of the heater or ground wires. Start the engine and let it idle. Typically, the meter will respond with a fluctuating millivolt reading when connected properly. Also, be sure the engine is in closed loop (warmed-up to operating temperature).

Watch very carefully as the voltage oscillates. The display will flash values ranging from 100 mV to 900 mV (0.100 to 0.900 V). The numbers will flash very quickly, so be observant. Record the high and low values over a period of one minute. With the engine operating properly, the oxygen sensor should average approximately 500 mV (0.500 V) **(see illustration)**.

To further test the oxygen sensor, remove a vacuum line and observe the readings as the engine stumbles from the excessively LEAN mixture. The voltage should LOWER to an approximate value of 200 mV (0.200 V). Install the vacuum line. Now, obtain some propane gas mixture (bottled) and connect it to a vacuum port on the intake manifold. Start the engine and open the propane valve (open the propane valve only partially and do so a little at a time to prevent over-richening the mixture). This will create a RICH mixture. Watch carefully as the readings INCREASE. **Warning:** *Propane gas is highly flammable. Be sure there are no leaks in your connections or an explosion could result.* If the oxygen sensor responds correctly to the makeshift lean and rich conditions, the sensor is working properly.

EVP (EGR Valve Position) sensors

What it does and how it works

The EGR Valve Position (EVP) sensor **(see illustration)** monitors the position of the EGR valve and keeps the computer informed on the exact amount the valve is open or closed. From this data, the computer can calculate the optimum EGR flow for the lowest NOx emissions and the best driveability, then control the EGR valve to alter the EGR flow by means of the EGR solenoid.

The EVP sensor is a linear potentiometer that operates very much like a Throttle Position Sensor (TPS). Its electrical resistance changes in direct proportion to the movement of the EGR valve stem. When the EGR valve is closed, the EVP sensor registers maximum resistance. As the valve opens, resistance decreases until it finally reaches a minimum value when the EGR valve is fully open.

Checking

Typical symptoms of a malfunctioning EGR valve position sensor include hesitating during acceleration, rough idling and hard starting. Be sure to distinguish between an EGR valve problem and an EGR valve position sensor problem. Consult the section on EGR valves for additional information on testing the EGR valve itself.

Generally, the EGR valve position sensor should change resistance smoothly as the EGR valve is opened and closed. Be sure to check the appropriate factory service manual to determine which terminals of the sensor to hook the ohmmeter up to (there's often more than two). A typical Ford EVP sensor should have no more than 5,500 ohms resistance when the EGR valve is closed and no less than 100 ohms when the valve is fully open.

Crankshaft position sensor

What it is and how it works

A crankshaft position sensor works very similarly to an ignition pick-up coil or trigger wheel in an electronic distributor **(see illustration)**. The crankshaft position sensor provides an ignition timing signal to the computer based on the position of the crankshaft. The difference between a crankshaft position sensor and a pick-up coil or trigger wheel is that the crankshaft position sensor reads the ignition timing signal directly off the crankshaft or harmonic balancer instead of from the distributor. This eliminates timing variations from backlash in the timing chain or distributor shaft. Crankshaft position sensors are necessary in most modern distributorless ignition (DIS) systems. Basically, the sensor reads the position of the crankshaft by detecting when pulse rings on the crankshaft or harmonic balancer pass by it **(see illustrations)**.

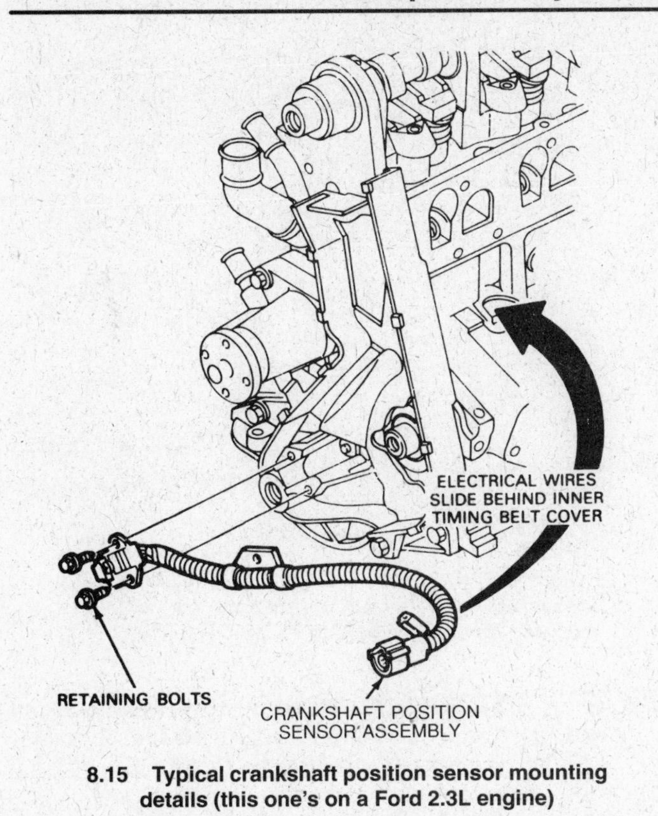

8.15 Typical crankshaft position sensor mounting details (this one's on a Ford 2.3L engine)

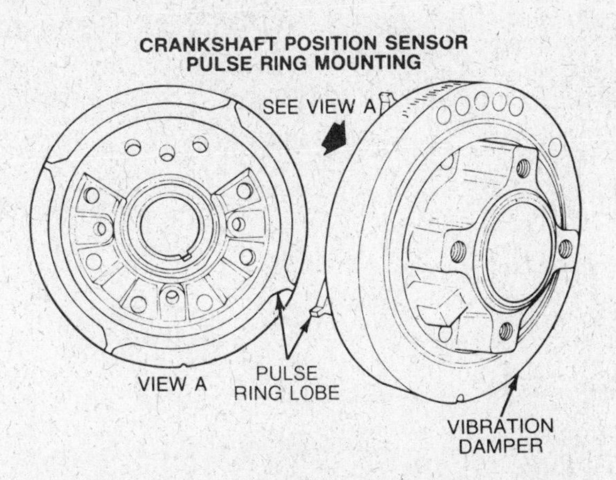

8.16 The crankshaft position sensor pulse rings are mounted on the harmonic balancer (vibration damper) on Ford V8 engines

8.17 On Ford V6 engines, the pulse rings are directly behind the crankshaft pulley, easily detected by the sensor

Checking

Most crankshaft position sensor problems can be traced to a fault in the wiring harness or connectors. These problems can cause a loss of the timing signal and consequently the engine will not start. When troubleshooting a crankshaft position sensor problem, it is advisable to follow the diagnostic flow chart in a factory service manual to isolate the faulty component. The problem could be in the ignition module, computer, wiring harness or crankshaft position sensor. Be aware of the interrelationship of these components.

If it is necessary to replace the sensor, be sure to install it correctly, paying attention to the alignment. Any rubbing or interference will cause driveability problems. Also, on variable reluctance type crankshaft position sensors, be sure to adjust the air gap properly. Consult a factory service manual for the correct specification.

VSS (Vehicle Speed Sensor)

What it is and how it works

Vehicle Speed Sensors (VSS) are used in modern vehicles for a number of different purposes. One purpose is to monitor the vehicle speed so the computer can determine the correct time for torque converter clutch (TCC) lock-up. The sensor may also provide input to the computer to control the function of various other emissions systems components based on vehicle speed. On some GM vehicles, the signal from the VSS is used by the computer to reset the Idle Air Control valve as well as the canister purge valve. Another purpose is to assist with the power steering. Here, the sensor input is used by the electronic controller to vary the amount of power assist according to the vehicle speed. The lower the speed, the greater the assist for easier maneuverability for parking. The higher the speed, the less the assist for better road feel. Another purpose is to change the position of electronically adjustable shock absorbers used in ride control systems. The ride control systems in Mazda 626's and Ford Probes automatically switch the shocks to a "firm" setting above 50 mph in the AUTO mode and "extra firm" in the SPORT mode. Also, vehicle speed sensors replace the mechanical speedometer cable in some modern vehicles.

Checking

The driveability symptoms of a faulty vehicle speed sensor depend on what control functions require an accurate speed input. For example, on some GM vehicles, the idle quality may be affected by a faulty sensor. Other symptoms include hard steering with increased speed, premature torque converter lock-up or fluctuating or inaccurate speedometer readings.

Different vehicles require different testing techniques. It is best to consult a factory service manual for the specific test procedures for your particular vehicle. Also, it is rare, but sometimes the input shaft on the transmission has broken or missing teeth that affect the accuracy of the sensor's reading.

8.18 Here's a typical knock sensor (this one's on a Corvette), mounted low on the side of the engine block

8.19 On many GM models, information from the knock sensor is sent to the Electronic Spark Control (ESC) module (arrow), which retards ignition timing if detonation is evident

Knock sensor

What it is and how it works

The knock sensor (sometimes called an Electronic Spark Control [ESC] sensor) is an auxiliary sensor that is used to detect the onset of detonation **(see illustrations)**. Although the knock sensor influences ignition timing, it doesn't have direct impact on the fuel and emission systems. It affects ignition timing only.

The sensor, which is usually mounted on the intake manifold or engine block, generates a voltage signal when the engine vibrations are between 6 to 8 Hz. The location of the sensor is very critical because it must be positioned so it can detect any vibrations from the most detonation-prone cylinders. On some engines, it is necessary to install two knock sensors.

When the knock sensor detects a pinging or knocking vibration, it signals the computer to momentarily retard ignition timing. The computer then retards the timing a fixed number of degrees until the detonation stops.

This system is vital on turbocharged vehicles to achieve maximum performance. When the knock control system is working properly, the maximum timing advance for all driving conditions is achieved.

Checking

The most obvious symptom of knock sensor failure will be an audible pinging or knocking, especially during acceleration under a light load. Light detonation usually does not cause harm, but heavy detonation over a period of time will cause engine damage. Knock sensors sometimes are fooled by other sounds such as rod knocks or worn timing chains. Reduced fuel economy and poor performance result from the constantly retarded timing.

Another thing to keep in mind is that most engine detonation has other causes than the knock sensor. Some causes include:

1) Defective EGR valve
2) Too much compression due to accumulated carbon in the cylinders
3) Overadvanced ignition timing
4) Lean fuel/air mixture; Possible vacuum leak
5) Overheated engine
6) Low-octane fuel

To check the knock sensor, use a wrench to rap on the intake manifold (not too hard or you may damage the manifold!) near the sensor while the engine is idling. Never strike the sensor directly. Observe the timing mark with a timing light. The vibration from the wrench will produce enough of a shock to cause the knock sensor to signal the computer to back off the timing. The timing should retard momentarily. If nothing happens, check the wiring, electrical connector or computer for any obvious shorts or problems. If the wiring and connectors are OK, the sensor is probably faulty.

The knock sensor is a sealed unit. If it is defective, replace it with a new part.

Output actuators

The output actuators receive commands from the computer and actuate the correct engine response after all the data and parameters have been analyzed by the computer. Output devices can be divided into three categories: solenoids, electric motors and controller modules.

Solenoids include the EGR solenoid, Canister Purge (CANP) solenoid, carburetor feedback solenoid (FBC), Electronic Air Control Valve (EACV), Torque Converter Clutch (TCC) solenoid and fuel injectors.

Electric motors include the Idle Speed Control motor (ISC), the fuel pump and cooling fan.

Controller modules are used to control more than one device. Modules can control air conditioner and cooling fan response as well as ignition functions.

The following discusses operation and diagnosis of the most common types of output actuators.

Fuel injectors

What they are and how they work

Fuel injectors are electro-mechanical devices which both meter and atomize fuel delivered to the engine **(see illustration)**. The injectors on multi-point fuel injection systems are usually mounted in the lower intake manifold **(see illustration)** and positioned so that their tips are directing fuel in front of each engine intake valve. On vehicles equipped with Throttle Body Injection (TBI), the injector(s) are mounted in the throttle body (the carburetor-like device on the intake manifold) **(see illustration)**.

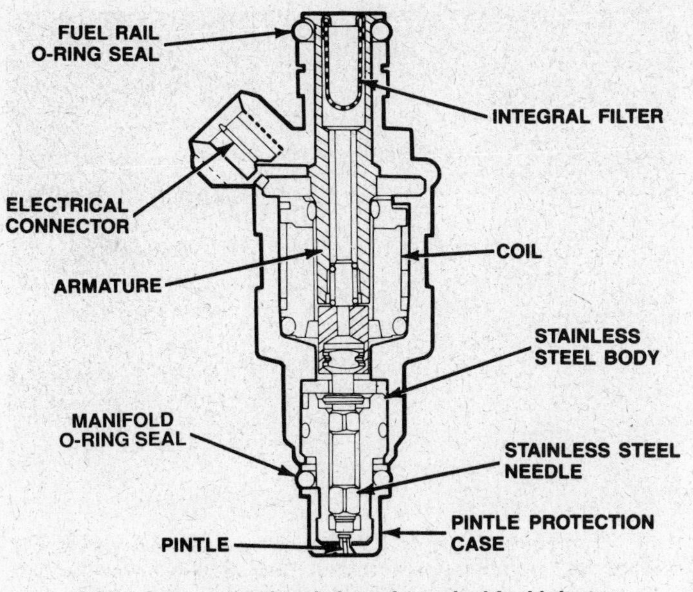

8.20 A cross-sectional view of a typical fuel injector (multi-point type shown)

8.21 Removing a fuel injector from a typical throttle-body-type fuel injection system (GM throttle body shown)

The injector bodies consist of a solenoid-actuated pintle and needle valve assembly. An electrical signal from the computer activates the solenoid, causing the pintle to move inward off the seat and allow fuel to inject into the engine.

Fuel flow to the engine is controlled by how long the solenoid is energized, since the injector flow orifice is fixed and the fuel pressure drop across the injector tip is constant. This duration can be measured electronically and is referred to as the injector pulse width.

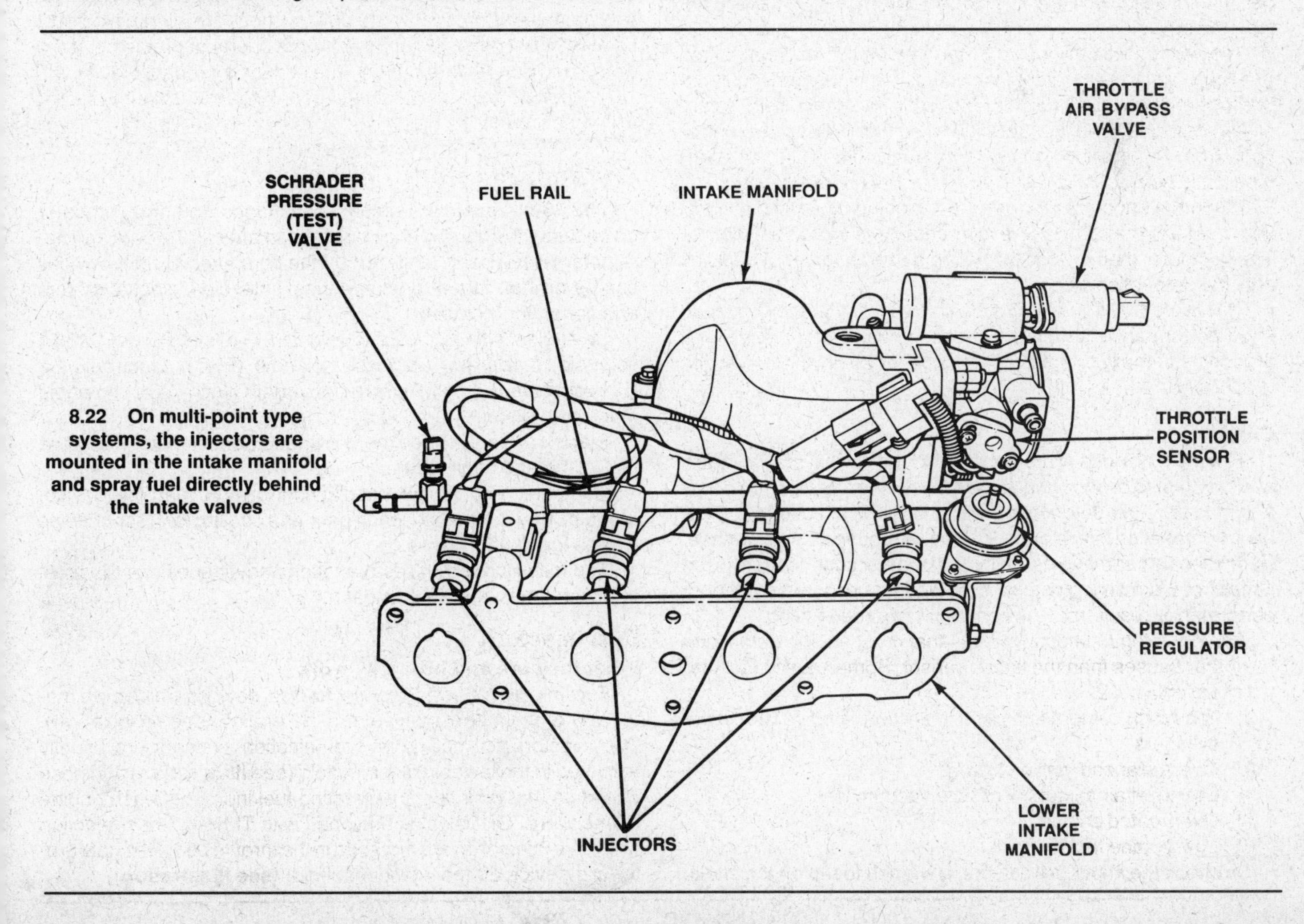

8.22 On multi-point type systems, the injectors are mounted in the intake manifold and spray fuel directly behind the intake valves

8.23 Hook up a vacuum gauge to the EGR-valve side of the solenoid – with the engine running at about 2,000 rpm, the gauge should register at least ten in-Hg of vacuum

8.24 Connect a test light across the EGR solenoid wire harness terminals with the ignition switch ON (engine off) – the light should come on

Checking

On multi-point fuel injectors, check to make sure the wiring and electrical connectors to each fuel injector is/are in good shape and securely connected. With the engine running, listen to each injector through a mechanic's stethoscope (a long screwdriver can also be used – place the tip of the screwdriver against each injector and place the handle against your ear). **Warning:** *be careful of rotating engine components when performing this check.* If you can hear a rapid clicking noise, the injector is functioning, although it could still be clogged or leaking. Many auto repair shops can clean the injectors for a reasonable charge. Also, fuel injector cleaners that can be added to the fuel tank are available at auto parts stores and are often very effective at cleaning injectors. Further diagnosis of multi-point injectors should be left to a dealer service department or other qualified auto repair shop.

On throttle body injectors, check to make sure the wiring and electrical connectors are secure and in good shape. With the engine running, remove the air cleaner or air intake duct to expose the top of the throttle body. Look at the spray pattern coming from the injector(s). The pattern should be even, conical in shape and reaching all the way to the throttle bore walls. Also, the injectors should not be dripping, both with the engine running and with it shut off. If the spray pattern is not as described, have the injector(s) cleaned by an auto repair shop or try some fuel injector cleaner in the fuel tank (available at auto parts stores). If cleaning does not correct the injector problem, it may need to be replaced (refer to the *Haynes Automotive Repair Manual* for your particular vehicle or the vehicle's factory service manual).

EGR valve solenoid

What it is and how it works

On computer-controlled vehicles, the action of the EGR valve is usually controlled by commanding the EGR control solenoid(s). Refer to the information earlier in this Chapter on EGR valve position sensors for additional information concerning these systems. The EGR valve solenoid is computer controlled and located in the vacuum line between the EGR valve and vacuum source. It opens and closes electrically to maintain finer control of EGR flow than is possible with ported-vacuum-type systems. The computer uses information from the coolant temperature, throttle position and manifold pressure sensors to regulate the EGR valve solenoid.

During cold operation and at idle, the solenoid circuit is grounded by the computer to block vacuum to the EGR valve. When the solenoid circuit is not grounded by the computer, vacuum is allowed to the EGR valve.

Checking

Note: *For further information and checking procedures for the EGR system, see Chapter 1 and Section 1 of this Chapter.*

First, inspect all vacuum hoses, wires and electrical connectors associated with the EGR solenoid and system. Make sure nothing is damaged, loose or disconnected.

Locate the vacuum line that runs from the vacuum source to the EGR solenoid. Disconnect it at the solenoid and hook up a vacuum gauge to the hose. Start the engine, bring it to normal operating temperature and observe the vacuum reading. There should be at least ten in-Hg of vacuum. If not, repair the hose to the vacuum source. Disconnect the gauge and re-connect the hose.

If there is at least ten in-Hg vacuum to the EGR solenoid, locate the vacuum hose running from the EGR solenoid to the EGR valve. Disconnect and plug the hose at the solenoid and hook up a vacuum gauge to the solenoid. Open the EGR solenoid by starting the engine and raising the speed to about 2,000 rpm. With the solenoid open, the vacuum gauge should read at least ten in-Hg **(see illustration)**. If there is no vacuum, either the solenoid valve is defective or there is a problem in the wiring circuit or computer.

To check the wiring circuit, disconnect the electrical connector from the EGR solenoid. With the ignition on and the engine off, connect a test light across the two terminals of the connector **(see illustration)**. The test light should come on. If it does not, there is a problem in the wiring or computer. If the light does come on, but there is no vacuum from the EGR solenoid to the EGR valve, the solenoid is probably defective. Check the solenoid's resistance. Normally, it should not be less than about 20 ohms.

ISC (Idle Speed Control) motor

See the heading in Section 6, *Carburetor controls* for information on the ISC motor.

EACV (Electronic Air Control Valve)

What it is and how it works

The EACV (sometimes called an Idle Air Control [IAC] valve) changes the amount of air bypassed (not flowing through the throttle valve) into the intake manifold in response to the changes in the electrical signals from the computer. EACV's are usually located on the throttle body, although some are mounted remotely. After the engine starts, the EACV opens, allowing air to bypass the throttle and thus increase idle speed. While the coolant temperature is low, the EACV remains open to obtain the proper fast idle speed. As the engine warms up, the amount of bypassed air is controlled in relation to the coolant temperature. After the engine reaches normal operating temperature, the EACV is opened, as necessary, to maintain the correct idle speed.

Checking

To check the EACV valve circuit, connect the positive probe of a voltmeter to the signal wire in the EACV connector and the negative probe to the ground **(see illustration)**. Check the voltage as the engine starts to warm up from cold to warm conditions. Most EACV valves will indicate an increase in voltage as the system warms and the valve slowly cuts off the additional air. Consult a factory service manual for the correct voltage specifications for your vehicle. If the voltage is correct, but the EACV valve is not opening or closing to provide the correct airflow, replace the valve.

8.25 Observe the voltage reading at the EACV connector

TCC (Torque Converter Clutch) solenoid

What it is and how it works

Lock-up torque converters are installed on newer vehicles to help eliminate torque converter slippage and thus reduce power loss and poor fuel economy. The torque converter is equipped with a clutch that is activated by a solenoid valve. The computer determines the best time to lock up the clutch device based on data it receives from various sensors and switches.

When the vehicle speed sensor indicates speed above a certain range and the coolant temperature sensor is warm, the Throttle Position Sensor (TPS) determines the position of the throttle (acceleration or deceleration) and the transmission sensor relays the particular gear the transmission is operating in to the computer for a complete analysis of operating parameters. If all parameters are within a certain range, the computer sends an electrical signal to the clutch, telling it to lock up. Needless to say, diagnosing a problem in this system can become complicated.

Checking

One symptom of TCC failure is a clutch that will not disengage, causing the engine to stall when slowing to a stop. Another symptom is an increase in engine rpm at cruising speed, resulting in decreased fuel economy (this usually means the converter is not locking up). If the converter is not locking up, the driver might not notice any differences unless he/she checks fuel consumption and the increase in tachometer readings. Without the TCC operating, the engine will turn an additional 300 to 500 rpm at cruising speed to maintain the same speed. Also, when the converter is not locking up, there is a chance the transmission will overheat and become damaged due to the higher operating temperatures.

Before diagnosing the TCC system as defective, make some preliminary checks. Check the transmission fluid level, linkage adjustment and the condition of the vacuum lines. After you've checked that all the basics are in order, check for any trouble codes (see Chapter 2). Further diagnosis should be referred to a dealer service department or other qualified repair shop.

Vacuum diagrams and VECI labels

General Information

The illustrations in this Chapter are intended to help you further your understanding of emissions control systems by studying the way components are connected on various types of vehicles. They do not attempt to cover all vehicles, but rather are a representative sampling.

To better understand your vehicle, year and model you need to study the Vehicle Emission Control Information (VECI) label located under the hood (for more information on VECI labels, see Chapter 1). The VECI label contains tune-up specifications and vital information regarding the emission control devices and vacuum hose routing.

If your vehicle does not have a VECI label, it's possible that it never came with one or that it's missing. Sometimes when body panels are replaced with used ones, the VECI labels get replaced along with them, and often this results in the wrong VECI label being displayed under the hood. Check with the dealer parts department if your VECI label is missing or you think you have the wrong one. The dealer parts department can get the right one for you.

One more thing you could do if the VECI label is missing is take a photo or make a drawing of the vacuum hoses and connections before any work is started. Reassembly will then be simpler.

Chrysler

(including Dodge and Plymouth)

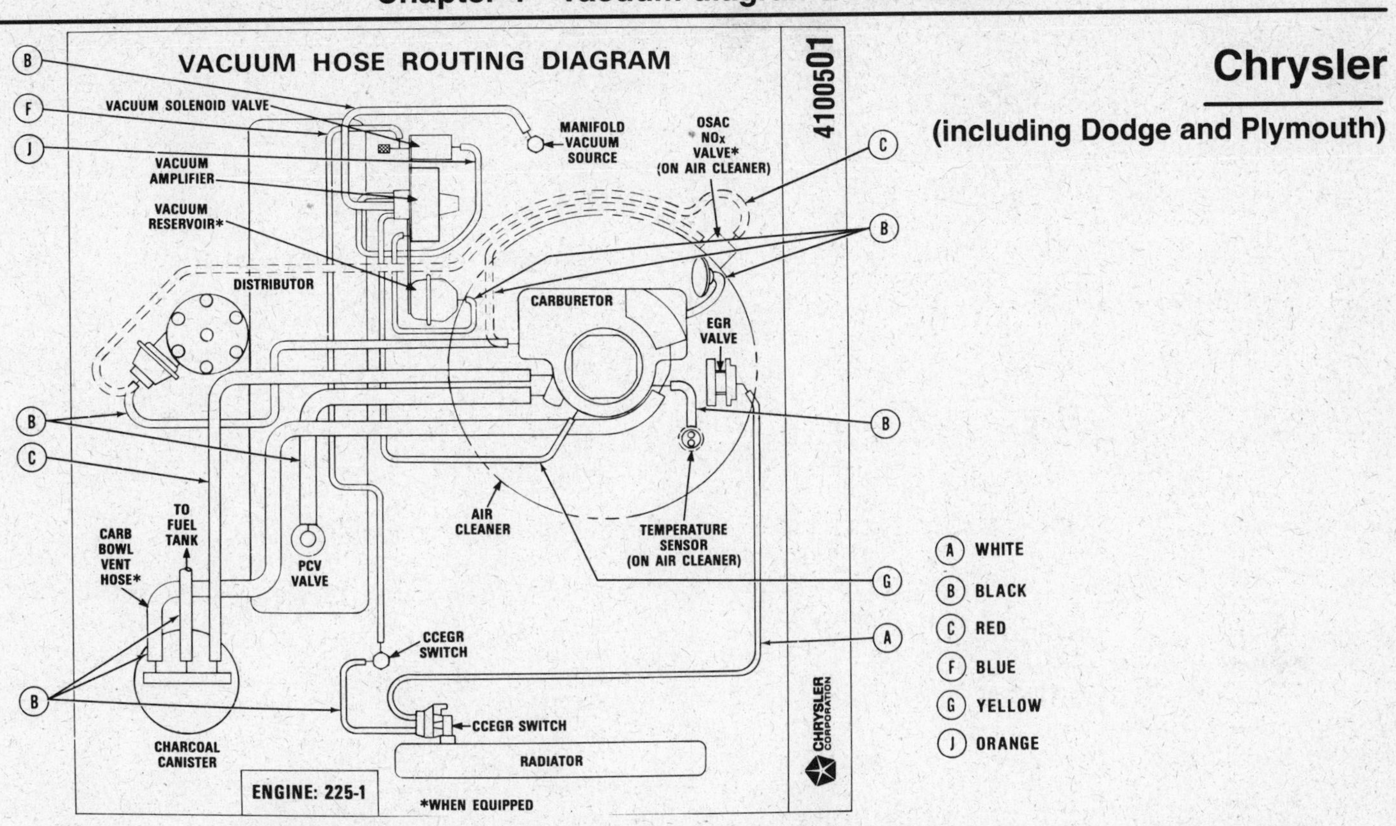

Vacuum hose routing diagram for the 1978 225 slant six-cylinder engine with 1 BBL carburetor (Federal and Canada, manual and automatic transmission)

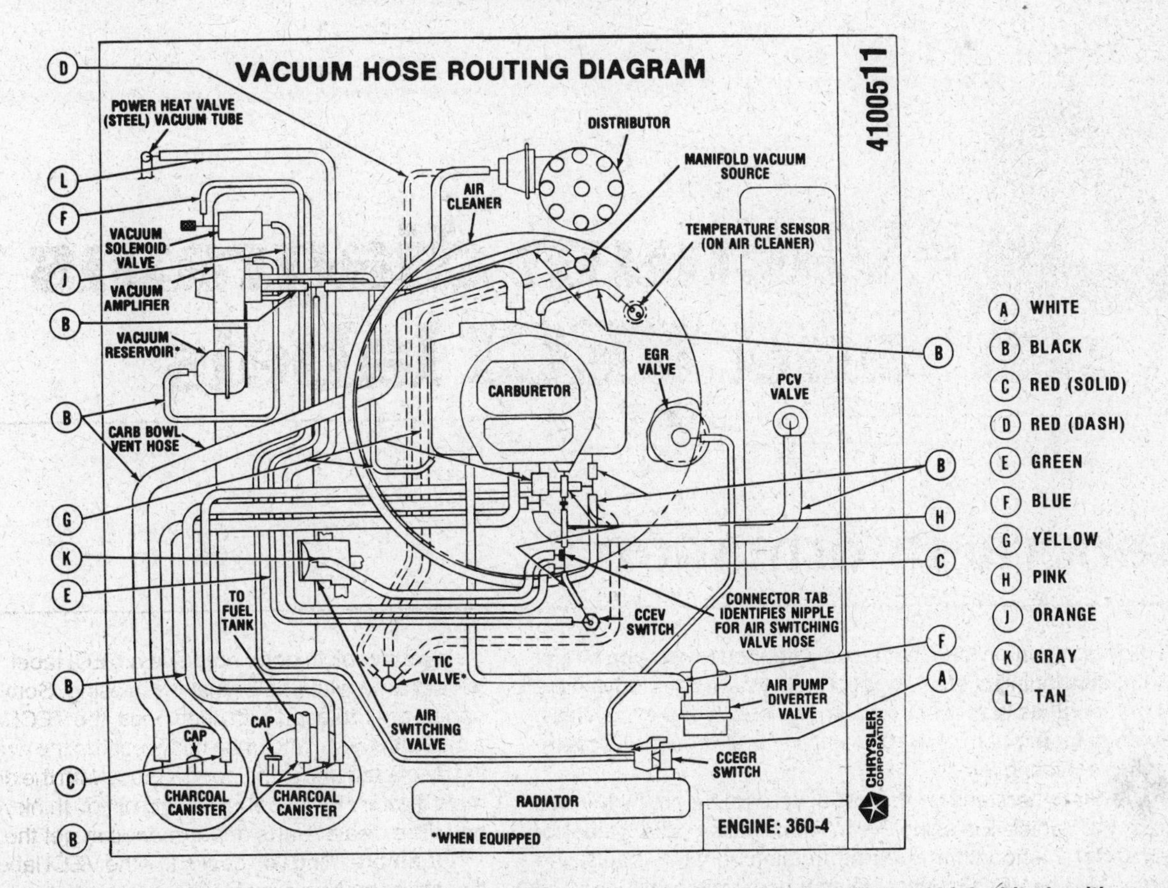

Vacuum hose routing diagram for the 1978 360 (V-8) engine, 4 BBL carburetor, B and C bodies (Federal high altitude with automatic transmission)

Chrysler

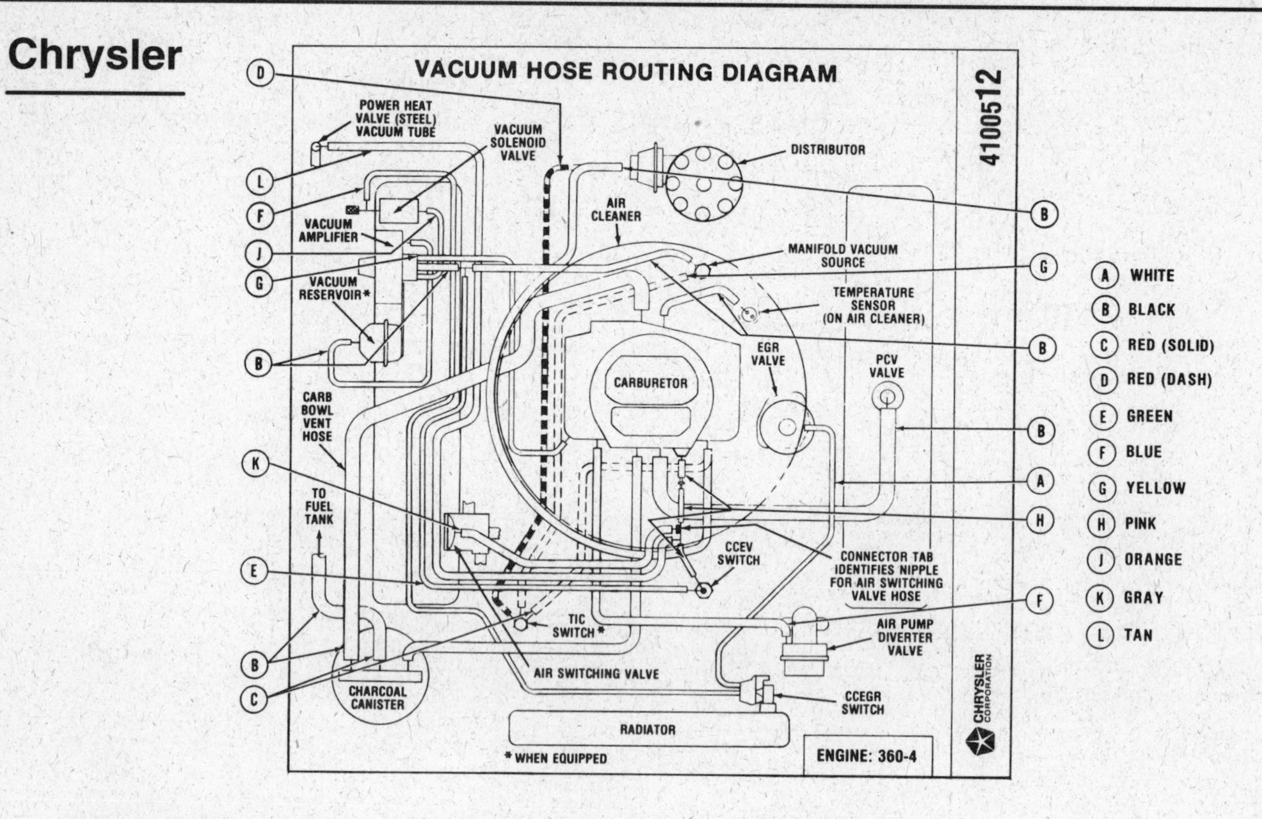

Vacuum hose routing diagram for the 1978 360 (V-8) engine, 2 and 4 BBL carburetor (Federal high altitude and California F and M bodies with automatic transmission)

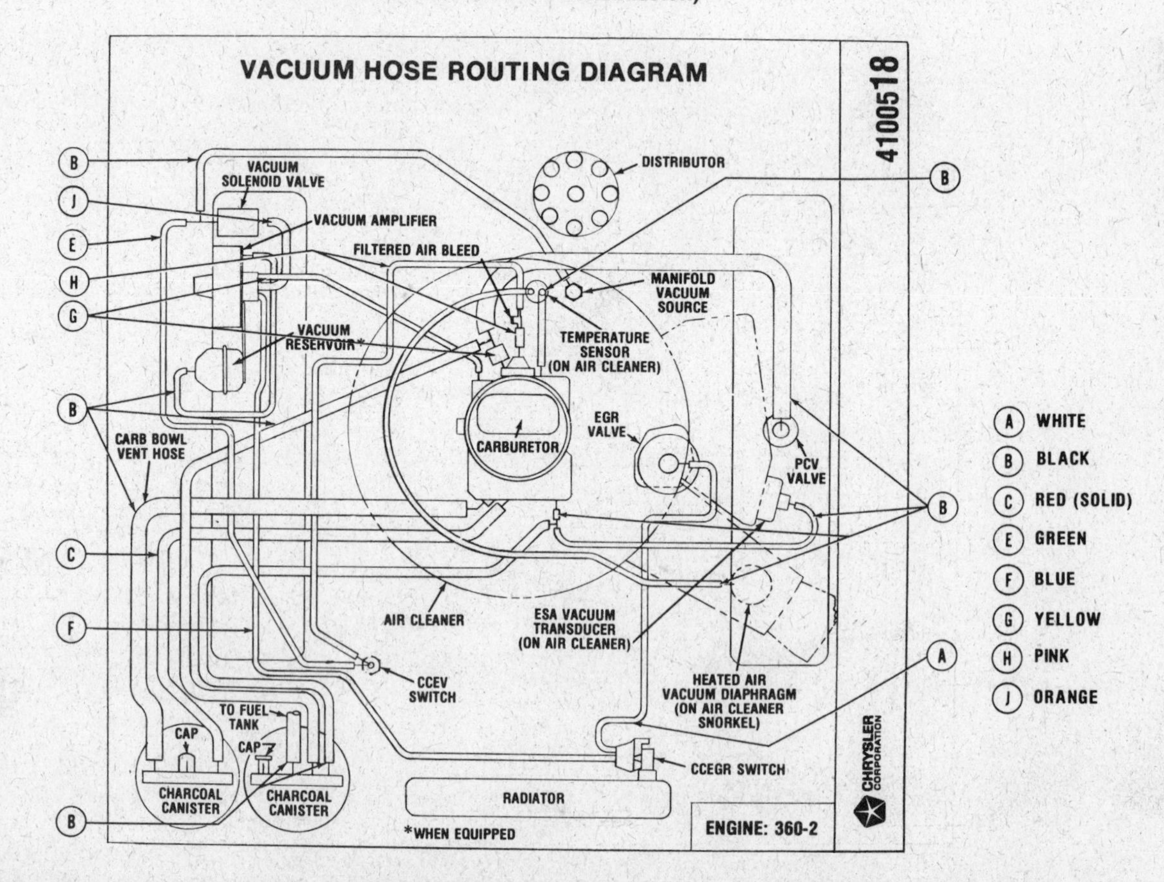

Vacuum hose routing diagram for the 1978 360 (V-8) engine, 2 BBl carburetor (Federal with ELB and catalyst, C body, automatic transmission)

Chrysler

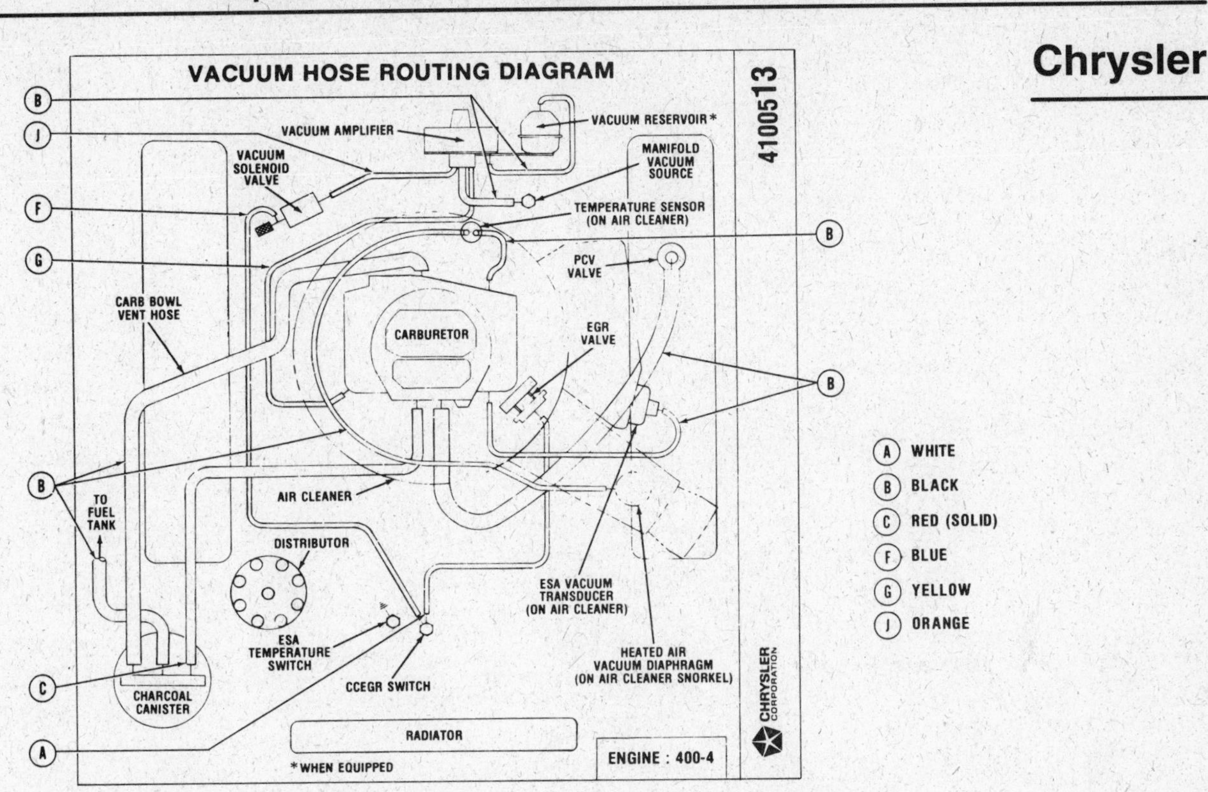

Vacuum hose routing diagram for the 1978 400 (V-8) engine, 4 BBL carburetor (Federal with ELB and catalyst, all except C body with automatic transmission)

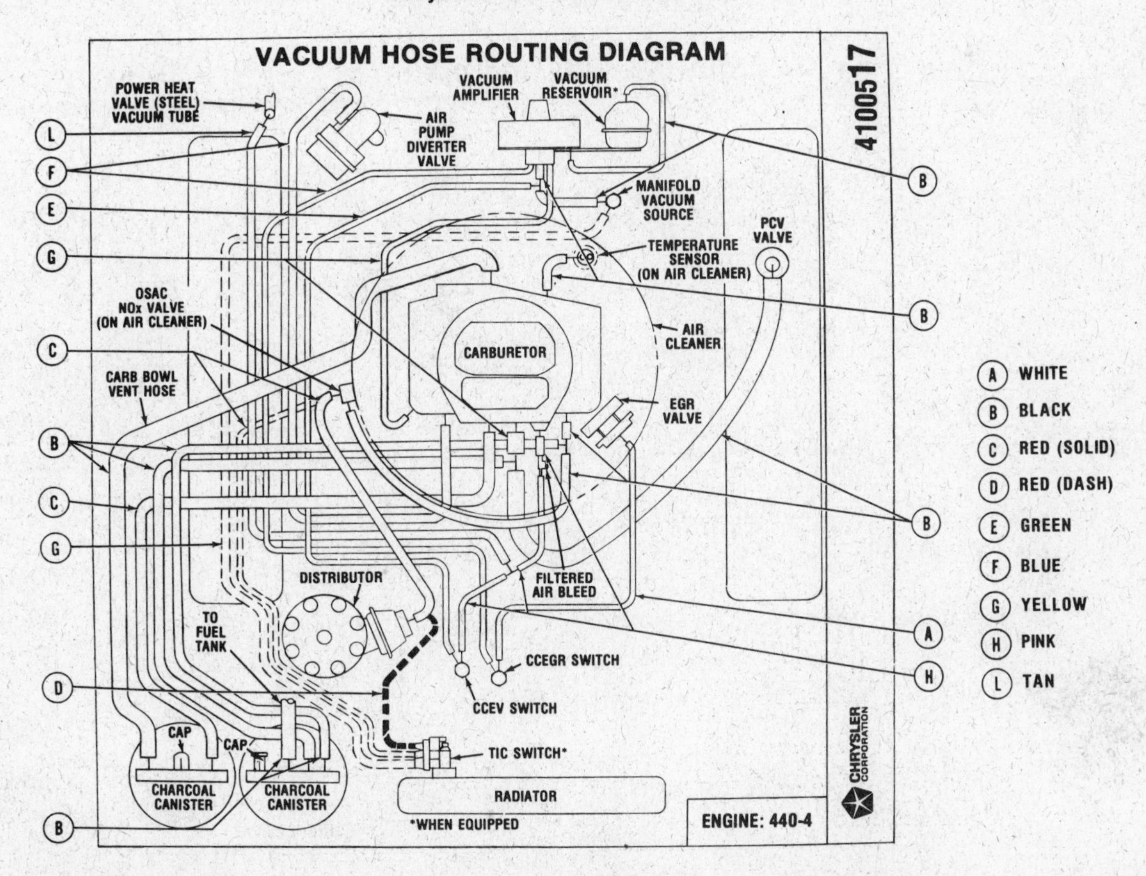

Vacuum hose routing diagram for the 1978 440 (V-8) engine, 4 BBL carburetor (California with catalyst and air pump and automatic transmission)

Chrysler

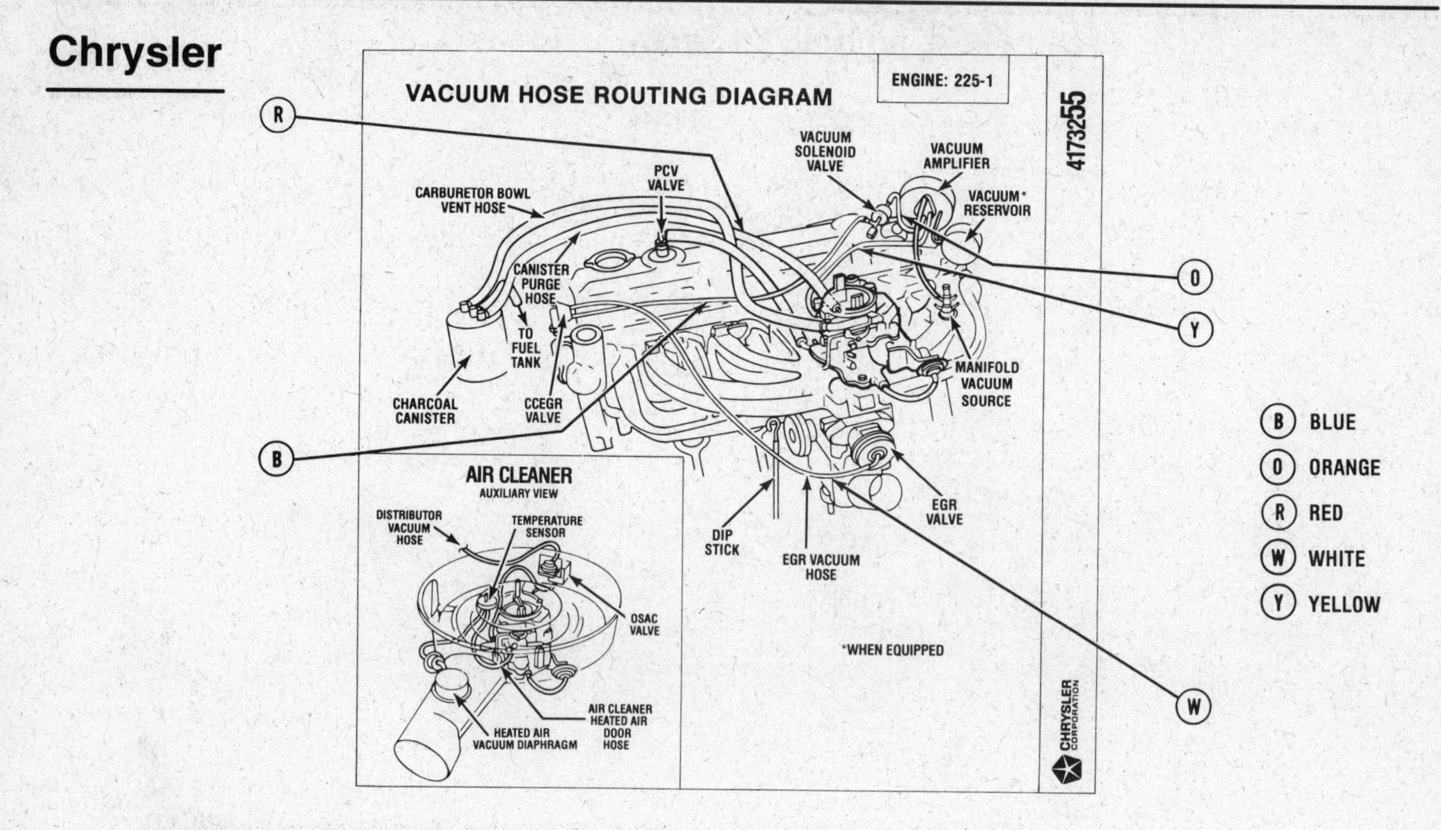

Vacuum hose routing diagram for the 1979 225 slant six-cylinder engine, 1 BBL carburetor (Federal and Canada)

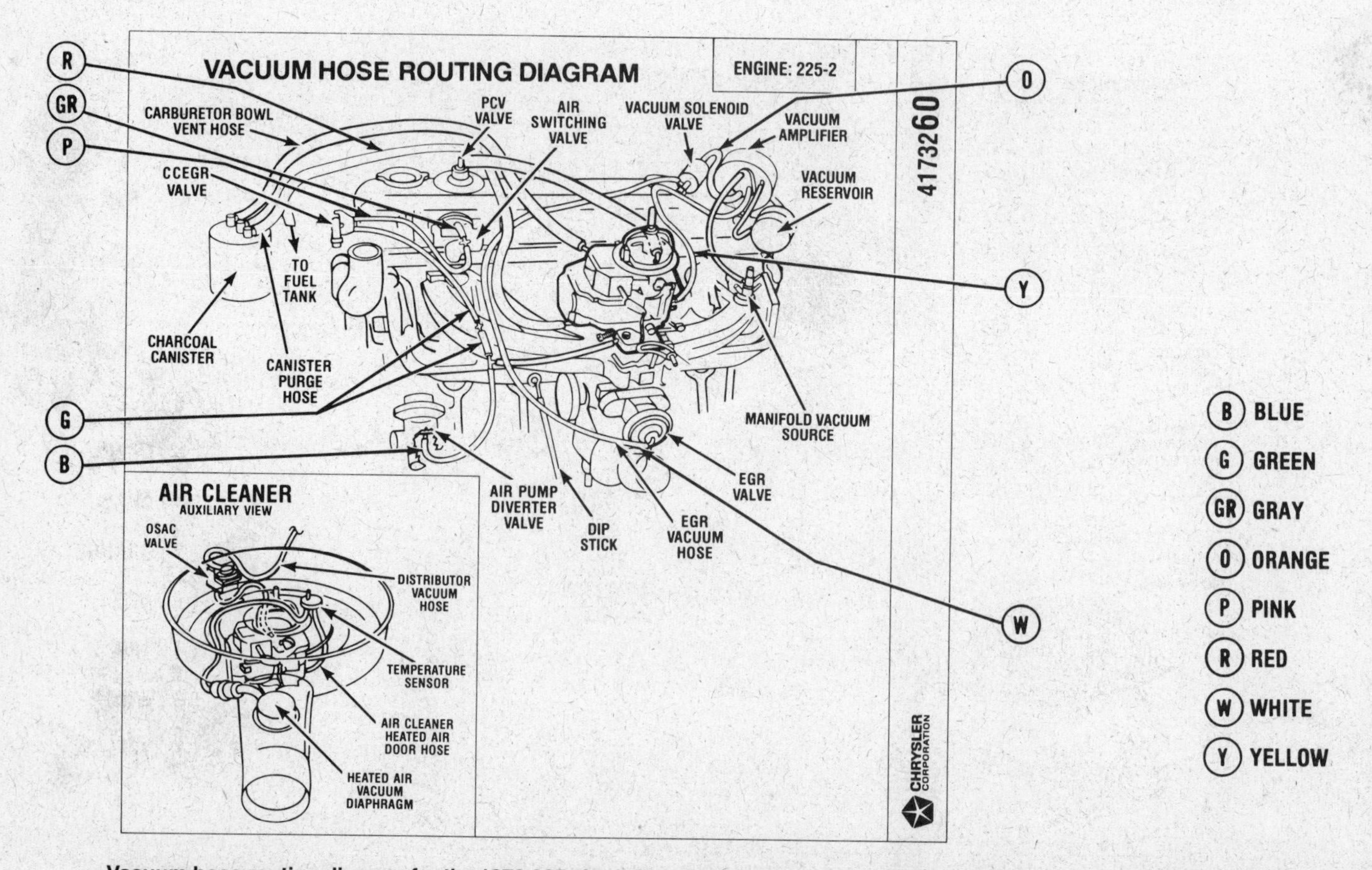

Vacuum hose routing diagram for the 1979 225 slant six-cylinder engine with the #4000 2 BBL carburetor (Federal with automatic transmission)

Chrysler

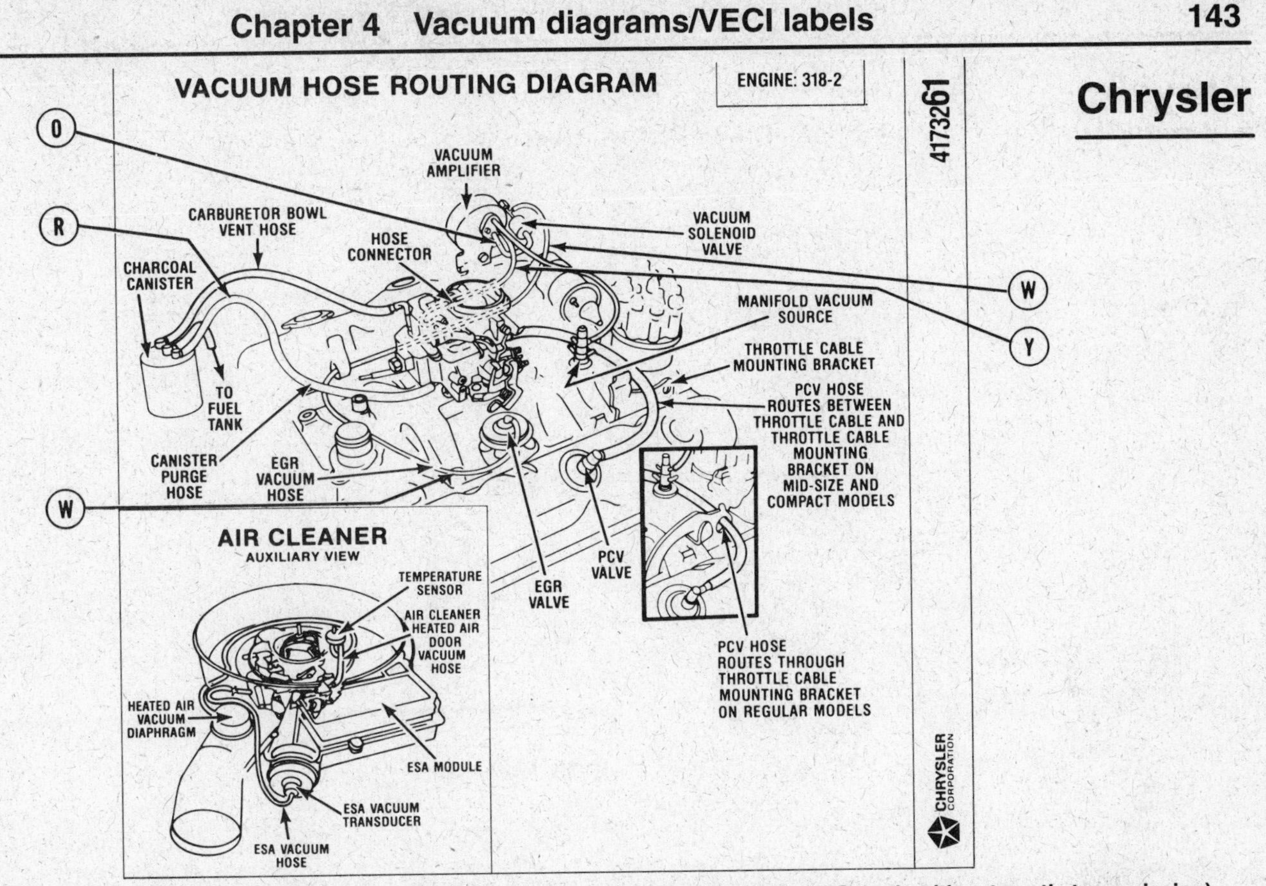

Vacuum hose routing diagram for the 1979 318 (V-8) engine, 2 BBL carburetor (Federal and Canada with automatic transmission)

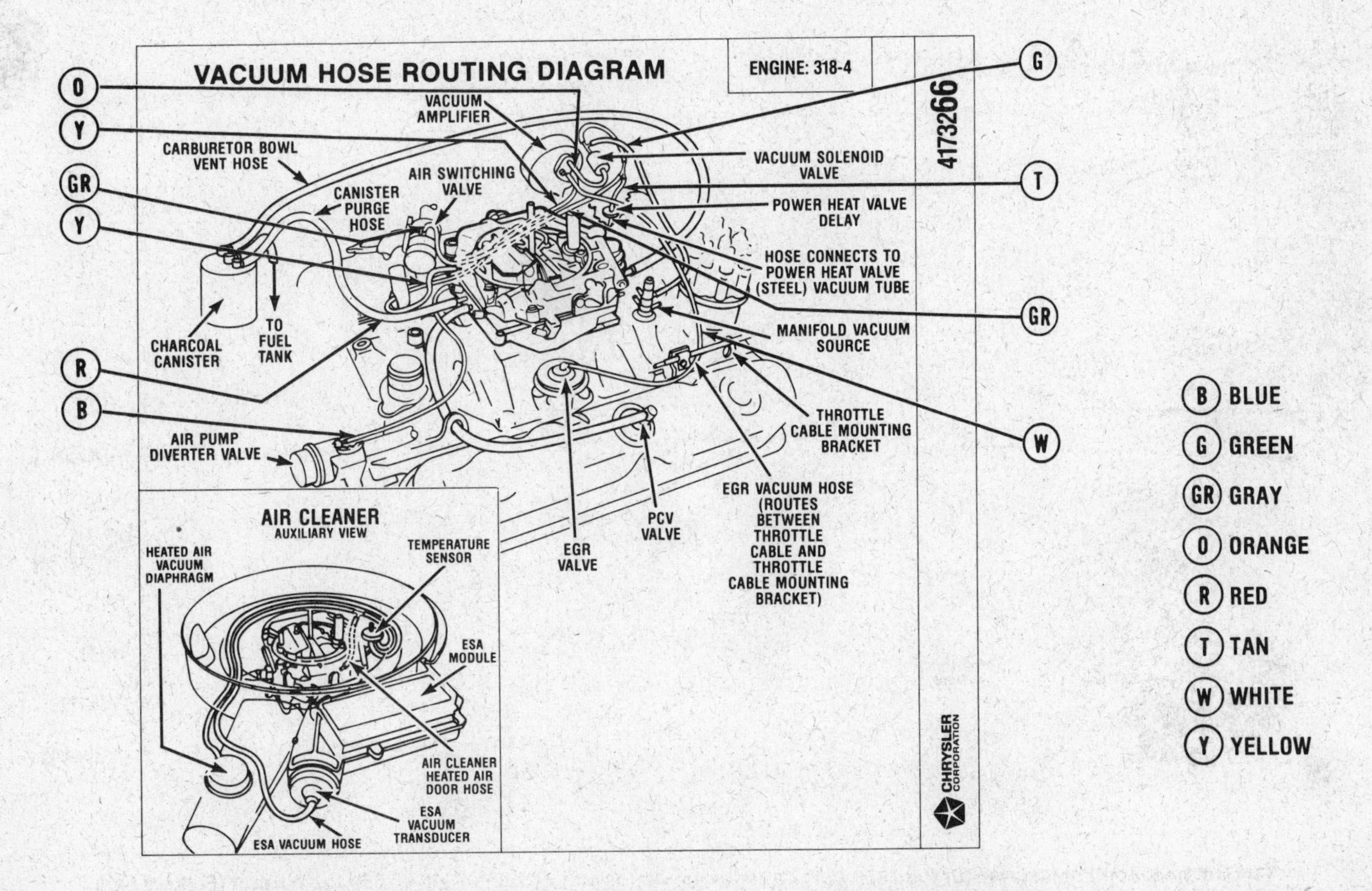

Vacuum hose routing diagram for the 1979 318 (V-8) engine, 4 BBL carburetor (California with automatic transmission)

Chrysler

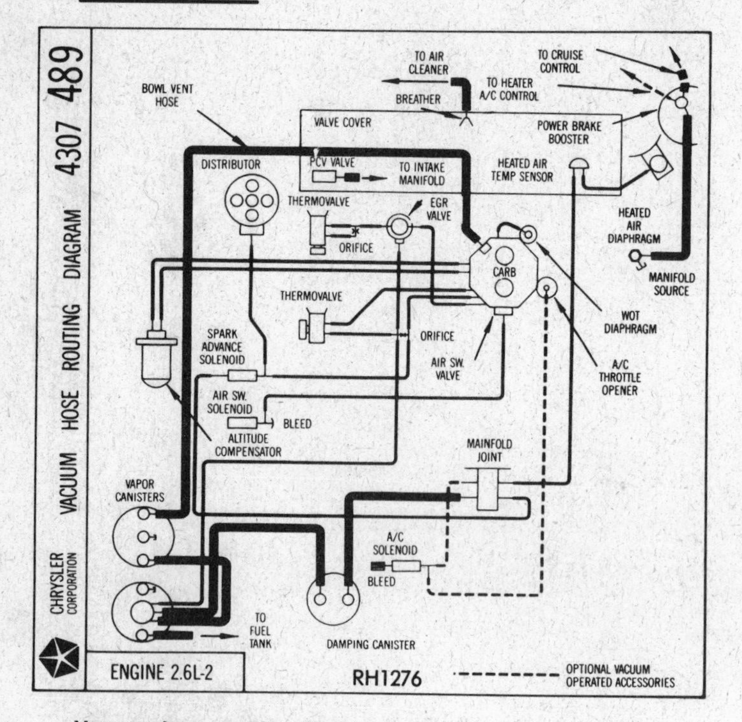

Vacuum hose routing diagram for the Chrysler, Dodge and Plymouth Mini vans with the 2.6L engine, 2 BBL carburetor (for California)

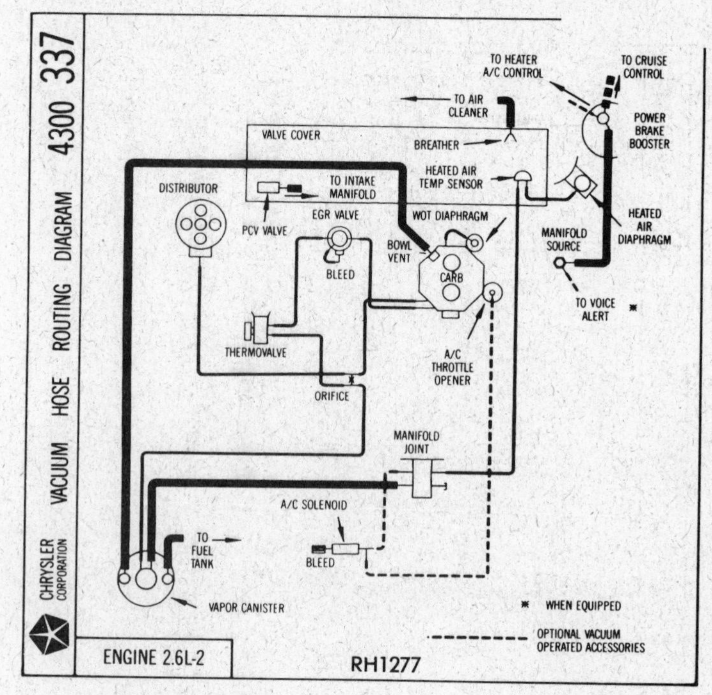

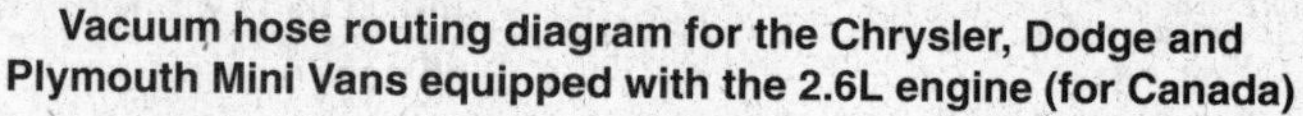

Vacuum hose routing diagram for the Chrysler, Dodge and Plymouth Mini Vans equipped with the 2.6L engine (for Canada)

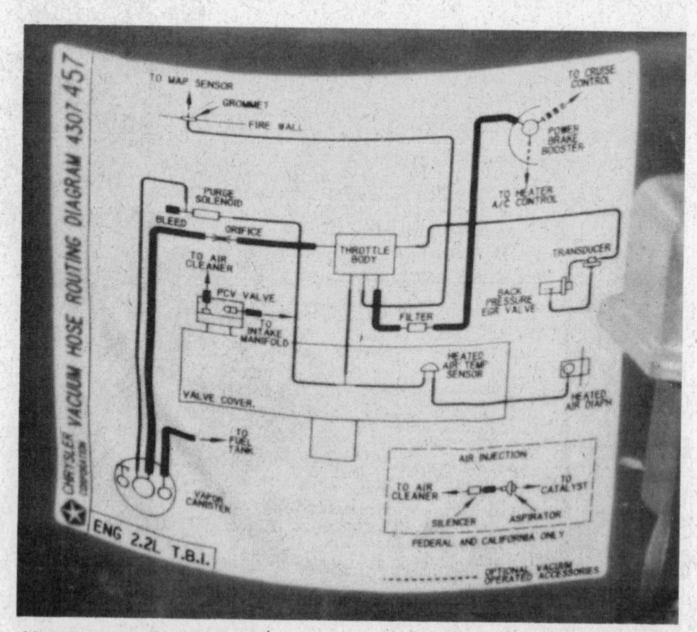

Vacuum hose routing diagram for the Chrysler Laser and Dodge Charger/Daytona equipped with the 2.2L engine with TBI (for California)

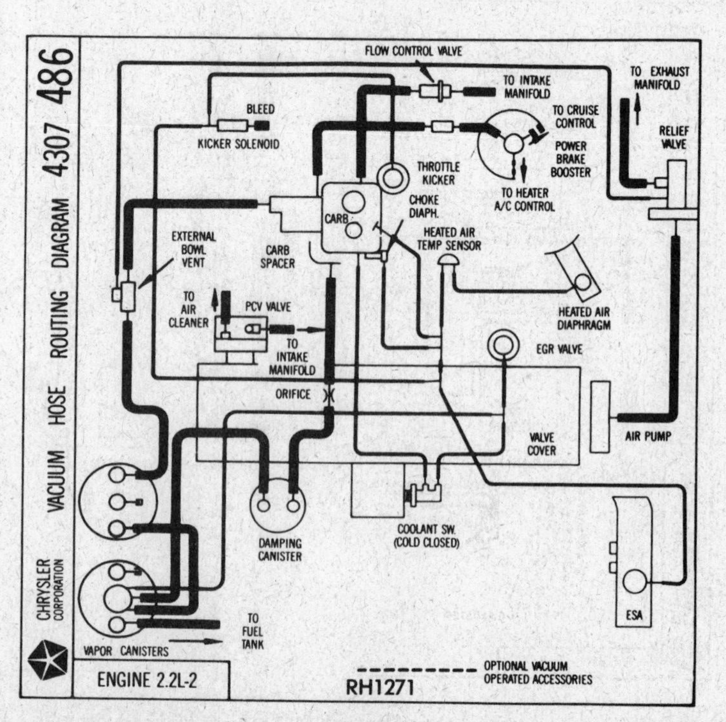

Typical vacuum hose routing diagram for the 2.2L engine, 2 BBL carburetor (Canada)

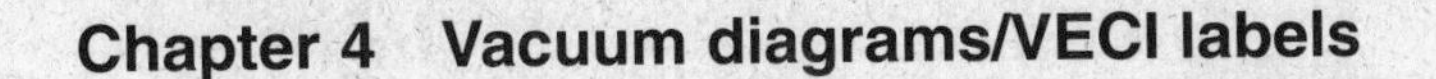

Chrysler

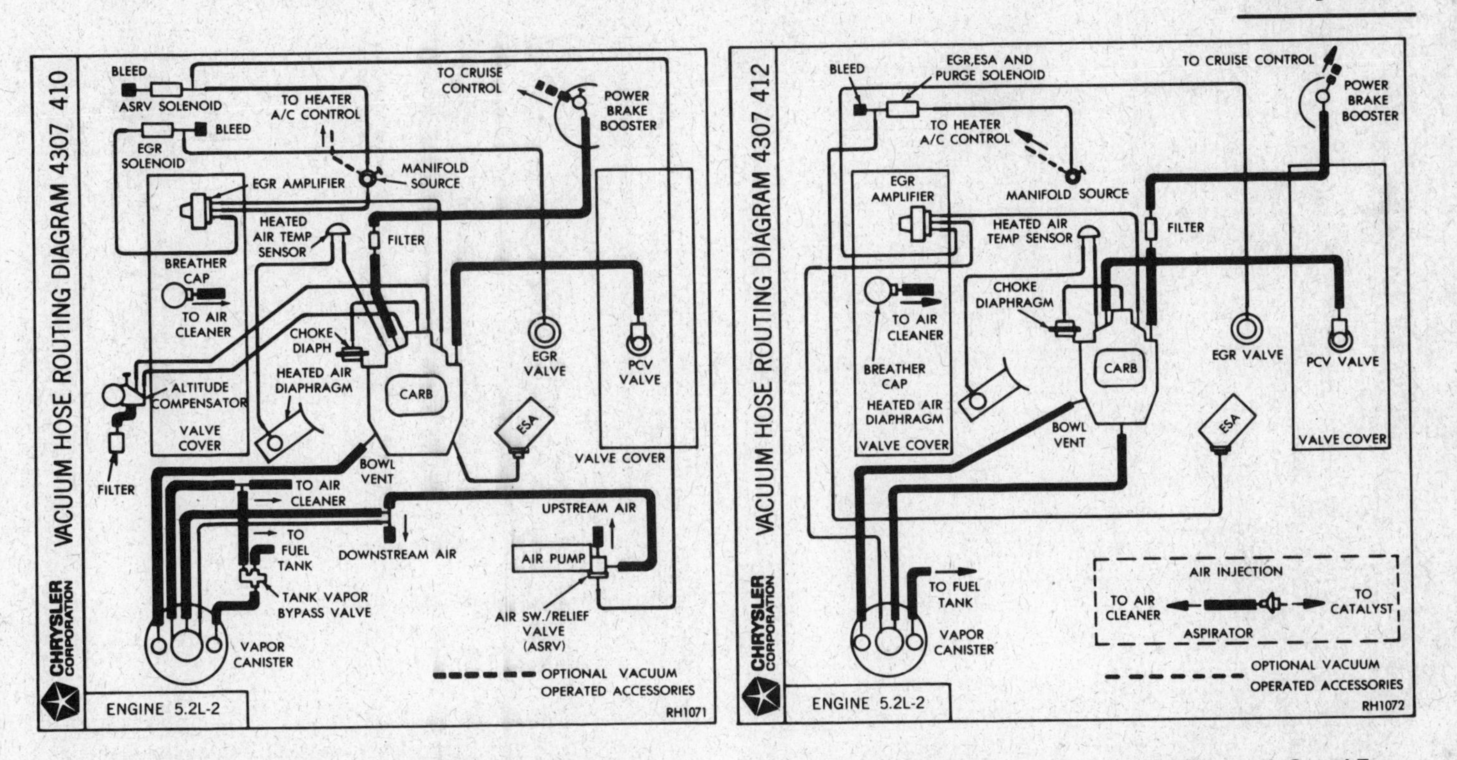

Vacuum hose routing diagram for the 1984 Dodge Diplomat with the 5.2L (318 V-8) engine, 2 BBL carburetor (Federal and California)

Vacuum hose routing diagram for the 1984 Plymouth Grand Fury with the 5.2L (318 V-8) engine, 2 BBL carburetor (Canada)

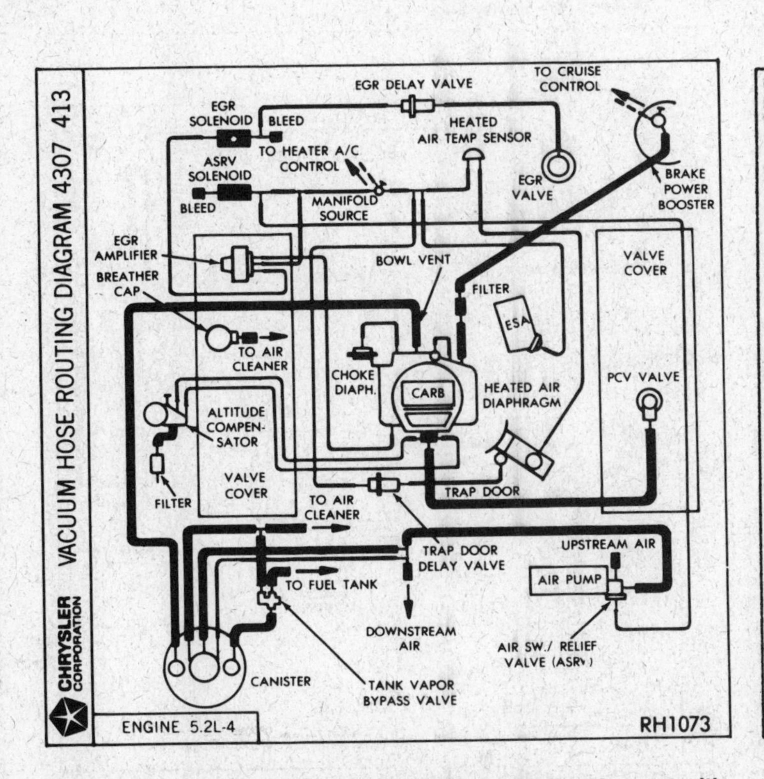

Vacuum hose routing diagram for the 1984 Chrysler Newport with the 5.2L (318 V-8) engine, 4 BBL carburetor (Federal and Canada)

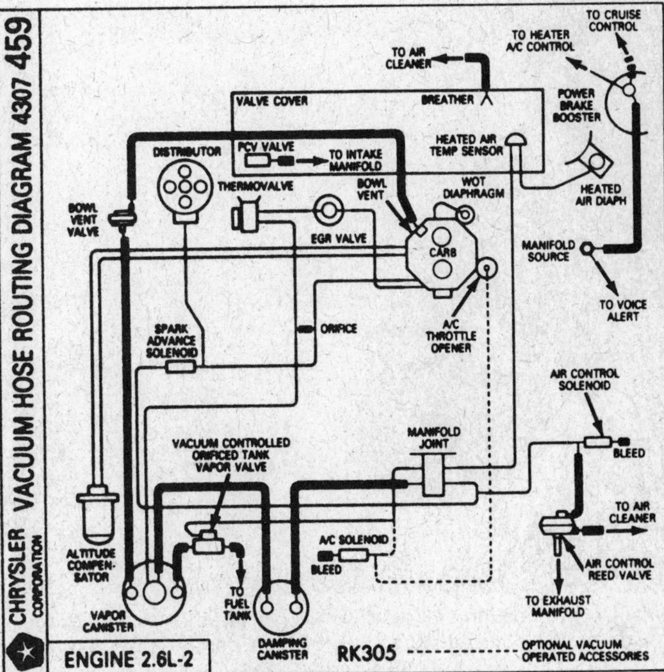

Vacuum hose routing diagram for the 1985 K and E cars with the 2.6L engine, 2 BBL carburetor (California and Canada)

Chrysler

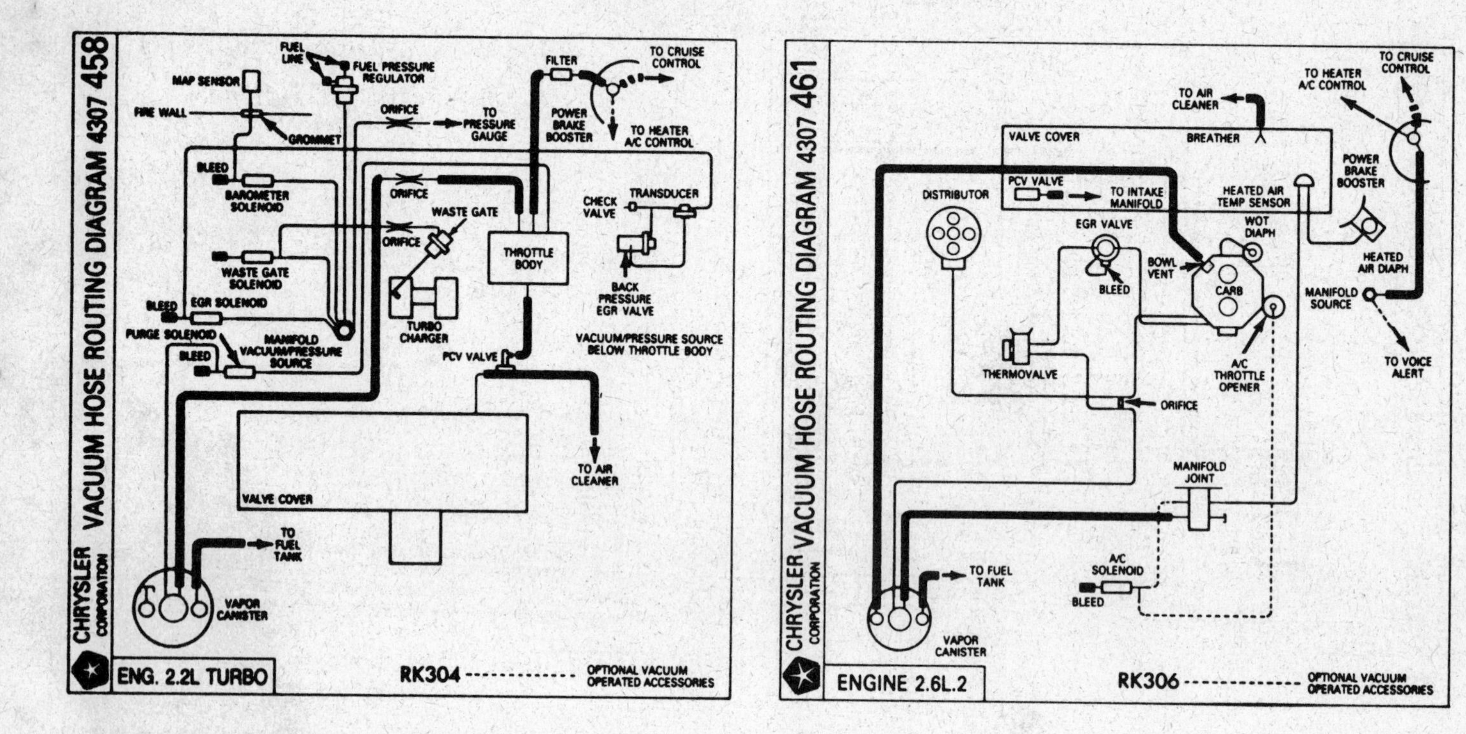

Vacuum hose routing diagram for the 1985 2.2L engine with TBI and turbo

Vacuum hose routing diagram for the 1985 2.6L engine with 2 BBL carburetor (Canada)

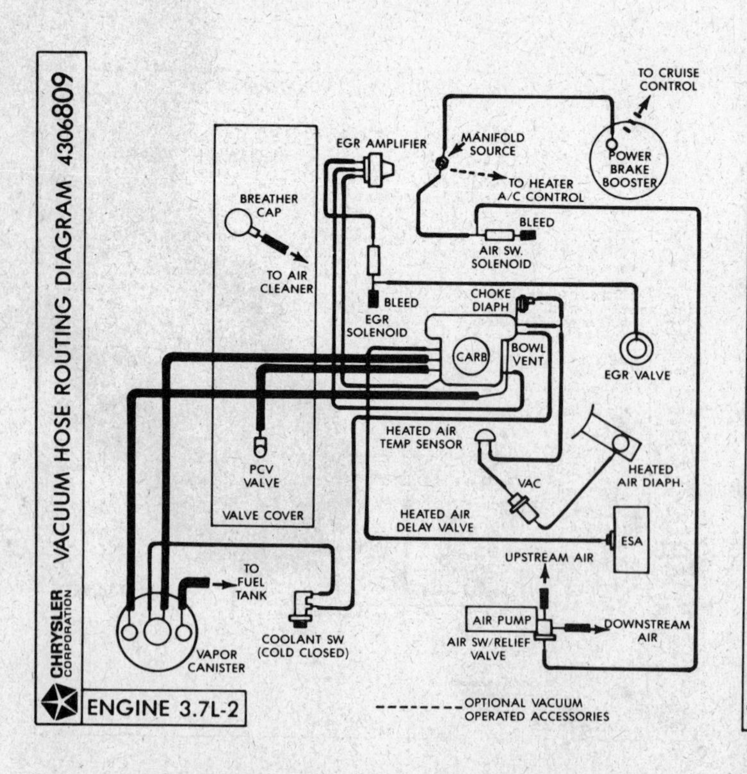

Vacuum hose routing diagram for the 1986 225 slant six- cylinder engine, 2 BBL carburetor, ESA with catalyst (Federal)

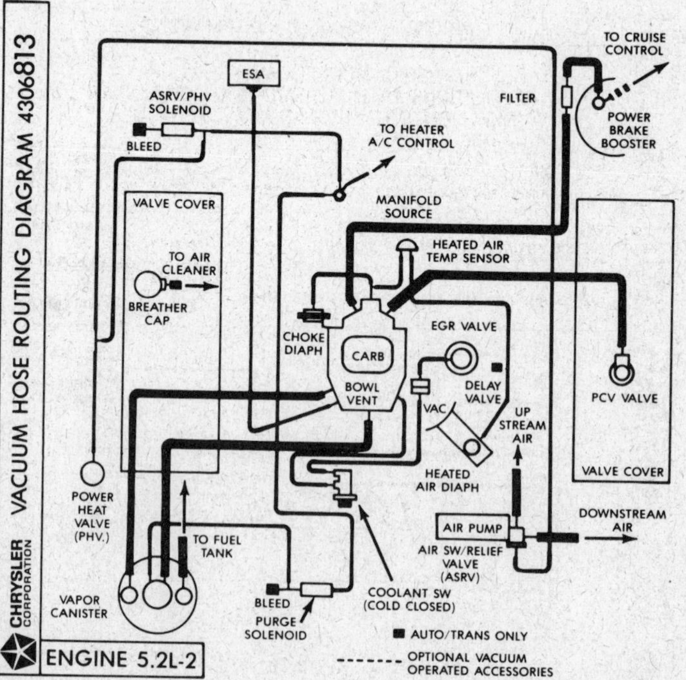

Vacuum hose routing diagram for the 1986 5.2L (318 V-8) engine, 2 BBL carburetor (Federal and Canada)

Chrysler

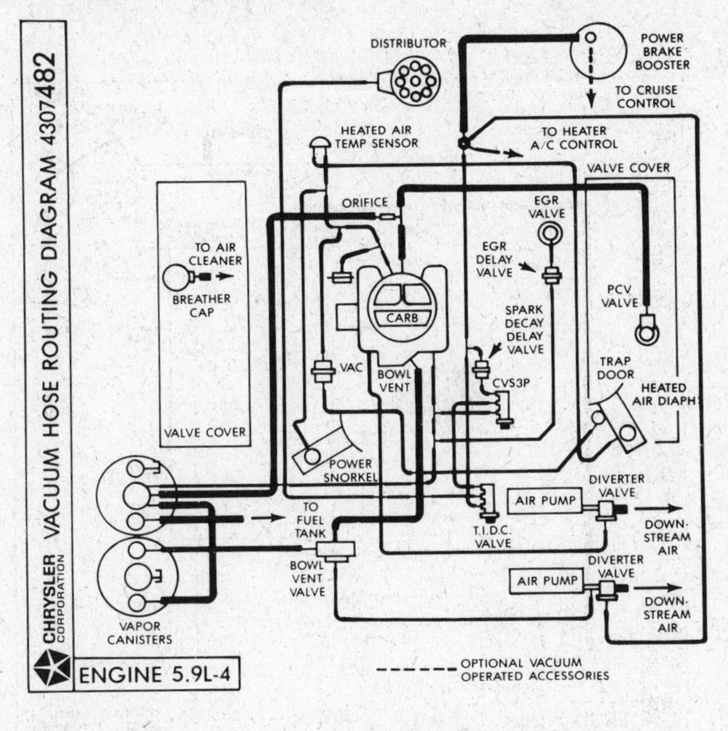

Vacuum hose routing diagram for the 1986 5.9L (360 V-8) engine, 4 BBL carburetor (California with air conditioning)

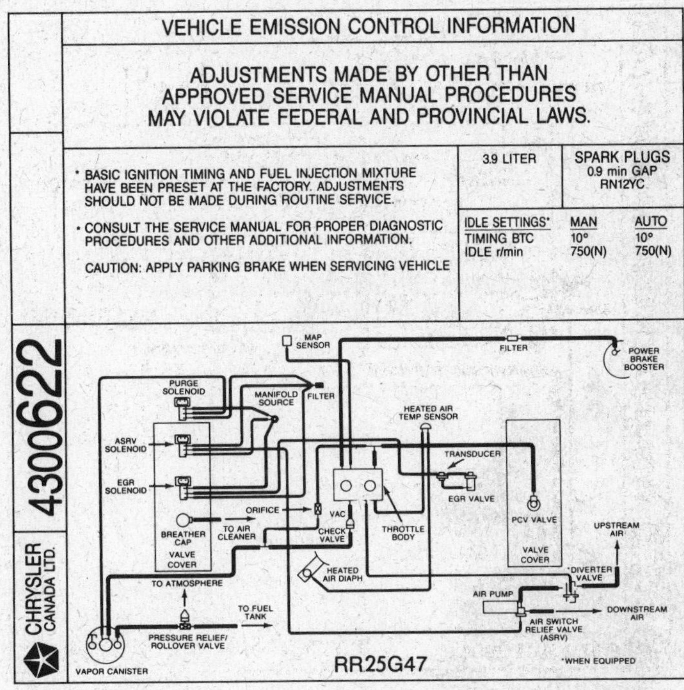

Vacuum hose routing diagram for the 1988 3.9L (V-6) engine with TBI (Canada)

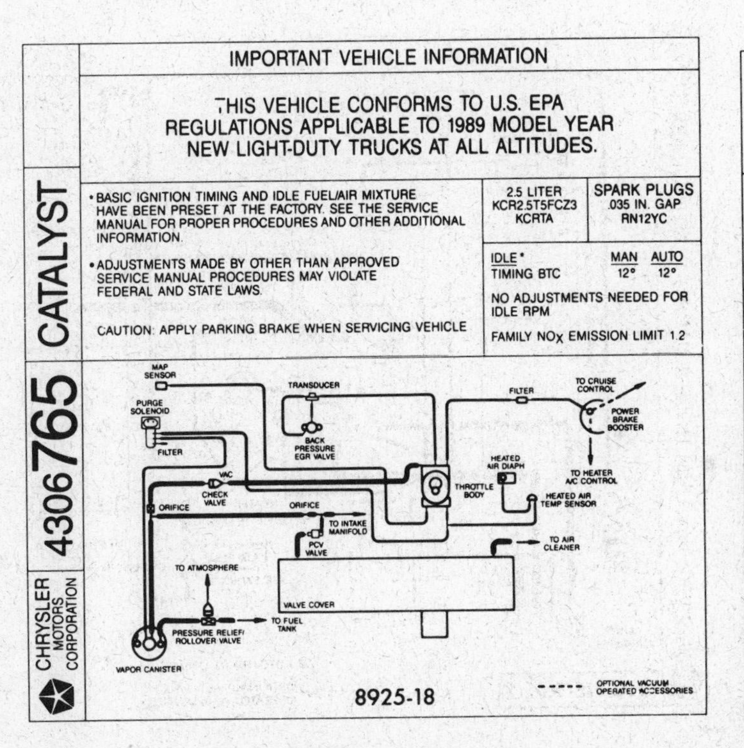

Vacuum hose routing diagram for the 1989 2.5L engine with TBI (California, Canada and Federal high altitude)

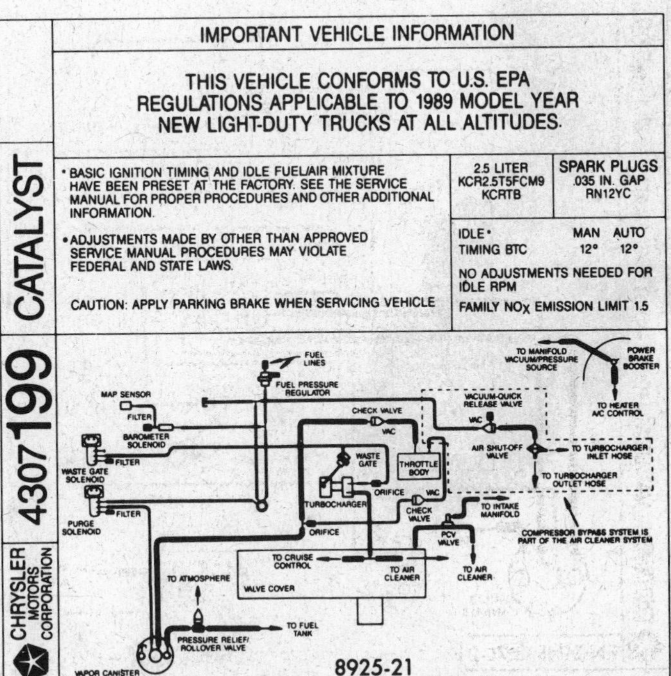

Vacuum hose routing diagram for the 1989 2.5L engine with TBI and turbo (California, Canada and Federal)

Chrysler

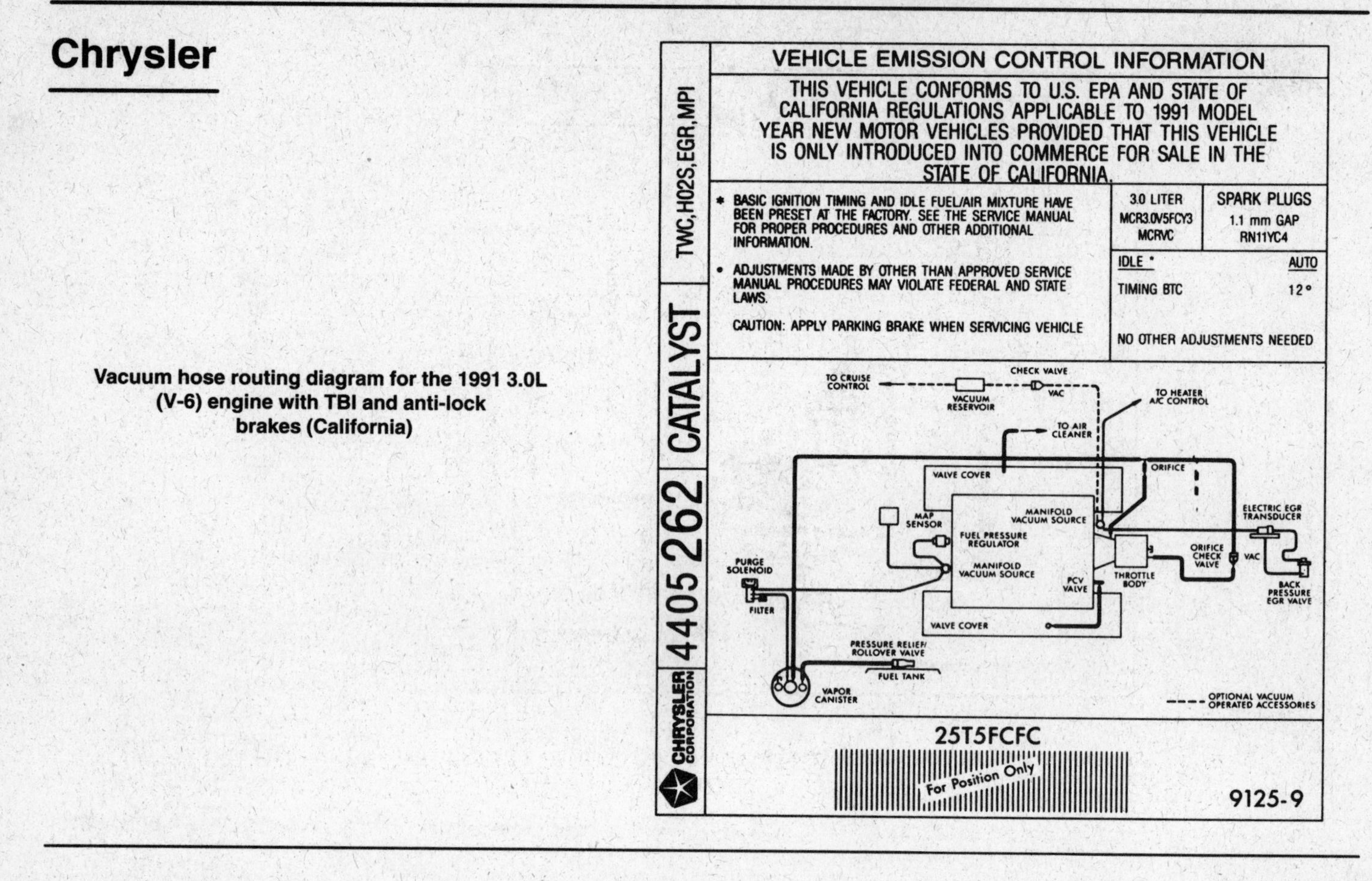

VEHICLE EMISSION CONTROL INFORMATION

THIS VEHICLE CONFORMS TO U.S. EPA AND STATE OF CALIFORNIA REGULATIONS APPLICABLE TO 1991 MODEL YEAR NEW MOTOR VEHICLES PROVIDED THAT THIS VEHICLE IS ONLY INTRODUCED INTO COMMERCE FOR SALE IN THE STATE OF CALIFORNIA.

* BASIC IGNITION TIMING AND IDLE FUEL/AIR MIXTURE HAVE BEEN PRESET AT THE FACTORY. SEE THE SERVICE MANUAL FOR PROPER PROCEDURES AND OTHER ADDITIONAL INFORMATION.
- ADJUSTMENTS MADE BY OTHER THAN APPROVED SERVICE MANUAL PROCEDURES MAY VIOLATE FEDERAL AND STATE LAWS.

CAUTION: APPLY PARKING BRAKE WHEN SERVICING VEHICLE

3.0 LITER	SPARK PLUGS
MCR3.0V5FCY3 MCRVC	1.1 mm GAP RN11YC4

IDLE *	AUTO
TIMING BTC	12°

NO OTHER ADJUSTMENTS NEEDED

25T5FCFC

For Position Only

9125-9

Vacuum hose routing diagram for the 1991 3.0L (V-6) engine with TBI and anti-lock brakes (California)

Ford

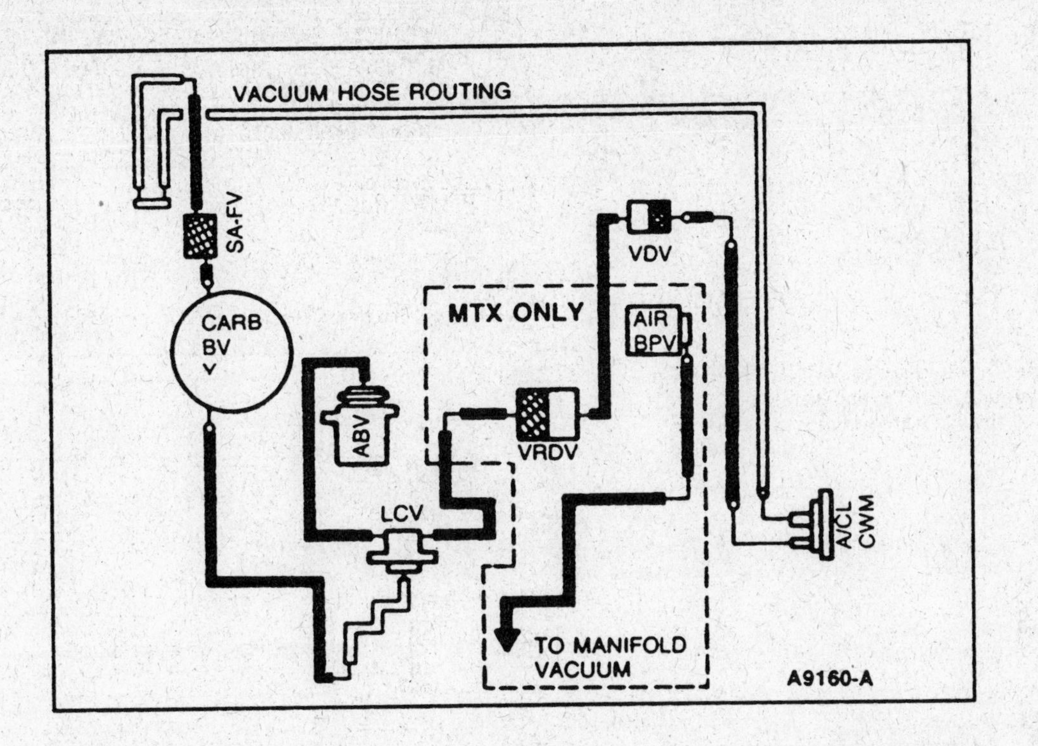

Vacuum hose routing diagram for the idle air bypass system on a typical carbureted 2.3L HSC engine

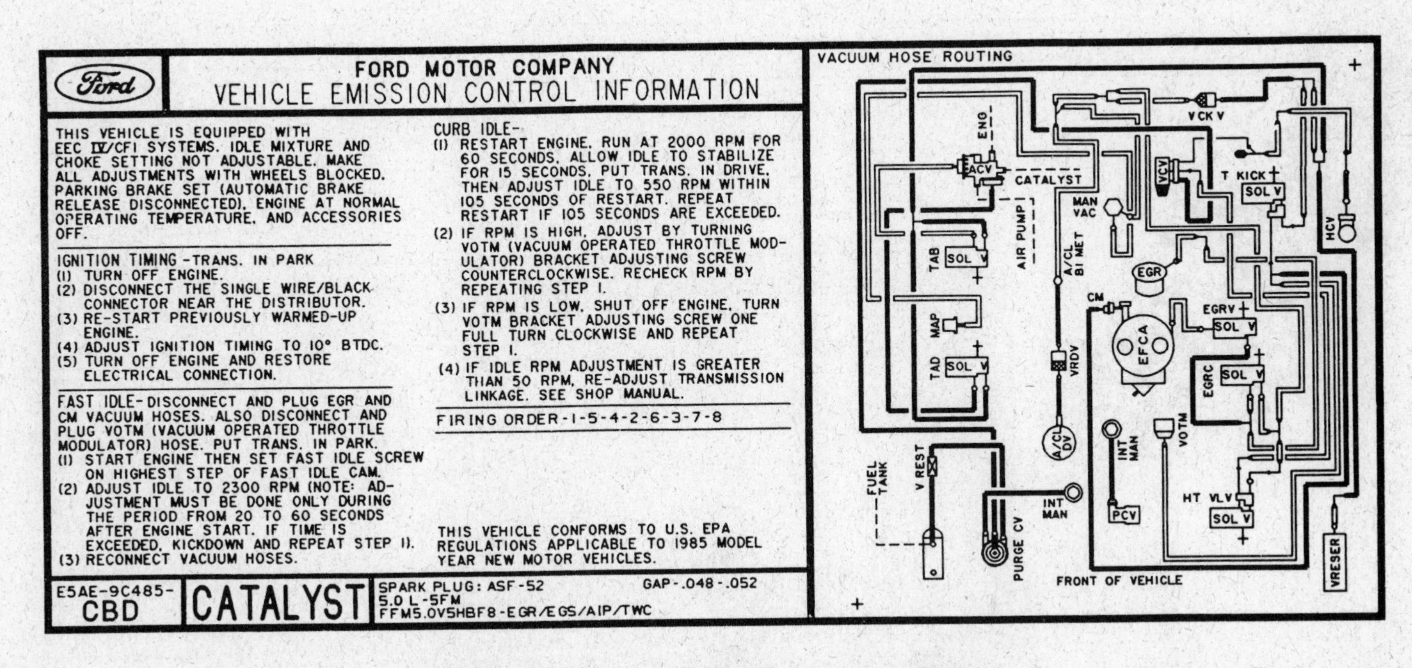

FORD MOTOR COMPANY
VEHICLE EMISSION CONTROL INFORMATION

THIS VEHICLE IS EQUIPPED WITH EEC IV/CFI SYSTEMS. IDLE MIXTURE AND CHOKE SETTING NOT ADJUSTABLE. MAKE ALL ADJUSTMENTS WITH WHEELS BLOCKED, PARKING BRAKE SET (AUTOMATIC BRAKE RELEASE DISCONNECTED), ENGINE AT NORMAL OPERATING TEMPERATURE, AND ACCESSORIES OFF.

IGNITION TIMING - TRANS. IN PARK
(1) TURN OFF ENGINE.
(2) DISCONNECT THE SINGLE WIRE/BLACK CONNECTOR NEAR THE DISTRIBUTOR.
(3) RE-START PREVIOUSLY WARMED-UP ENGINE.
(4) ADJUST IGNITION TIMING TO 10° BTDC.
(5) TURN OFF ENGINE AND RESTORE ELECTRICAL CONNECTION.

FAST IDLE - DISCONNECT AND PLUG EGR AND CM VACUUM HOSES. ALSO DISCONNECT AND PLUG VOTM (VACUUM OPERATED THROTTLE MODULATOR) HOSE. PUT TRANS. IN PARK.
(1) START ENGINE THEN SET FAST IDLE SCREW ON HIGHEST STEP OF FAST IDLE CAM.
(2) ADJUST IDLE TO 2300 RPM (NOTE: ADJUSTMENT MUST BE DONE ONLY DURING THE PERIOD FROM 20 TO 60 SECONDS AFTER ENGINE START. IF TIME IS EXCEEDED, KICKDOWN AND REPEAT STEP 1).
(3) RECONNECT VACUUM HOSES.

CURB IDLE-
(1) RESTART ENGINE. RUN AT 2000 RPM FOR 60 SECONDS, ALLOW IDLE TO STABILIZE FOR 15 SECONDS, PUT TRANS. IN DRIVE, THEN ADJUST IDLE TO 550 RPM WITHIN 105 SECONDS OF RESTART. REPEAT RESTART IF 105 SECONDS ARE EXCEEDED.
(2) IF RPM IS HIGH, ADJUST BY TURNING VOTM (VACUUM OPERATED THROTTLE MODULATOR) BRACKET ADJUSTING SCREW COUNTERCLOCKWISE. RECHECK RPM BY REPEATING STEP 1.
(3) IF RPM IS LOW, SHUT OFF ENGINE, TURN VOTM BRACKET ADJUSTING SCREW ONE FULL TURN CLOCKWISE AND REPEAT STEP 1.
(4) IF IDLE RPM ADJUSTMENT IS GREATER THAN 50 RPM, RE-ADJUST TRANSMISSION LINKAGE. SEE SHOP MANUAL.

FIRING ORDER - 1-5-4-2-6-3-7-8

THIS VEHICLE CONFORMS TO U.S. EPA REGULATIONS APPLICABLE TO 1985 MODEL YEAR NEW MOTOR VEHICLES.

E5AE-9C485-CBD | CATALYST | SPARK PLUG: ASF-52 GAP-.048-.052
5.0 L-5FM
FFM5.0V5HBF8-EGR/EGS/AIP/TWC

Typical VECI label for the 1985 5.0L (V-8) engine (Federal)

Ford

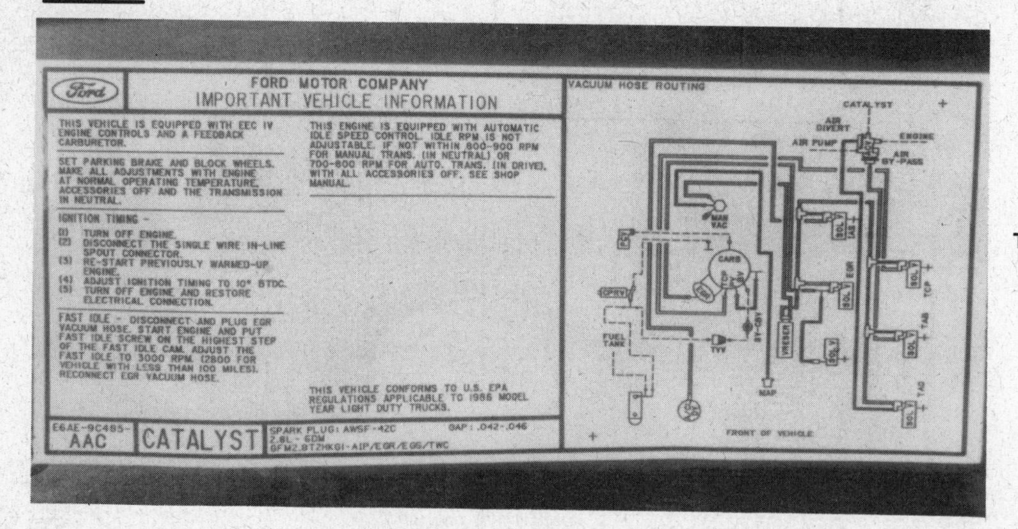

Typical VECI label for a 1986 2.8L engine in an Astro Van

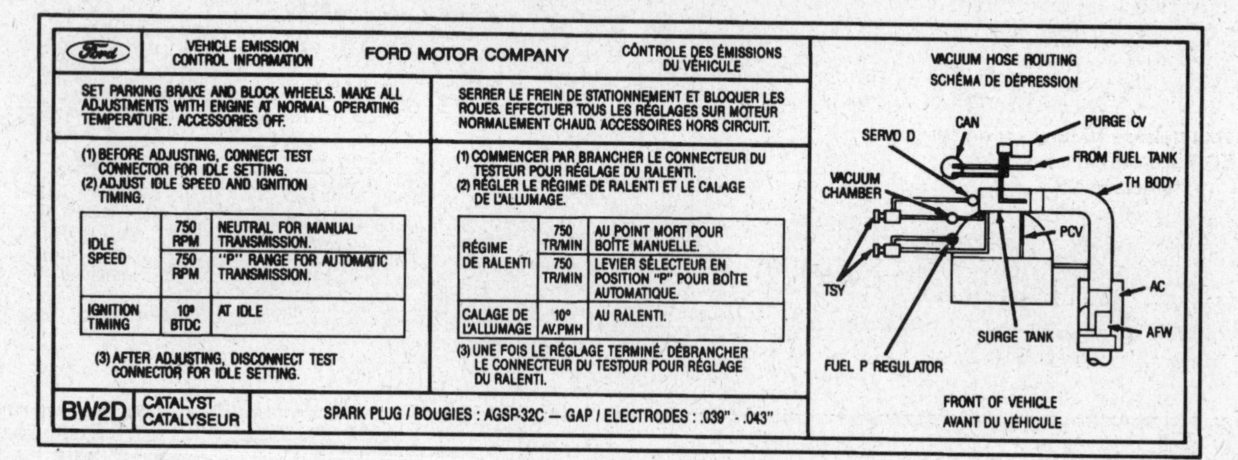

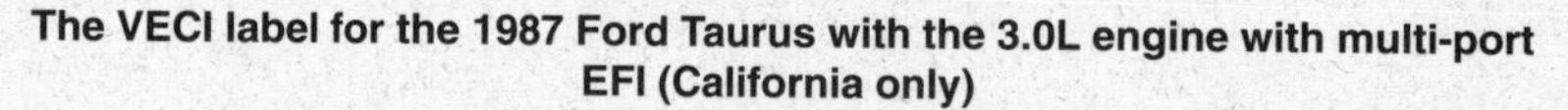

The VECI label for the 1987 Ford Taurus with the 3.0L engine with multi-port EFI (California only)

The VECI label for the 1987 Ford Thunderbird with the 3.8L engine

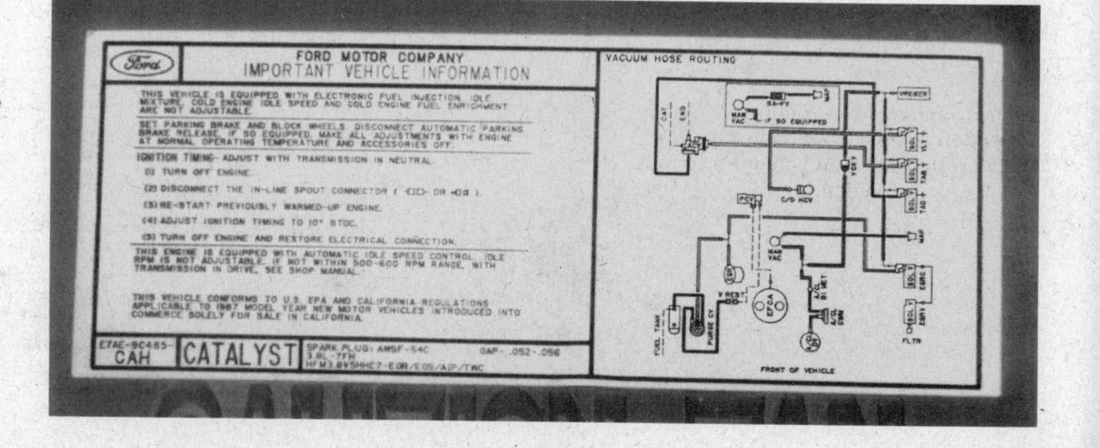

Ford

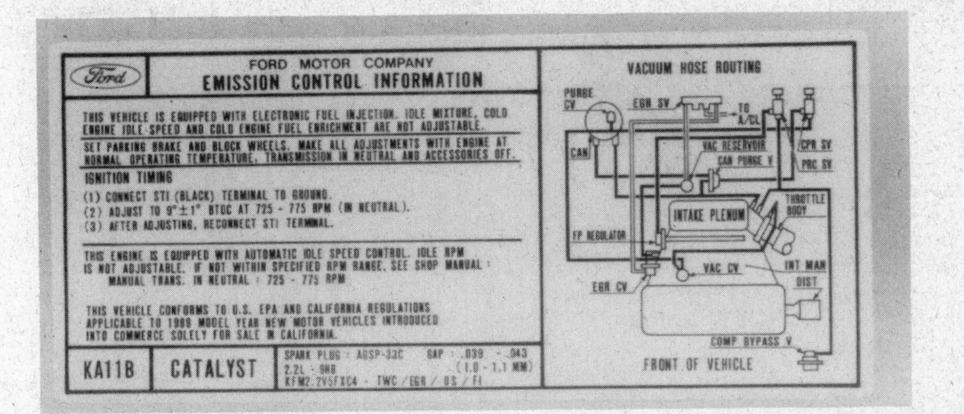
Ford

FORD MOTOR COMPANY

EMISSION CONTROL INFORMATION

THIS VEHICLE IS EQUIPPED WITH ELECTRONIC FUEL INJECTION. IDLE MIXTURE, COLD ENGINE IDLE SPEED AND COLD ENGINE FUEL ENRICHMENT ARE NOT ADJUSTABLE.

SET PARKING BRAKE AND BLOCK WHEELS. MAKE ALL ADJUSTMENTS WITH ENGINE AT NORMAL OPERATING TEMPERATURE, TRANSMISSION IN NEUTRAL AND ACCESSORIES OFF.

IGNITION TIMING

(1) CONNECT STI (BLACK) TERMINAL TO GROUND.
(2) ADJUST TO 9°±1° BTDC AT 725 - 775 RPM (IN NEUTRAL).
(3) AFTER ADJUSTING, RECONNECT STI TERMINAL.

THIS ENGINE IS EQUIPPED WITH AUTOMATIC IDLE SPEED CONTROL. IDLE RPM IS NOT ADJUSTABLE. IF NOT WITHIN SPECIFIED RPM RANGE, SEE SHOP MANUAL :
MANUAL TRANS. IN NEUTRAL : 725 - 775 RPM

THIS VEHICLE CONFORMS TO U.S. EPA AND CALIFORNIA REGULATIONS APPLICABLE TO 1989 MODEL YEAR NEW MOTOR VEHICLES INTRODUCED INTO COMMERCE SOLELY FOR SALE IN CALIFORNIA.

KA11B | CATALYST | SPARK PLUG : AGSP-33C GAP : .039 - .043 (1.0 - 1.1 MM) 2.2L - 9H8 KFM2.2V5FXC4 - TWC / EGR / OS / FI

The VECI label is affixed to the underside of the hood; this one is from a 1989 Ford Probe with the 2.2L engine and multi-port EFI

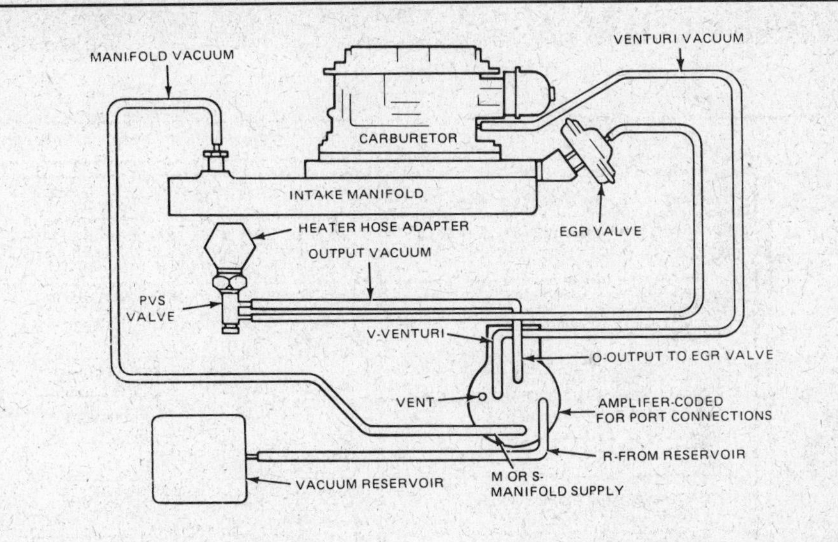

Typical vacuum hose routing for an early Ford EGR system (this one's for a full-size van)

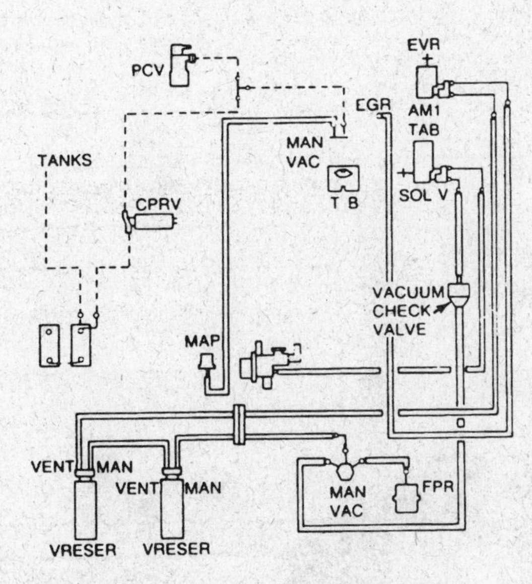

Typical vacuum hose routing for the air injection system used on heavy-duty trucks with a 460 (7.5L) engine and fuel injection

General Motors

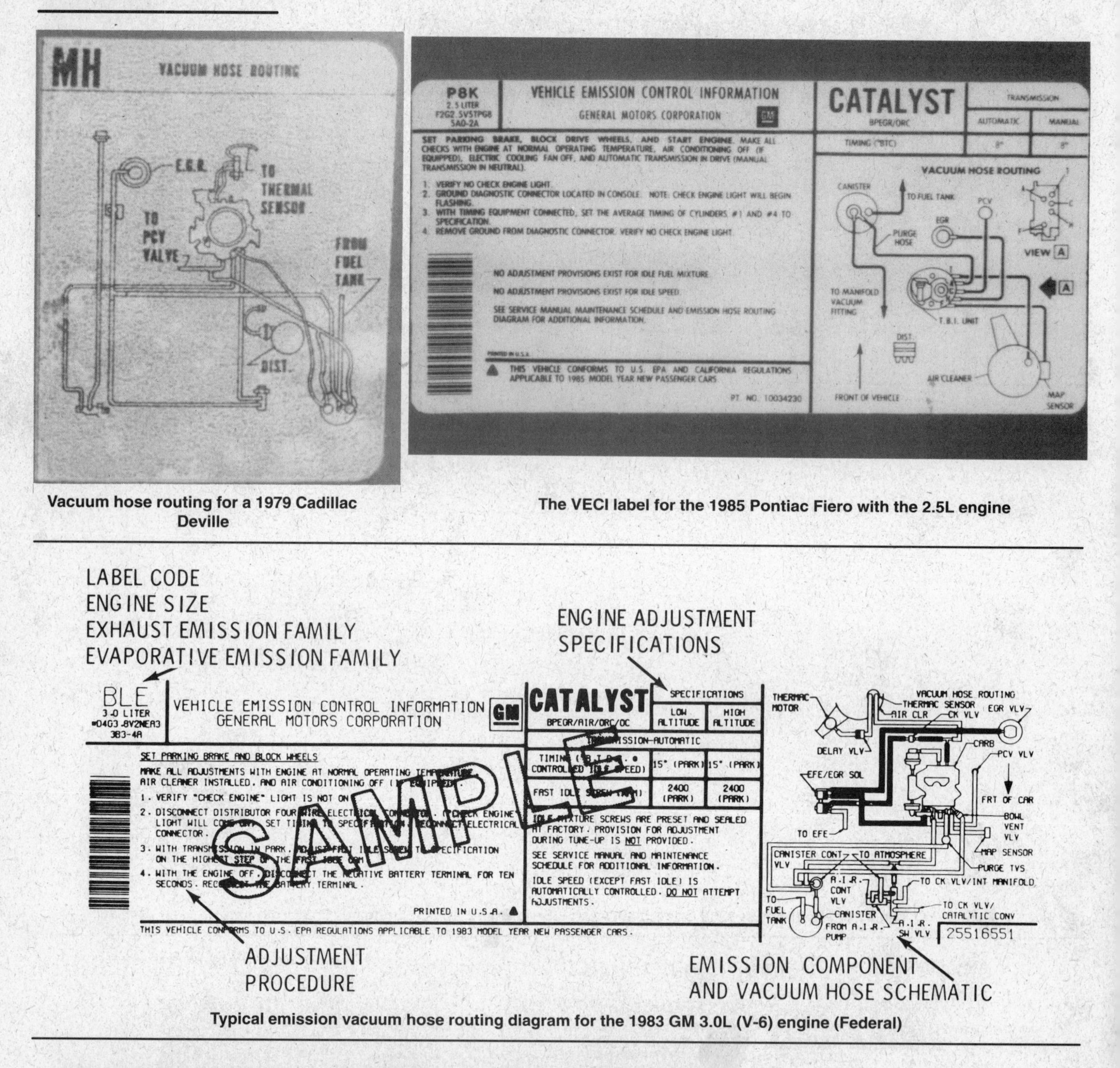

Vacuum hose routing for a 1979 Cadillac Deville

The VECI label for the 1985 Pontiac Fiero with the 2.5L engine

Typical emission vacuum hose routing diagram for the 1983 GM 3.0L (V-6) engine (Federal)

The VECI label for a GM N-car with the 3.0L (V-6) engine

General Motors

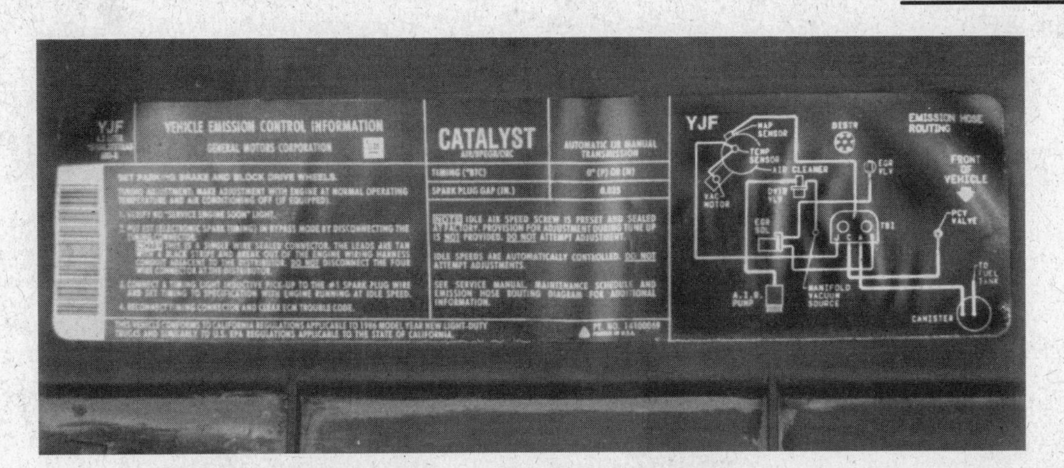

The VECI label for the 1986 Chevrolet Astro van or the GMC Safari van with the 4.3L engine is located on the fan shroud

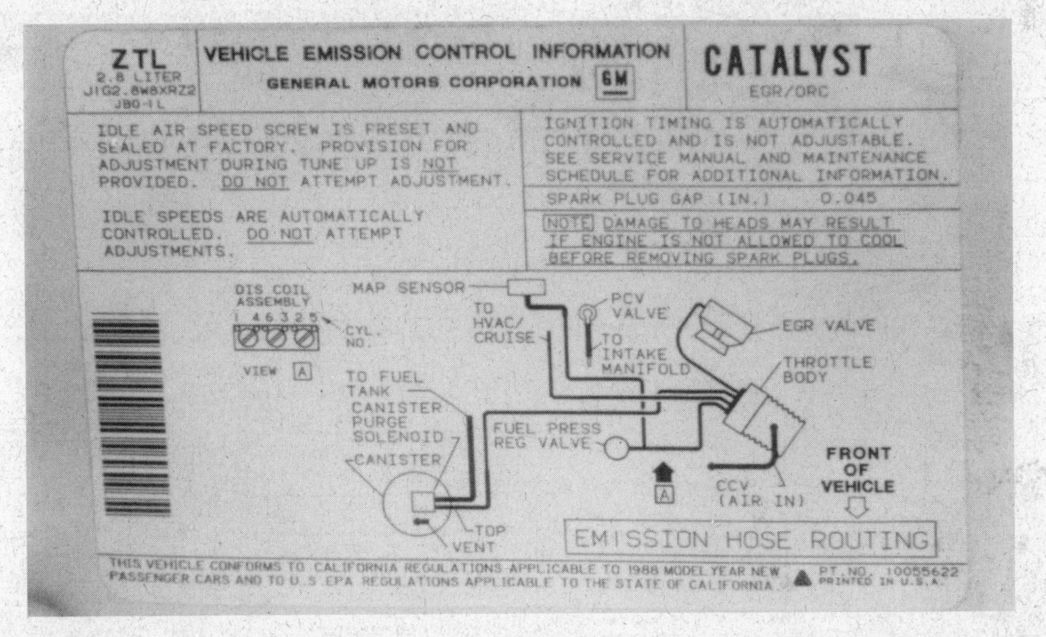

The VECI label for the 1988 Chevrolet Corsica/Beretta with the 2.8L engine (California only)

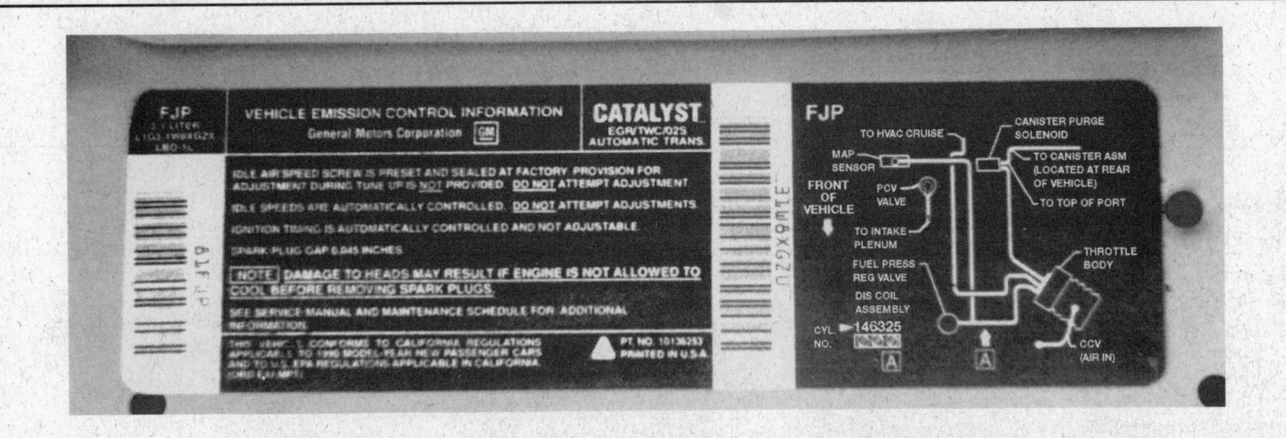

The VECI label for a California Oldsmobile Cutlass Supreme with a 3.1L engine

Honda/Isuzu/Jeep

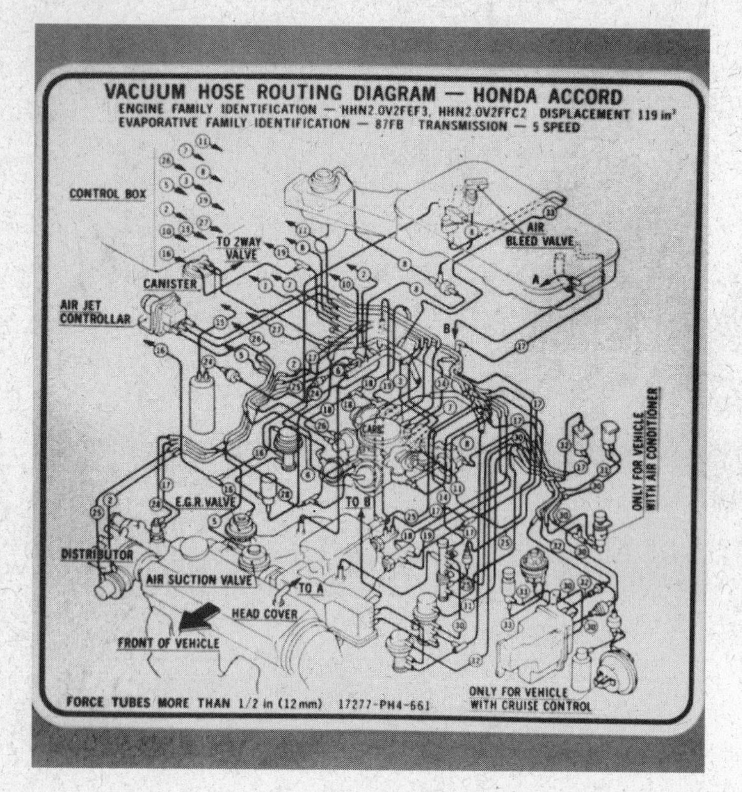

Typical vacuum hose routing diagram for a late-model Honda Accord with a 2000cc engine. Note that Honda vacuum hose routing often changes significantly from year to year

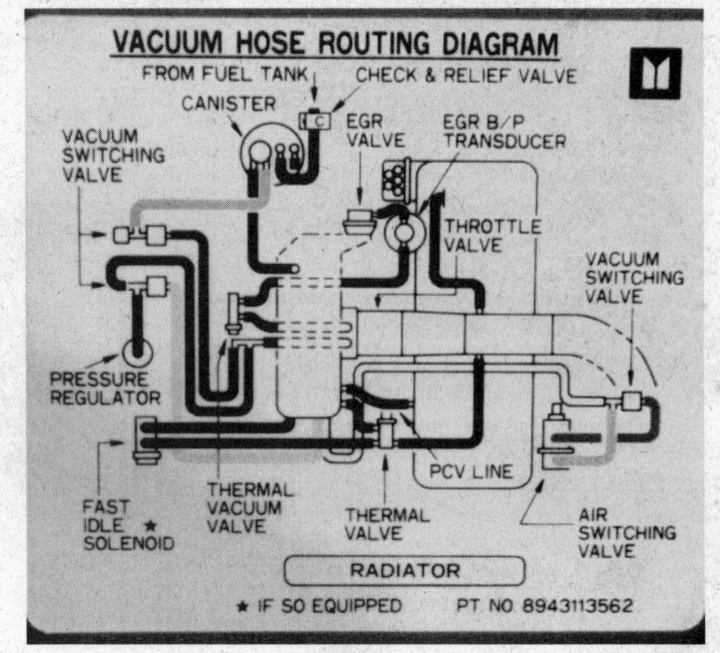

Typical vacuum hose routing diagram for a 1988 Isuzu Trooper with a four-cylinder engine

5277258

CHRYSLER CANADA

VEHICLE EMISSION CONTROL INFORMATION

THIS VEHICLE WAS BUILT FOR SALE IN CANADA AND WAS DESIGNED TO MEET THE EMISSION REQUIREMENTS OF THE CANADA MOTOR VEHICLE SAFETY ACT.

IT WAS NOT DESIGNED TO COMPLY WITH THE REQUIREMENTS OF OTHER COUNTRIES.

SEE SERVICE MANUAL FOR DETAILED INSTRUCTIONS. CAUTION: APPLY PARKING BRAKE WHEN SERVICING VEHICLE	2.5 LITRES	SPARK PLUGS 0.9 mm GAP RC-12LYC

RENSEIGNEMENTS RELATIFS AU SYSTEME ANTIPOLLUTION

LE PRÉSENT VÉHICULE A ÉTÉ FABRIQUÉ POUR ÊTRE VENDU AU CANADA ET IL A ÉTÉ CONCU DE MANIÈRE À SE CONFORMER AUX NORMES ANTIPOLLUTION DE LA LÔI SUR LA SÉCURITÉ DES VÉHICULES AUTOMOBILES DU CANADA.

IL N'EST PAS DESTINÉ À SE CONFORMER AUX NORMES D'AUTRES PAYS.

POUR PLUS DE RENSEIGNEMENTS, VEUILLEZ VOUS REPORTER AU MANUAL D'ENTRETIEN. AVERTISSEMENT: SERREZ LE FREIN DE STATIONNEMENT POUR FAIRE L'ENTRETIEN OU LA RÉPARATION DU VÉHICULE.	2.5 LITRES	BOUGIES ÉCARTEMENT 0,9 mm RC-12LYC

MAP SENSOR
CÂPTEUR DE PRESSION ABSOLUE DE LA TUBULURE
VAPOR CANISTER
FITRE À CHARBON ACTIF
ORIFICE AJUTAGE
TANK VENT
AÉRATEUR DU RÉSERVOIR
VALVE COVER
CACHE-SOUPAPES
VACUUM FITTING
RACCORD DE DÉPRESSION
THROTTLE BODY
CORPS DE PAPILLON
VACUUM FITTINGS
RACCORDS DE DÉPRESSION
ORIFICE AJUTAGE
FUEL PRESSURE REGULATOR
RÉGULATEUR DE LA PRESSION DE CARBURANT
FRONT OF VEHICLE
AVANT DU VÉHICULE
AIR CLEANER
FILTRE A AIR

Typical VECI label for a 1990 Jeep with the 2.5L engine

Jeep

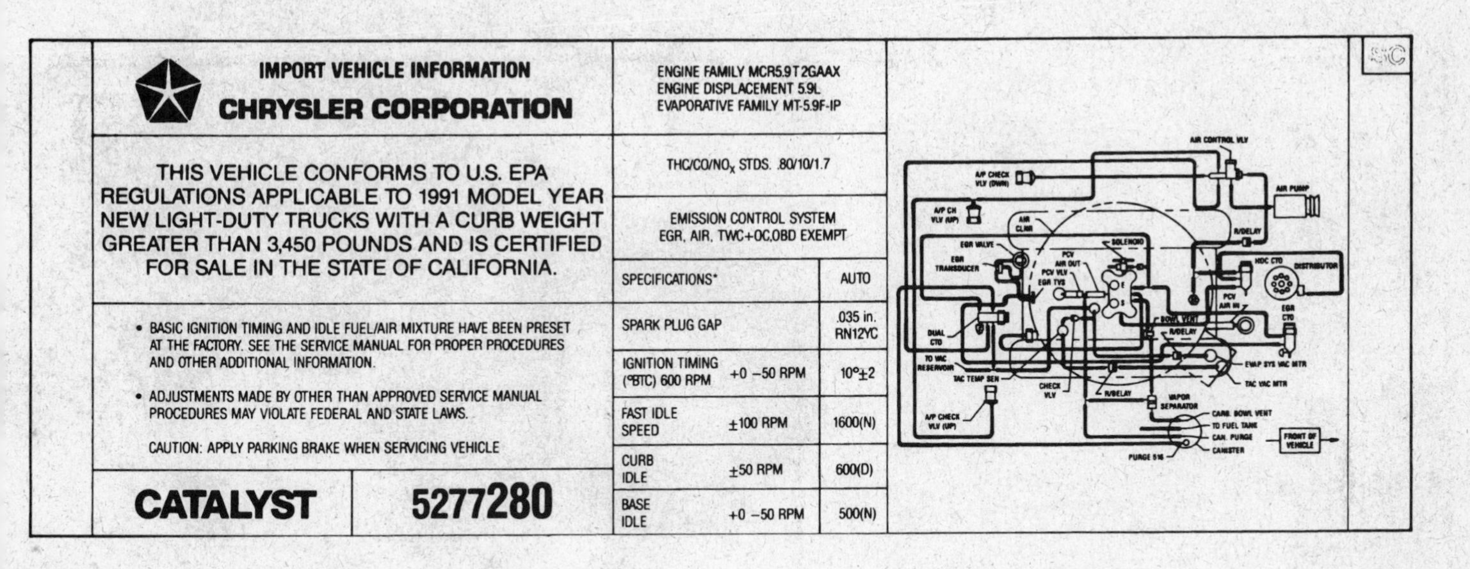

IMPORT VEHICLE INFORMATION

CHRYSLER CORPORATION

THIS VEHICLE CONFORMS TO U.S. EPA REGULATIONS APPLICABLE TO 1991 MODEL YEAR NEW LIGHT-DUTY TRUCKS WITH A CURB WEIGHT GREATER THAN 3,450 POUNDS AND IS CERTIFIED FOR SALE IN THE STATE OF CALIFORNIA.

- BASIC IGNITION TIMING AND IDLE FUEL/AIR MIXTURE HAVE BEEN PRESET AT THE FACTORY. SEE THE SERVICE MANUAL FOR PROPER PROCEDURES AND OTHER ADDITIONAL INFORMATION.
- ADJUSTMENTS MADE BY OTHER THAN APPROVED SERVICE MANUAL PROCEDURES MAY VIOLATE FEDERAL AND STATE LAWS.

CAUTION: APPLY PARKING BRAKE WHEN SERVICING VEHICLE

CATALYST 5277280

ENGINE FAMILY MCR5.9T2GAAX
ENGINE DISPLACEMENT 5.9L
EVAPORATIVE FAMILY MT-5.9F-IP

$THC/CO/NO_x$ STDS. .80/10/1.7

EMISSION CONTROL SYSTEM
EGR, AIR, TWC+OC,OBD EXEMPT

SPECIFICATIONS*		AUTO
SPARK PLUG GAP		.035 in. RN12YC
IGNITION TIMING (°BTC) 600 RPM	+0 –50 RPM	10°±2
FAST IDLE SPEED	±100 RPM	1600(N)
CURB IDLE	±50 RPM	600(D)
BASE IDLE	+0 –50 RPM	500(N)

Typical VECI label for the 1991 Jeep Grand Wagoneer with the 5.9L (360) V8 engine

Mazda

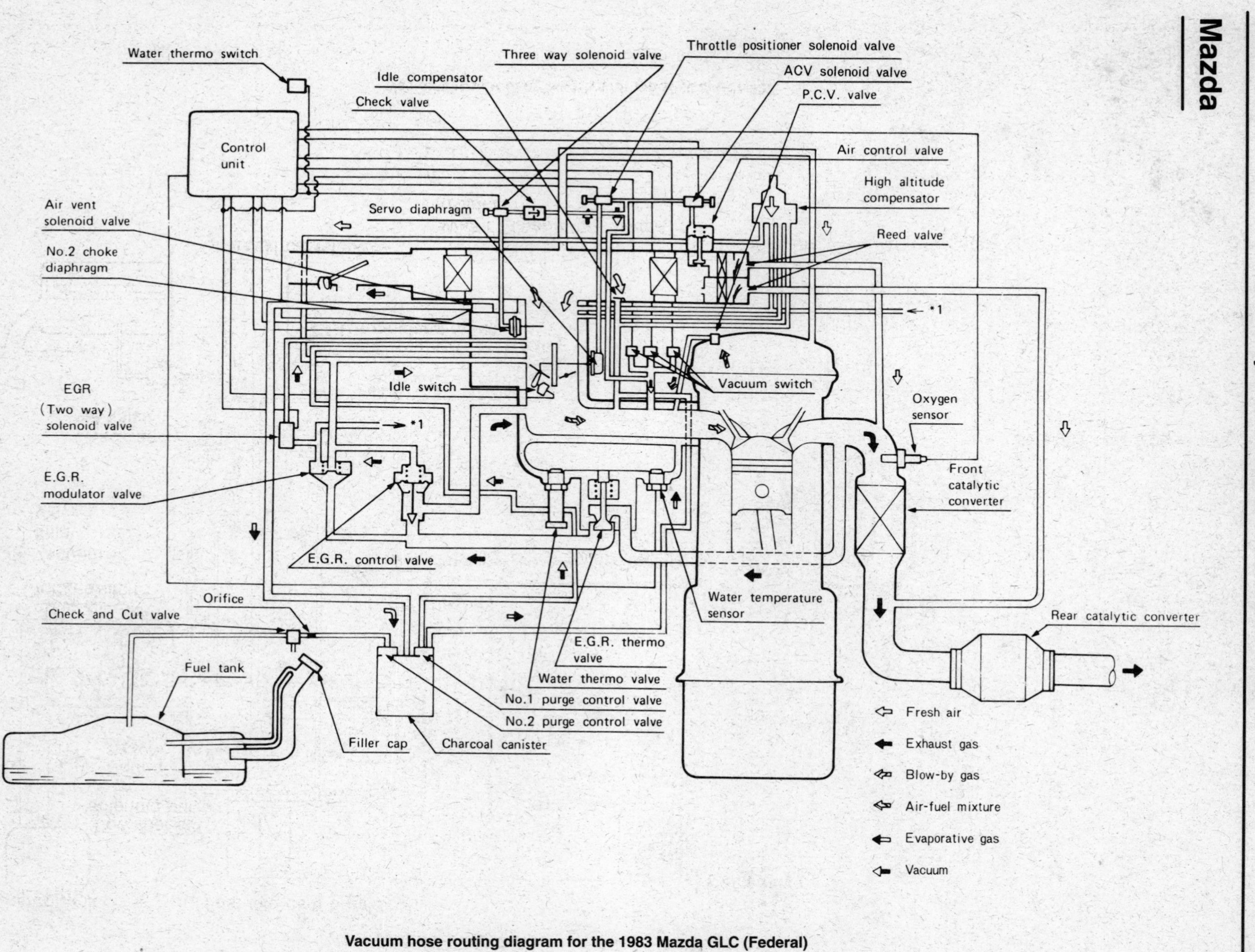

Vacuum hose routing diagram for the 1983 Mazda GLC (Federal)

Mazda

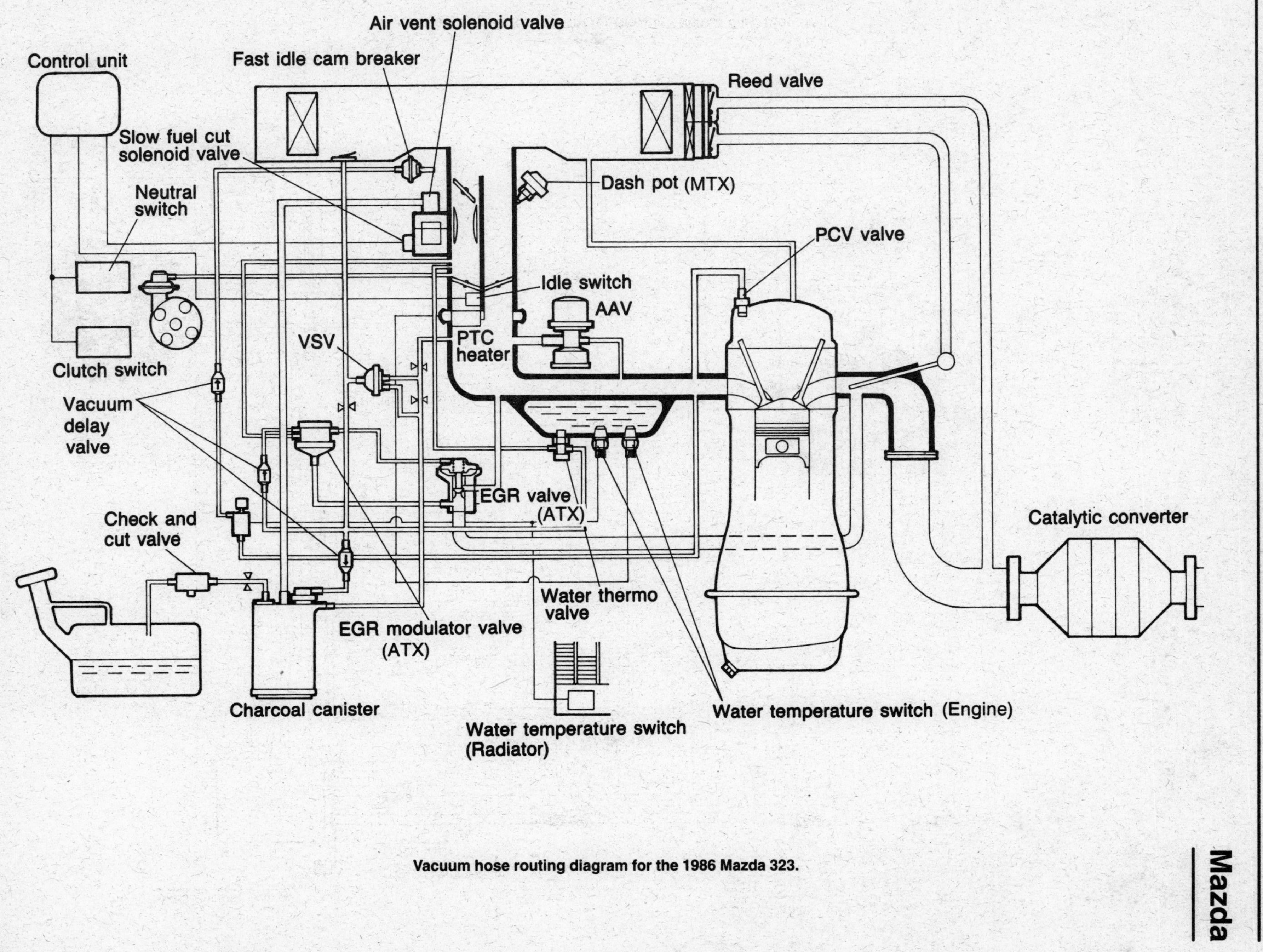

Vacuum hose routing diagram for the 1986 Mazda 323.

Mazda

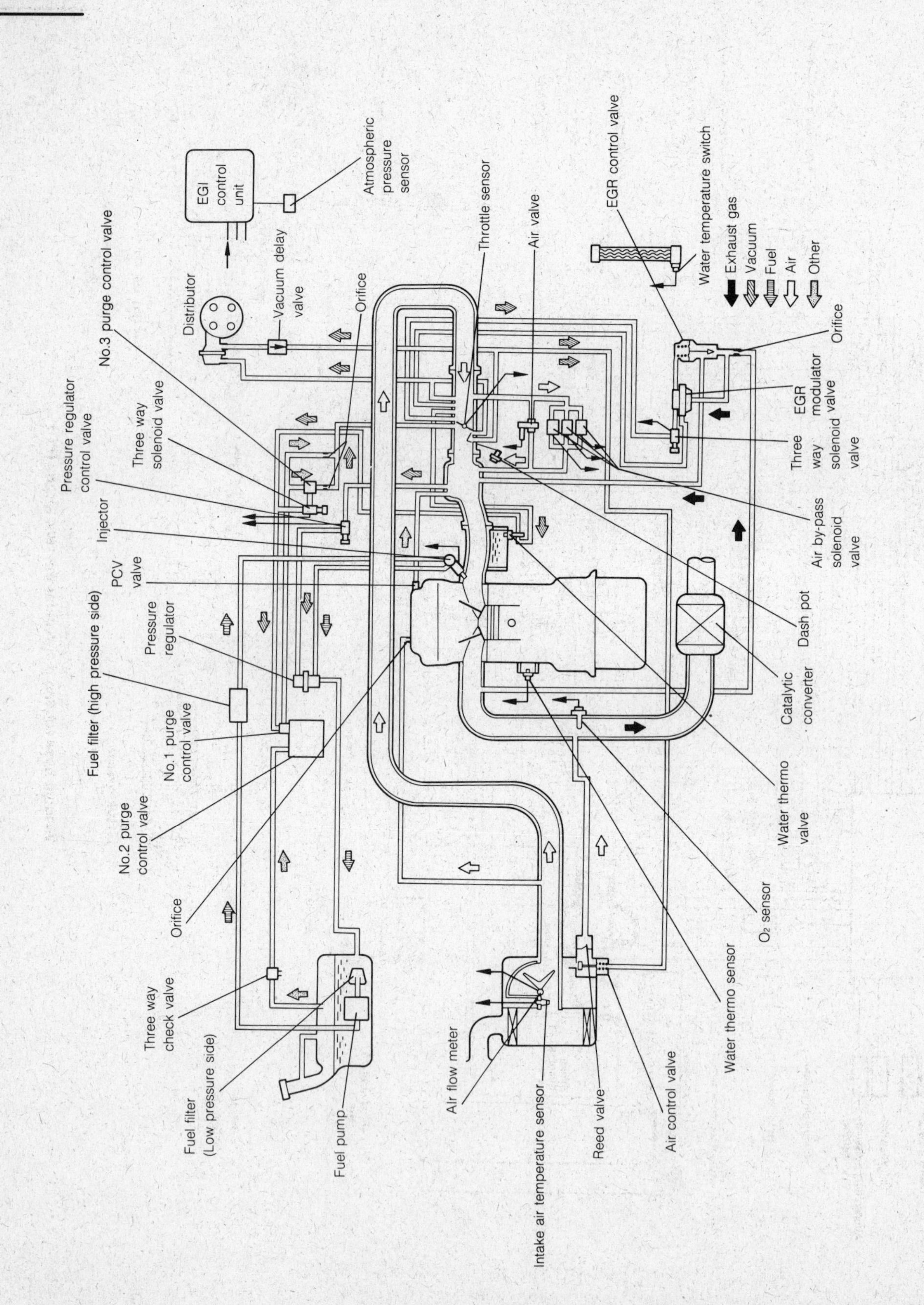

Vacuum hose routing diagram for the 1986 Mazda 626

Mazda

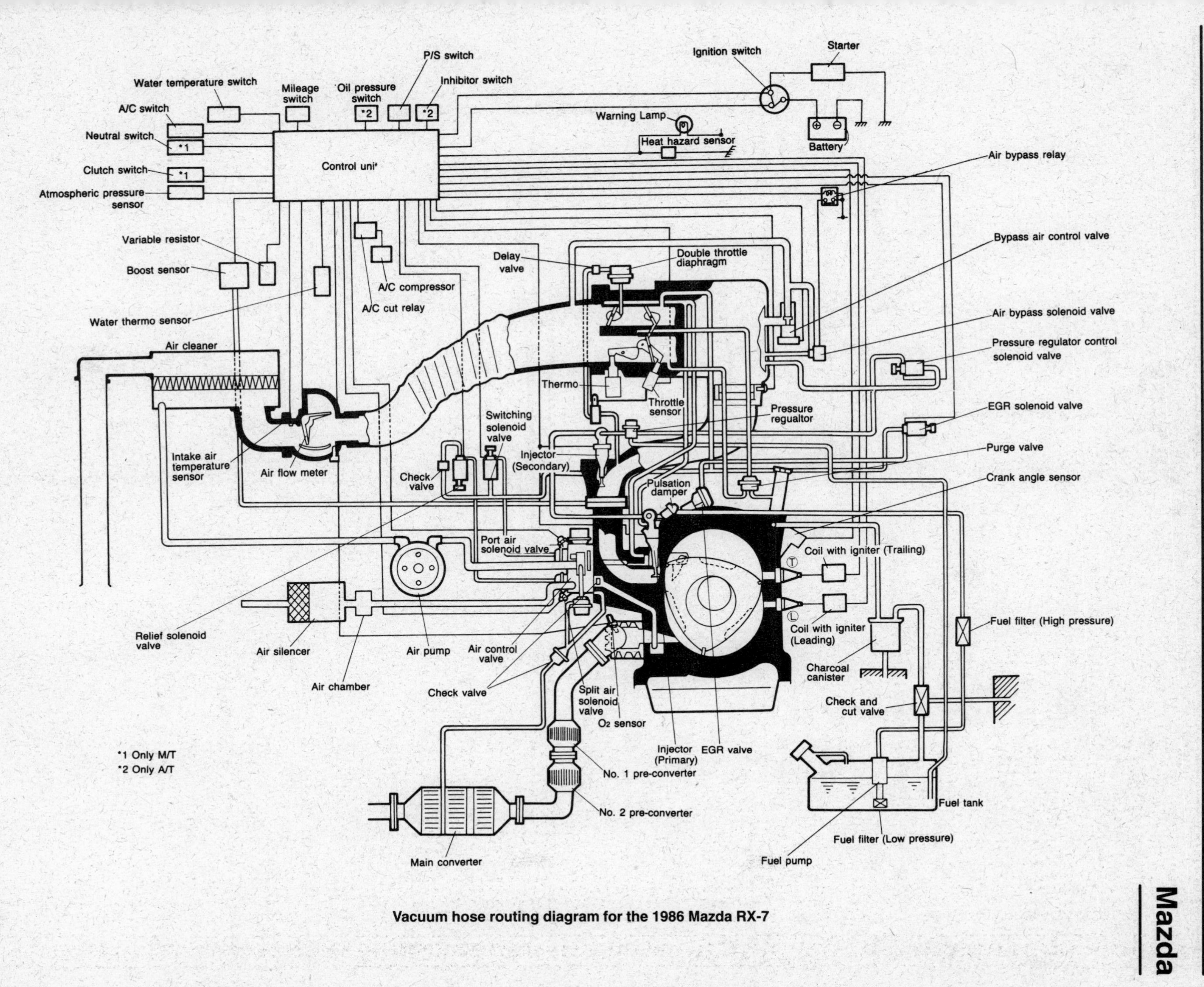

Vacuum hose routing diagram for the 1986 Mazda RX-7

Mazda

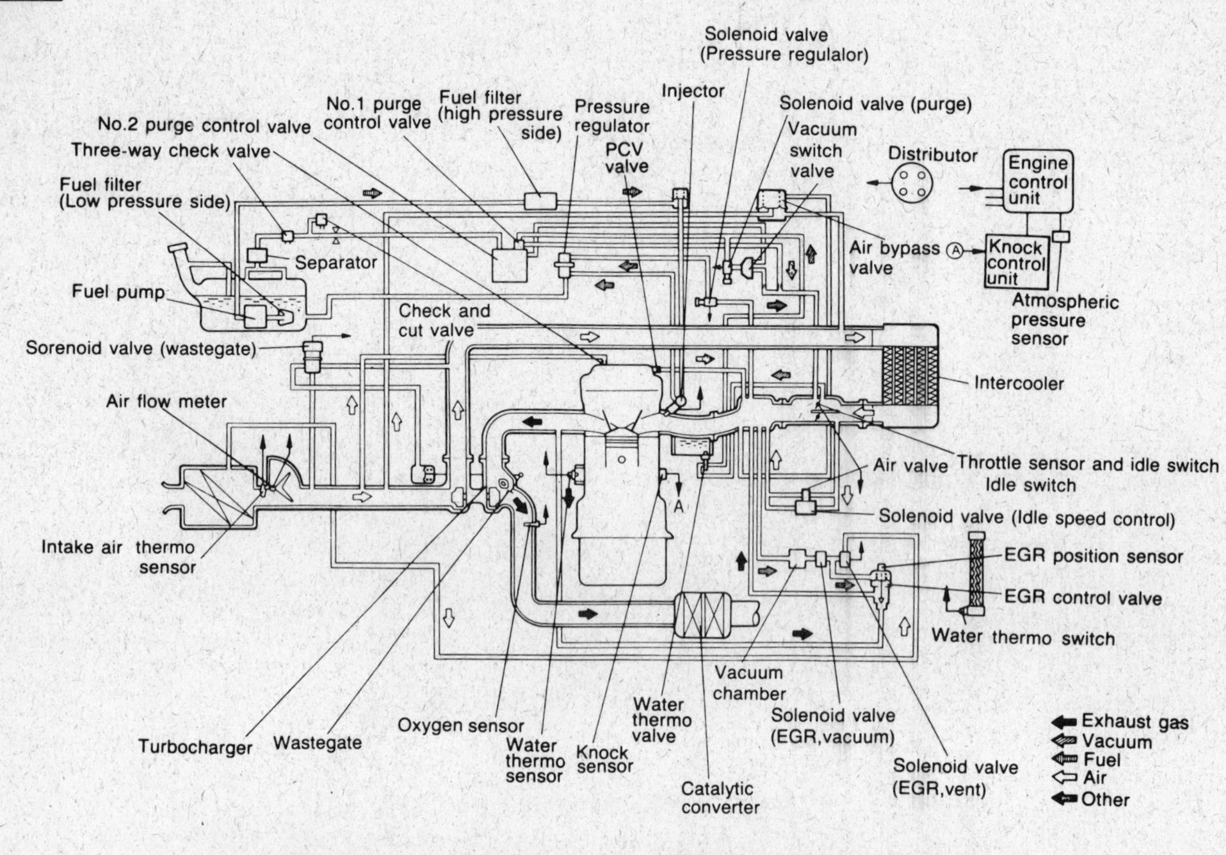

Vacuum hose routing diagram for the 1989 Mazda 626/MX-6 with turbo

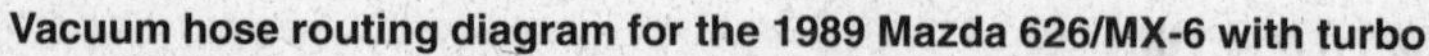

Mitsubishi

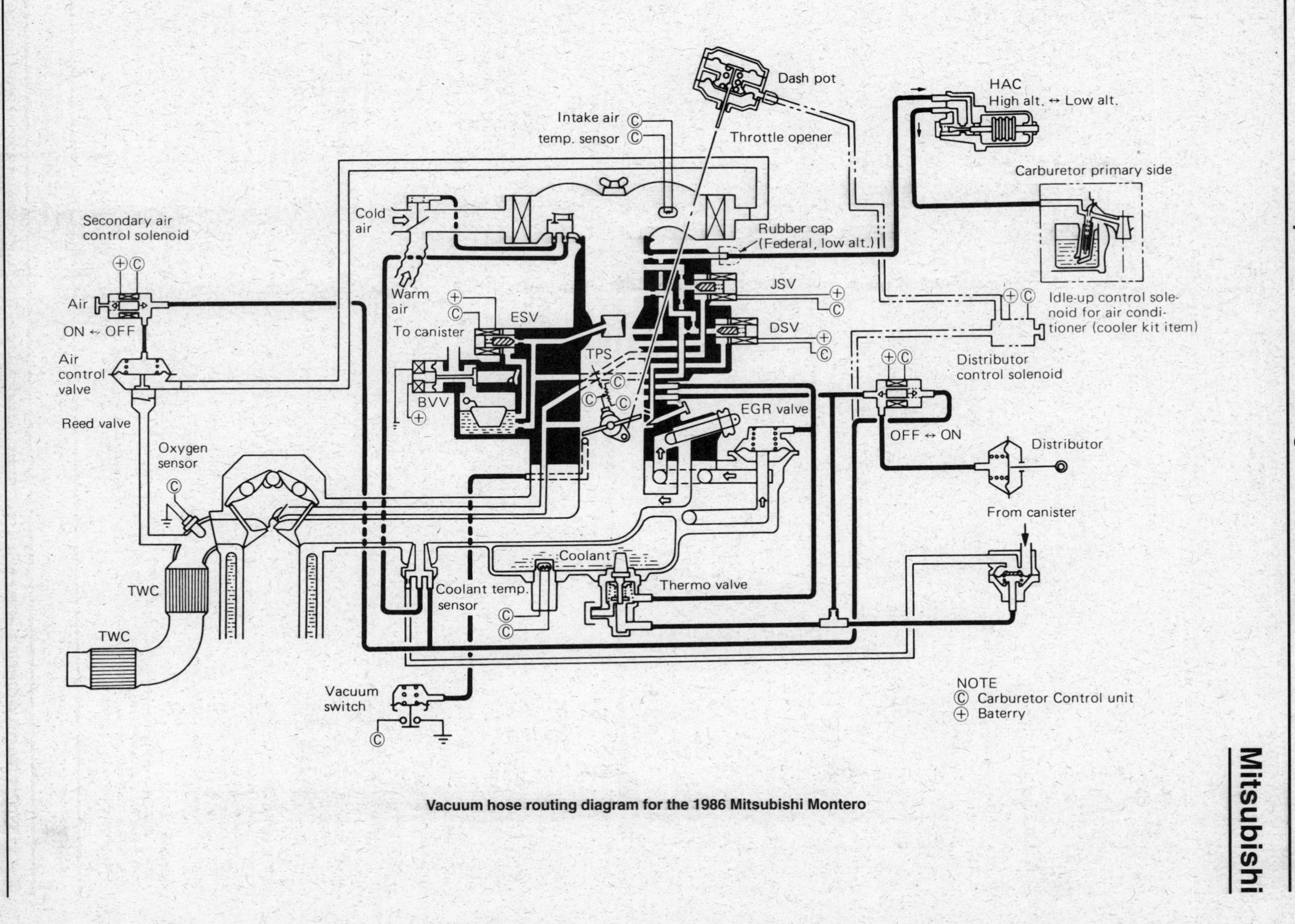

Vacuum hose routing diagram for the 1986 Mitsubishi Montero

Mitsubishi

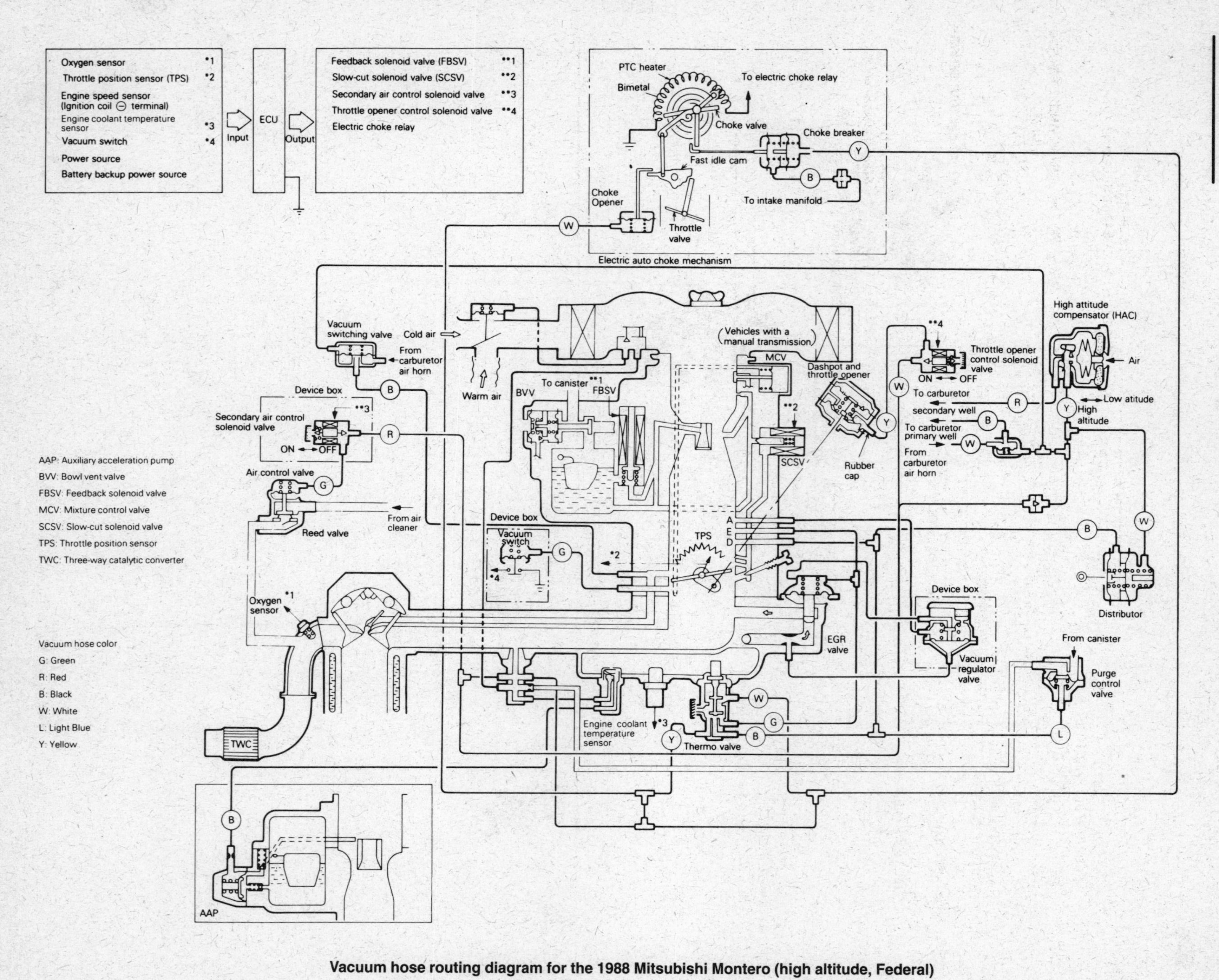

Vacuum hose routing diagram for the 1988 Mitsubishi Montero (high altitude, Federal)

Mitsubishi

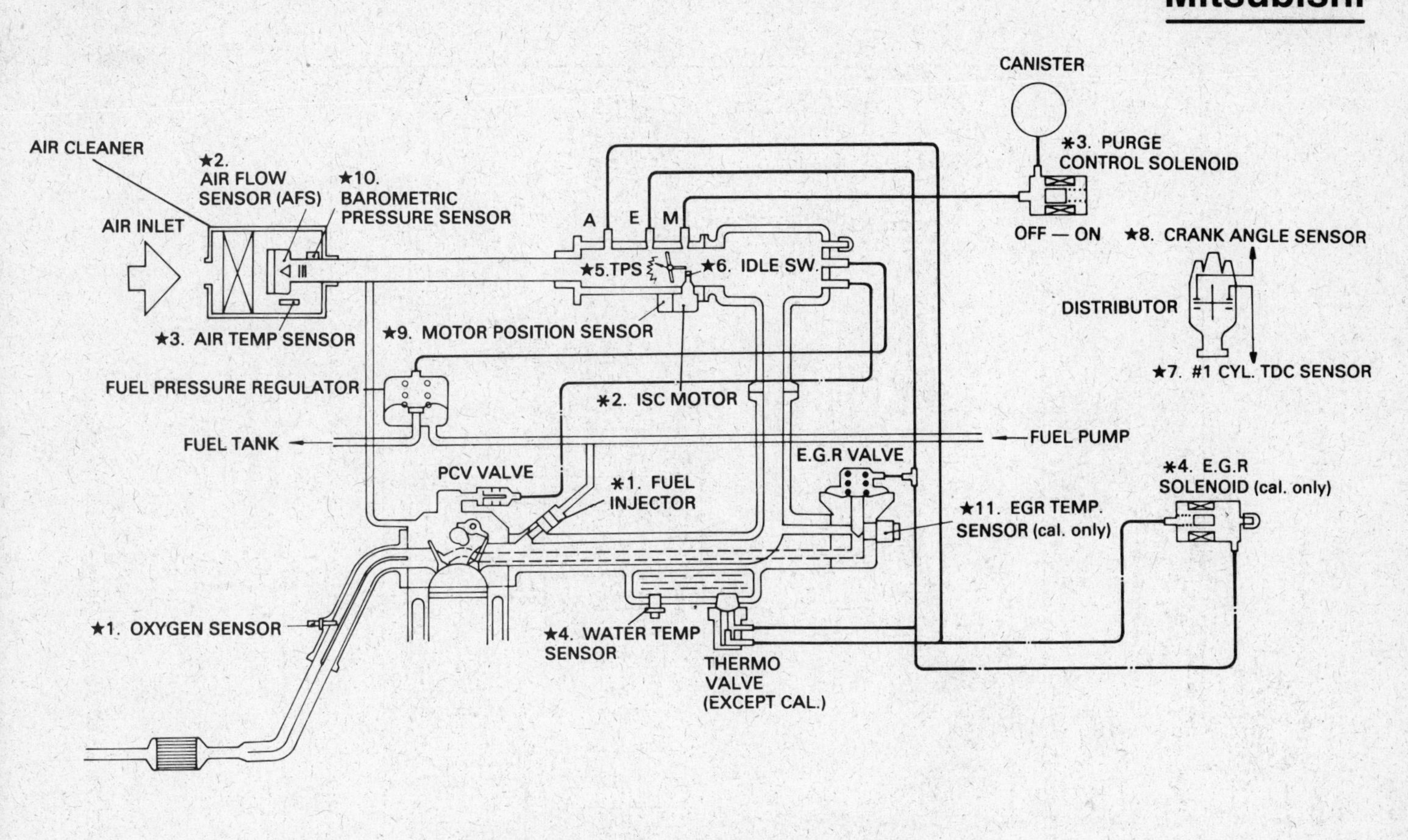

Vacuum hose routing diagram for the 1990 Mitsubishi Precis

Nissan/Datsun

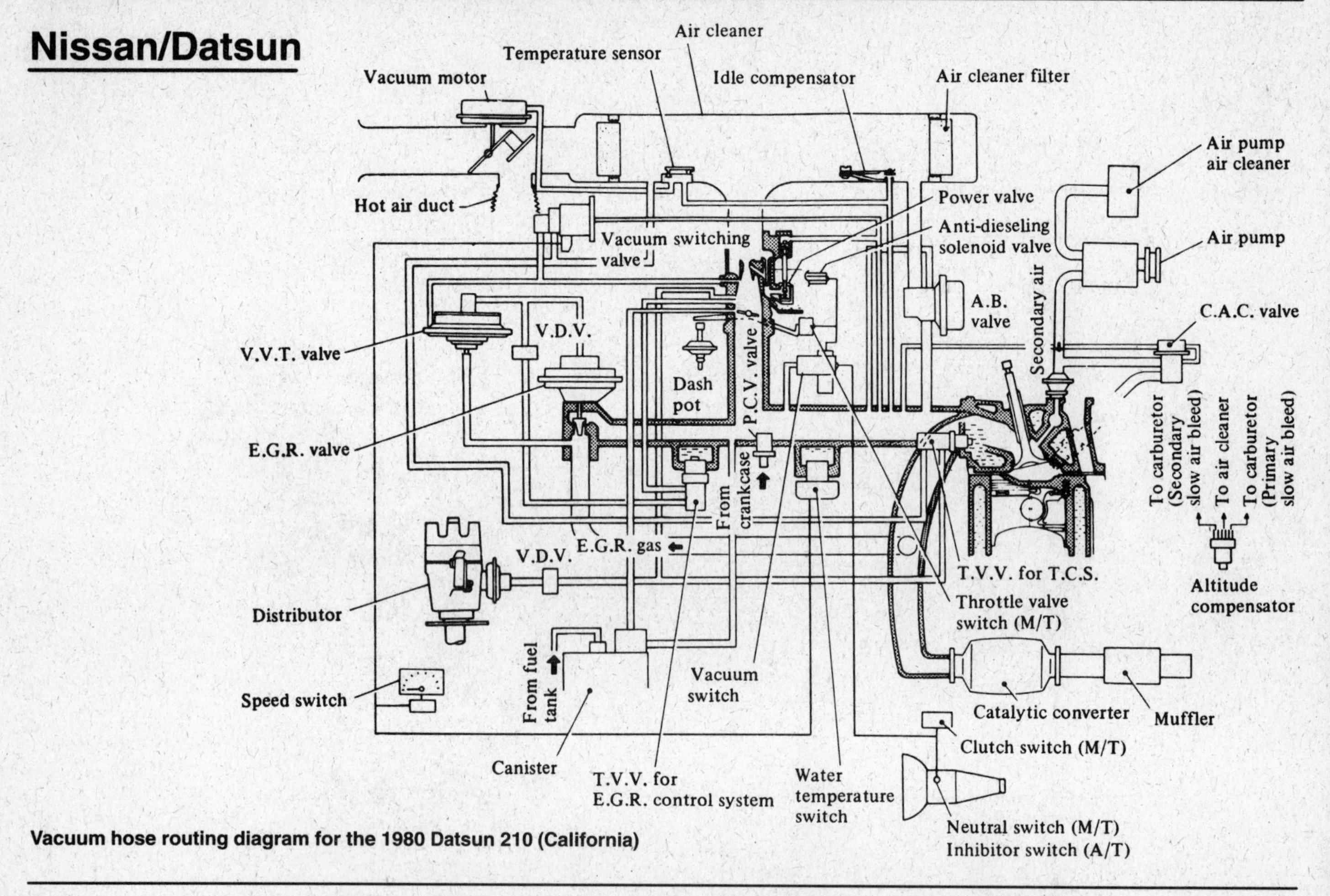

Vacuum hose routing diagram for the 1980 Datsun 210 (California)

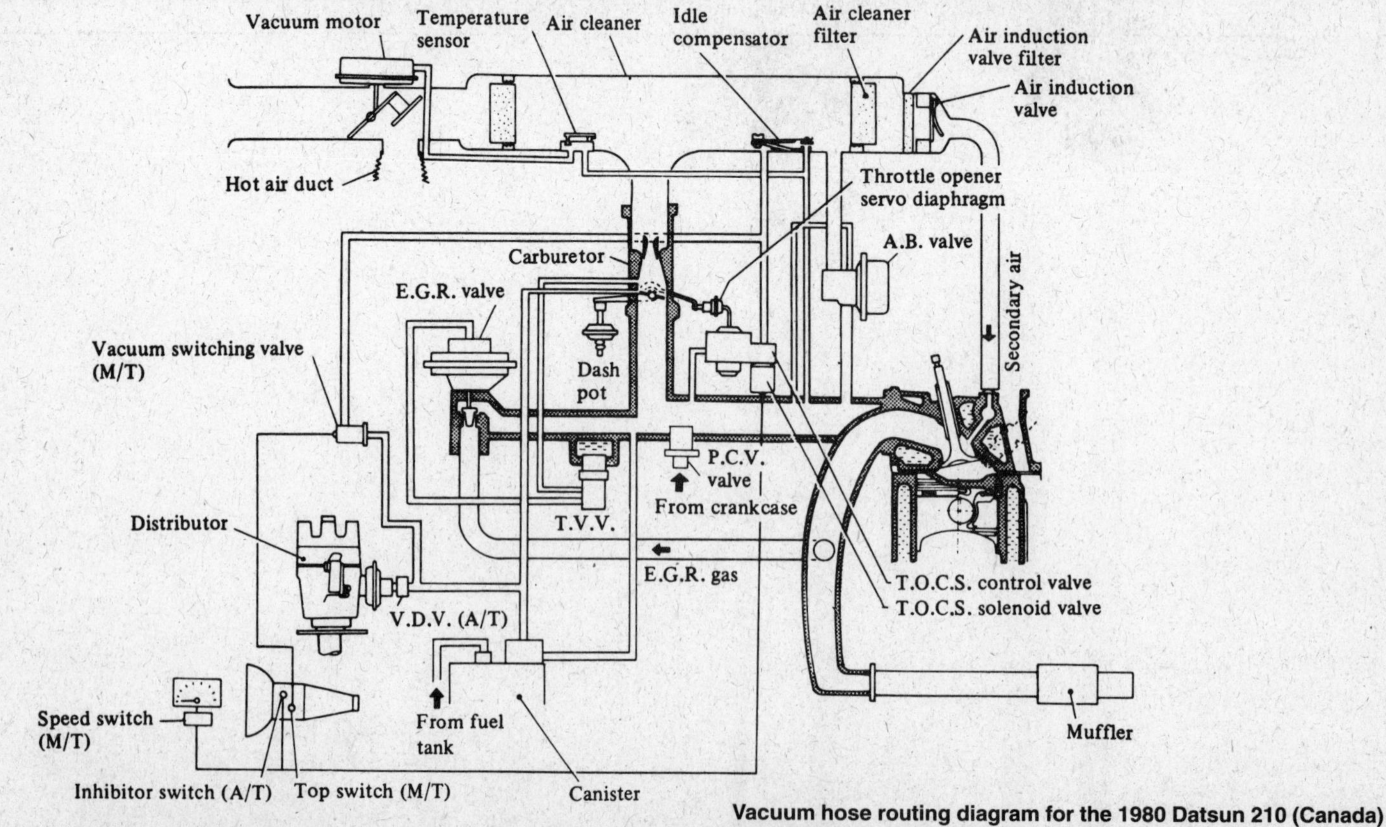

Vacuum hose routing diagram for the 1980 Datsun 210 (Canada)

Nissan/Datsun

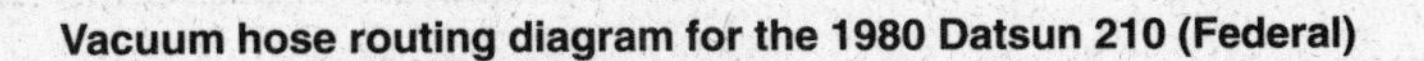

Vacuum hose routing diagram for the 1980 Datsun 210 (Federal)

Nissan/Datsun

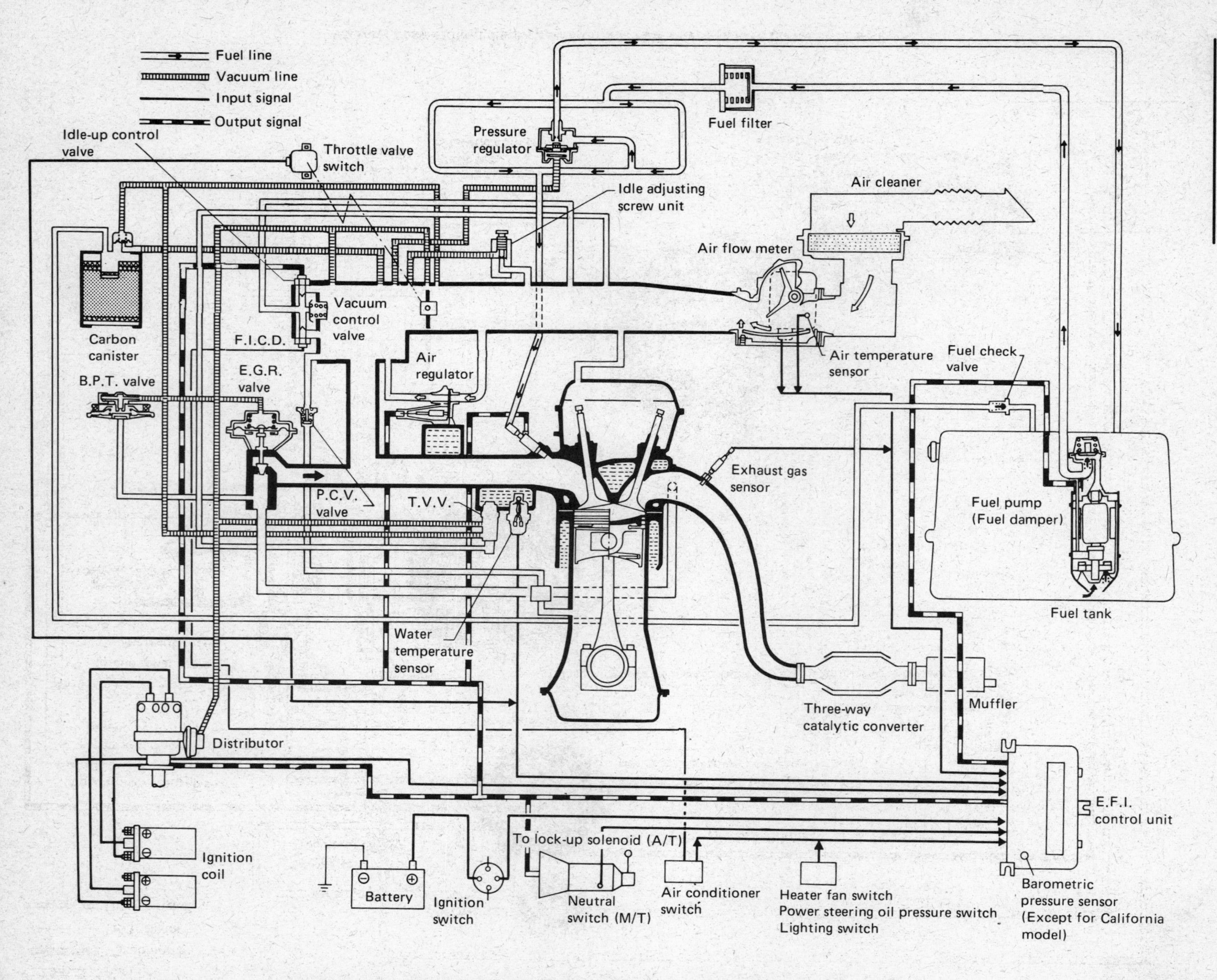

Vacuum hose routing diagram for the 1984 Nissan 200SX with the CA20E engine and EFI

Nissan/Datsun

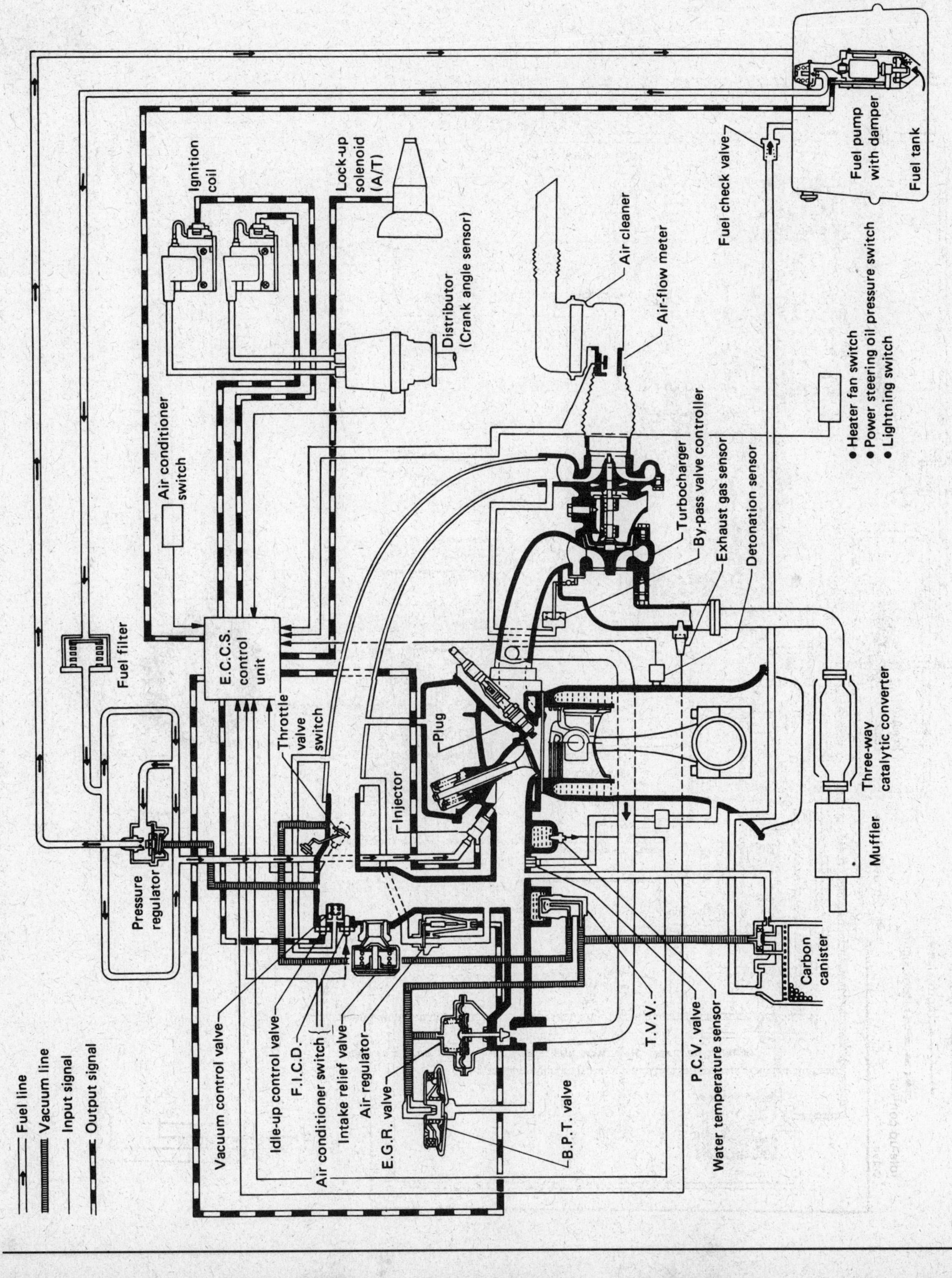

Fuel line
Vacuum line
Input signal
Output signal

Ignition coil
Lock-up solenoid (A/T)
Distributor (Crank angle sensor)
Air cleaner
Air-flow meter
Fuel check valve
Fuel pump with damper
Fuel tank
Air conditioner switch
E.C.C.S. control unit
Fuel filter
Pressure regulator
Throttle valve switch
Injector
Plug
Turbocharger
By-pass valve controller
Exhaust gas sensor
Detonation sensor
• Heater fan switch
• Power steering oil pressure switch
• Lightning switch
Three-way catalytic converter
Muffler
Carbon canister
Vacuum control valve
Idle-up control valve
F.I.C.D.
Air conditioner switch
Intake relief valve
Air regulator
E.G.R. valve
B.P.T. valve
T.V.V.
P.C.V. valve
Water temperature sensor

Vacuum hose routing diagram for the 1984 Nissan 200SX with the CA18ET engine

Nissan/Datsun

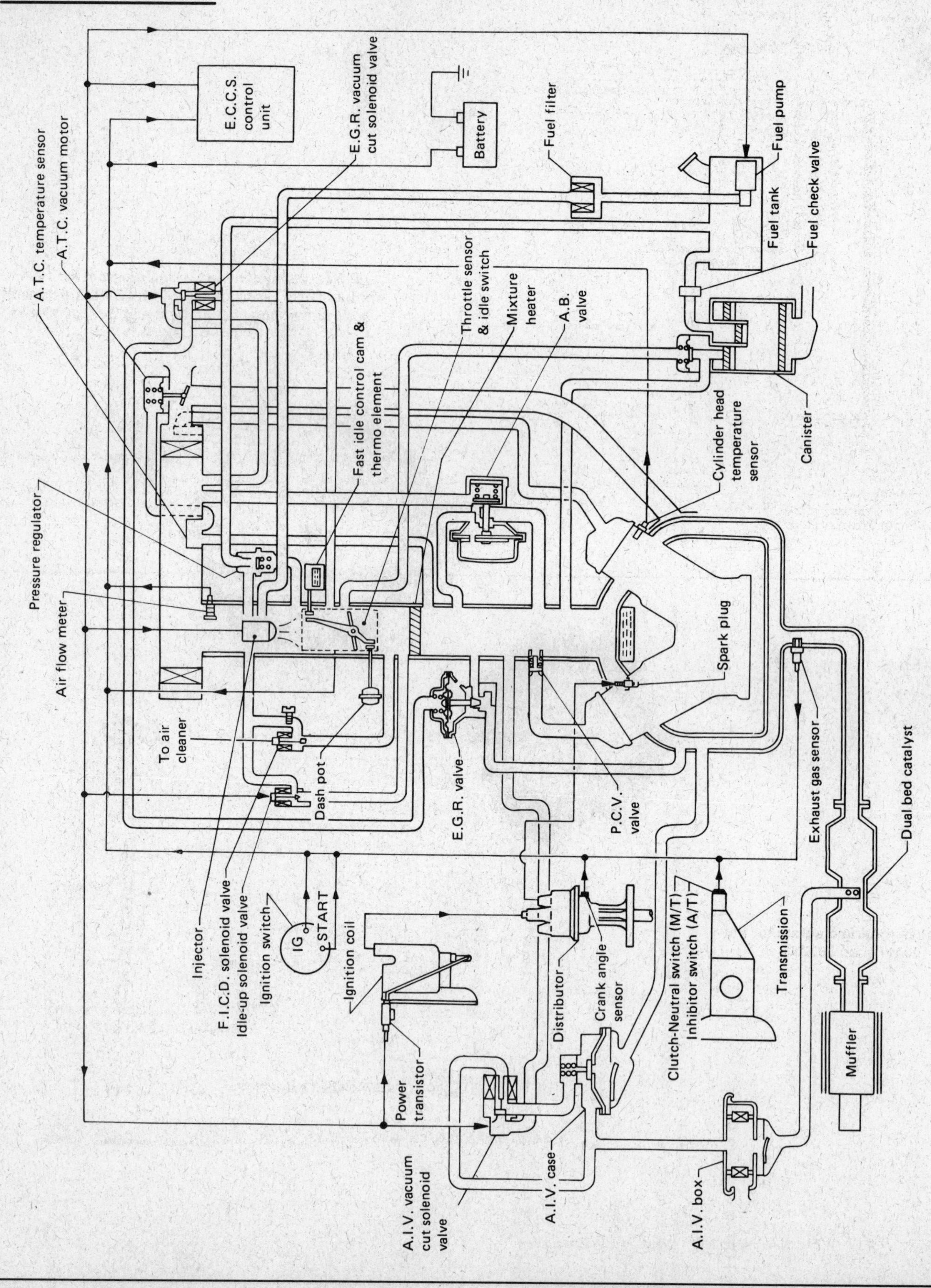

Vacuum hose routing diagram for the 1986 Nissan truck

Nissan/Datsun

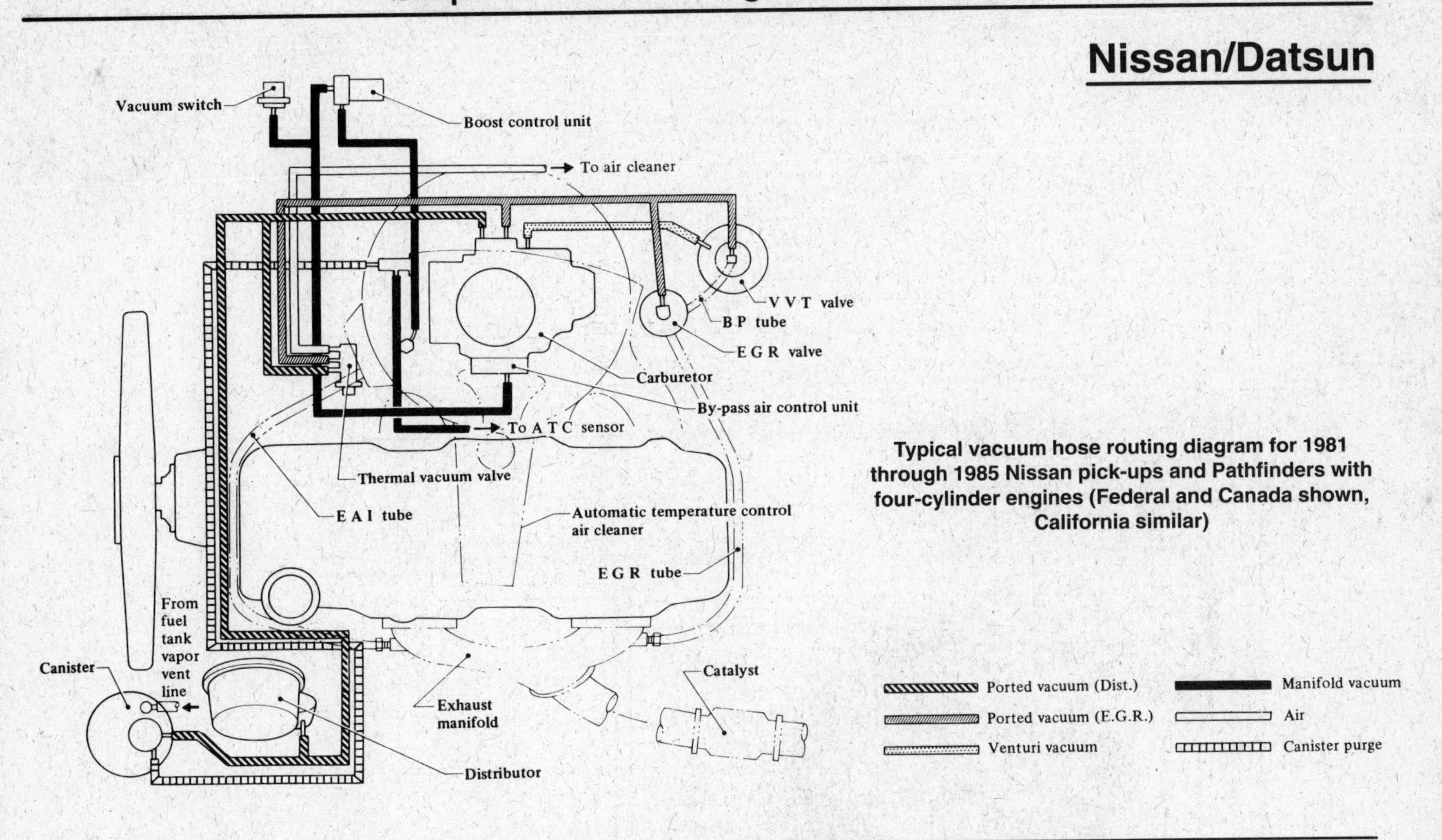

Typical vacuum hose routing diagram for 1981 through 1985 Nissan pick-ups and Pathfinders with four-cylinder engines (Federal and Canada shown, California similar)

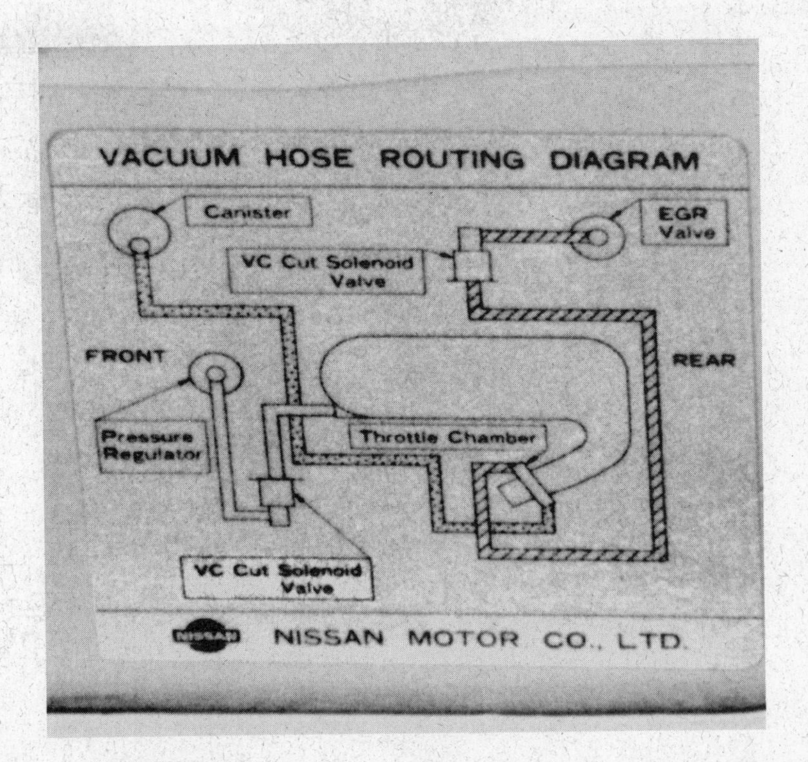

Vacuum hose routing diagram for the 1984 Nissan 300ZX

Nissan/Datsun

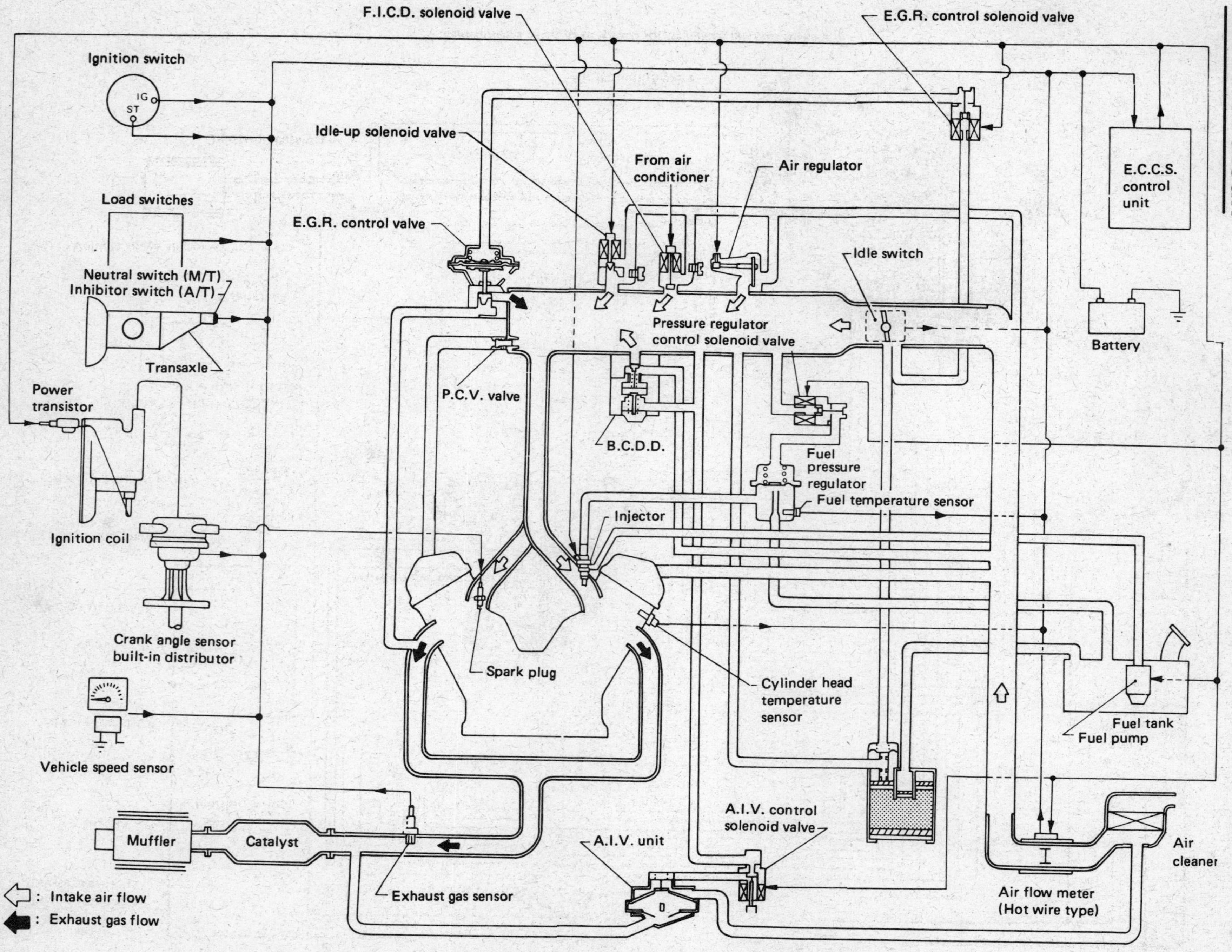

Vacuum hose routing diagram for the 1988 Nissan Maxima

Nissan/Datsun

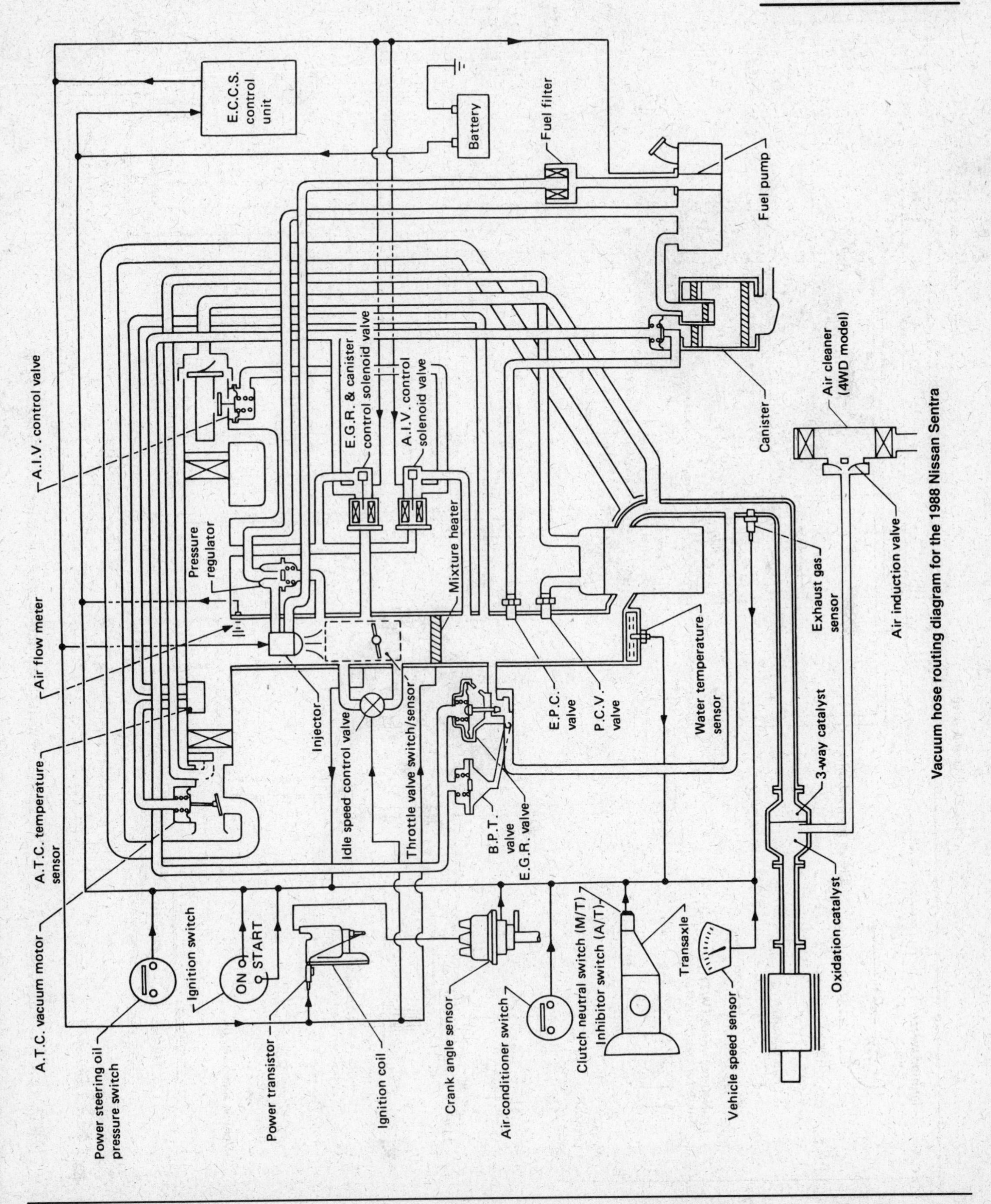

Vacuum hose routing diagram for the 1988 Nissan Sentra

Toyota

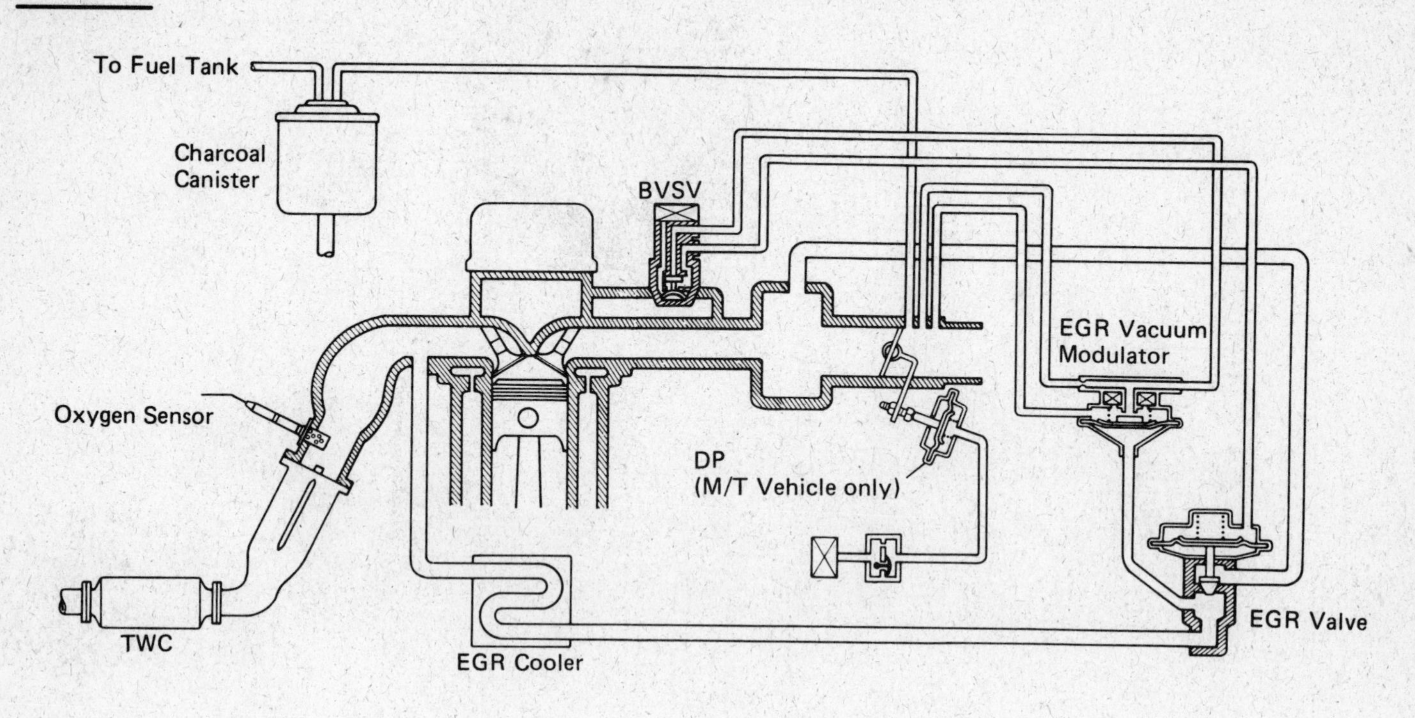

Typical vacuum hose routing diagram for Toyota pick-ups with the 22R engine (1986 shown)

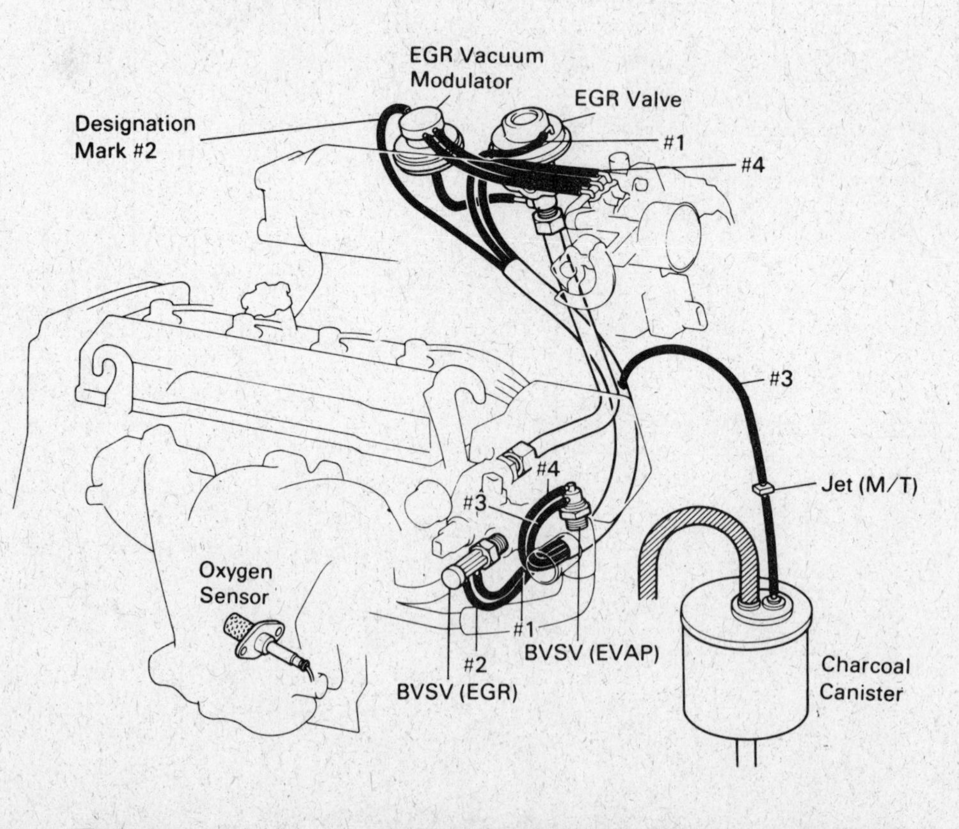

Typical vacuum hose routing diagram for the 1987 through 1990 Toyota Camry (four-cylinder models)

Toyota

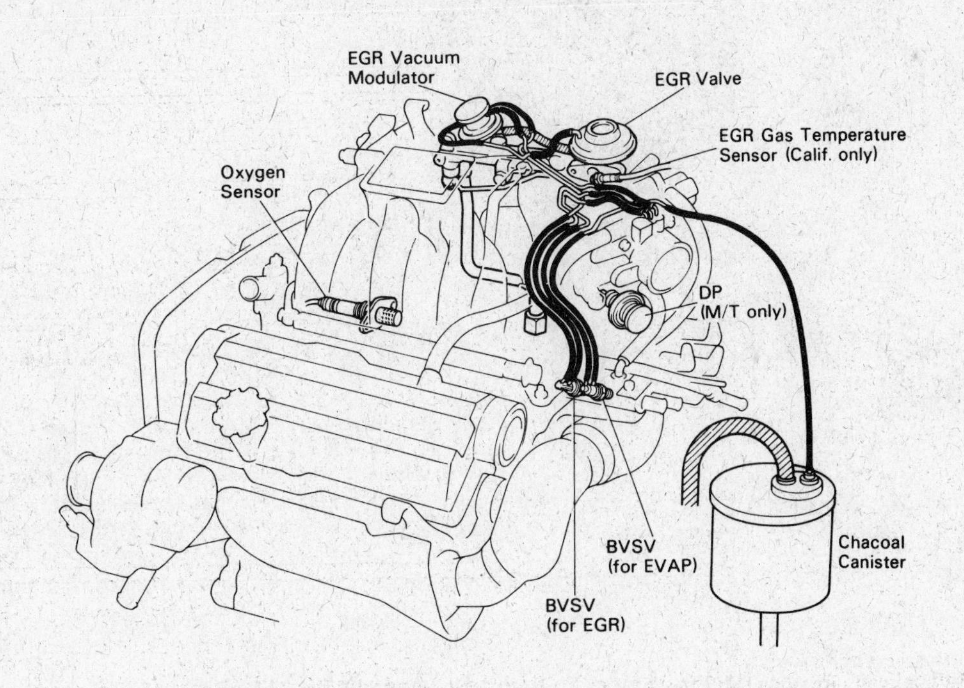

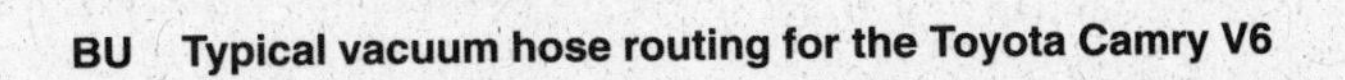

BU Typical vacuum hose routing for the Toyota Camry V6

Acronym list and glossary

A

AAS – Air Aspirator System or Aspirator Air System.

AAV – Anti-Afterburn Valve.

AB – Air Bleed.

ABPV or **ABV** – Air Bypass Valve.

A/C – Air Conditioner or Air Conditioning.

ACCC – A/C Compressor Clutch.

ACCS – A/C Clutch Cycling Switch.

ACL BI-Met – Air Cleaner Bimetal sensor.

ACL DV – Air Cleaner Duct and Valve Vacuum motor.

ACT – Air Charge Temperature sensor.

ACV – Air Control Valve.

Actuator – A device that delivers motion in response to an electrical signal; actuators alter the operation of the engine in response to commands from the engine management system computer. Examples are fuel injectors, Electronic Air Control Valves (EACV) and carburetor feedback solenoids.

AFC – AirFlow Controlled (usually refers to a type of fuel injector).

AFS – AirFlow Sensor.

AI – Air Injection. See "Air Injection System."

AICV – Air Injection Check Valve.

AIR – Air Injection Reactor system; typically the pump-operated air injection system used on GM vehicles. Also see "Air Injection System."

Air Aspirator System (AAS) – A passive air injection system that uses a one-way valve instead of an air pump to introduce extra air into the exhaust stream.

Air Bypass Valve (ABPV or ABV) – A backfire-suppressor valve used in air injection systems (also called an anti-backfire, diverter or gulp valve). During high engine vacuum conditions such as deceleration, it vents pressurized air from the air pump to the atmosphere in order to prevent backfiring. At other times, it sends air to the exhaust manifold; on vehicles with a three-way catalyst, it sends air to the oxidation catalyst only when the engine warms up.

Air Charge Temperature (ACT) sensor – A thermistor sensor that inputs the temperature of the incoming air stream in the air filter or intake manifold to the computer. It can be located in the intake manifold (EFI systems) or the air cleaner. On carbureted vehicles, if the air is cold, it signals the choke to let off slowly. It then alters engine speed after choke is off and, below a certain temperature, dumps air from the air injection system to the atmosphere for catalyst protection.

AIR-CHV – Air Check Valve.

Air Cleaner Bi-Metal (ACL BI-MET) sensor – A component of a thermostatic air cleaner system. It senses the temperature of incoming fresh air and bleeds off vacuum when the air is warm. When the air is cold, the sensor directs vacuum to the air cleaner vacuum motor.

Air Cleaner Duct and Valve (ACL DV) vacuum motor – A component of thermostatic air cleaner systems. It opens and closes the air duct valve to provide heated or unheated air to the engine in accordance with the temperature of the incoming air.

Air Conditioner Clutch Compressor (ACCC) signal – The input to the computer regarding the status of the A/C clutch (engaged or disengaged).

Air Control Valve (ACV) – A vacuum-controlled diverter valve (or a combination bypass/diverter valve) in an air injection system that diverts air pump air to either the upstream (exhaust manifold) or downstream (oxidation catalyst) air injection points as necessary.

AIR-DV – Air Diverter Valve. See "Diverter valve."

Air Guard (AG) – An American Motors air injection system that uses an air pump to supply air into the exhaust manifold to reduce HC and CO emissions.

Air Injection Reactor system – See "AIR" and "Air Injection System."

Air Injection System (AIS) – Any system that injects air into the exhaust stream to promote more complete oxidation of unburned exhaust gases.

AIR-IVV – Air Idle Vacuum Valve.

Air Management System (AMS) – Used to control the injected air to the exhaust manifold and catalytic converter. This improves the pollutant conversion efficiency in the converter.

Air Pump – A device which produces a slightly pressurized flow of air to the exhaust manifold and/or the catalytic converter. Also known as a Thermactor air supply pump.

Air Switching Valve (ASV) – A valve in an air injection system that senses intake manifold vacuum and, during heavy loads, dumps part of the air pump output to the air cleaner to reduce air injection system pressure.

AIS – Air Injection System.

AIS – Automatic Idle Speed motor.

AIV – Air Injection Valve.

ALCL – Assembly Line Communications Link.

ALDL – Assembly Line Diagnostic Link.

Altitude Compensation system – A barometric switch and solenoid used to provide better driveability over 4,000 ft. above sea level.

Ambient Temperature – The temperature of the air surrounding an object.

AMGV – Air Management Valve.

AMS – Air Management System.

ANTBV – Anti-Backfire valve.

Anti-Backfire Valve (ANTI-BFV) – See "Air Bypass Valve."

APDV – Air Pump Diverter Valve. See "Diverter valve."

AS – Airflow Sensor.

ASRV – Air Switching Relief Valve.

ASS – Air Switching Solenoid.

Assembly Line Communications Link (ALCL) or Assembly Line Diagnostic Link (ALDL) – On GM vehicles, the point where connector terminals are bridged to output trouble codes.

ASV – Air Switching Valve.

Atmospheric pressure – The weight of the air at sea level, which is about 14.7 pounds per square inch; decreases as the altitude increases.

ATS – Air Temperature Sensor.

Automotive emissions – Gaseous and particulate compounds (hydrocarbons, oxides of nitrogen and carbon monoxide) that are emitted from a vehicle's crankcase, exhaust, carburetor (or fuel injection system) and fuel tank.

Auxiliary Acceleration Pump (AAP) – A pump that increases driveability during cold engine operation by providing an extra amount of fuel to the acceleration nozzle to supplement the main acceleration pump.

B

Backfire Suppressor Valve – A device used in conjunction with earlier design air injection systems. Its primary function is to lean out the excessively rich fuel mixture which follows closing of the throttle after acceleration. It allows extra air into the induction system whenever the intake manifold vacuum decreases.

Backpressure Variable Transducer (BVT) – A system combining a ported EGR valve and a backpressure variable transducer to control emissions of NOx.

Barometric pressure – Atmospheric pressure.

BARO sensor – A barometric pressure sensor (see below).

Barometric and Manifold Absolute Pressure (BMAP) sensor – A housing containing both BP and MAP sensors.

Barometric Pressure (BP) sensor – Sends a variable frequency signal to the computer regarding the atmospheric pressure, allowing adjustment of the spark advance, EGR flow and air/fuel ratio as a function of altitude.

Base Idle – The idle speed determined by the throttle lever setting on the carburetor or throttle body while the Idle Speed Control (ISC) motor is fully retracted and disconnected.

BCDD – Boost-Controlled Deceleration Device.

BDC – Bottom Dead Center.

BHS – Bi-Metal Heat Sensor.

Bimetal Heat Sensor (BHS) – A strip (usually coiled) consisting of two metals with different expansion characteristics. Bimetal strips are used in thermostatically controlled devices because they move or bend toward the metal that expands least when heat is applied.

BMAP – Barometric and Manifold Absolute Pressure sensor.

Boost-Controlled Deceleration Device (BCDD) – A valve that, during deceleration, is triggered into action by high intake manifold vacuum. The BCDD valve allows an additional source of air and fuel to enter the intake manifold during deceleration to obtain a more burnable mixture.

Bowl Vent (BV) port – The port in the carburetor which vents fumes and excess pressure from the float bowl to maintain atmospheric pressure.

BP – Barometric Pressure sensor.

BPA – By-Pass Air solenoid.

BPEGR – Backpressure EGR.

BPS – Barometric Pressure Sensor.

Break-Out Box (BOB) – A service tool that tees-in between the computer and the multi-pin harness connector. Once connected in series with the computer and the harness, this test device permits measurements of computer inputs and outputs.

BV – Bowl Vent port.

BVT – Backpressure Variable Transducer.

Bypass Air (BPA) solenoid – Used to control the idle speed on some fuel-injected vehicles.

Bypass Valve (BPV) – A valve that opens under certain conditions to permit a flow of liquid or gas by some alternate to its normal route.

C

C-3 – Computer Command Control system (GM vehicles).

C-4 – Computer Controlled Catalytic Converter system (GM vehicles).

CAM – Choke Air Modulator switch.

Canister – A container in an evaporative emission control system; contains charcoal to trap vapors from the fuel system.

Canister Purge Shut-off Valve (CPSOV) – A vacuum-operated valve that shuts off canister purge when the air injection diverter valve dumps air downstream.

Canister purge solenoid – An electrical solenoid that opens the canister purge valve between the fuel vapor canister line and the intake manifold when energized.

Canister purge valve – Valve used to regulate the flow of vapors from the evaporative canister to the engine.

CANP – Canister Purge solenoid.

Carbon dioxide (CO_2) – A colorless gas that is a major constituent of automotive exhaust.

Carbon monoxide (CO) – A colorless, odorless and very poisonous gas that is a major product of incomplete combustion, especially at engine idle. As little as 0.3 percent by volume can be lethal within 30 minutes.

CAT – Catalytic Converter.

Catalyst – A compound or substance which can speed up or slow down the reaction of other substances without being consumed itself. In an automotive catalytic converter, special metals (i.e., platinum, palladium) are used to promote more complete combustion of unburned HC and CO.

Catalytic converter – A muffler-like device in the exhaust system that promotes a chemical reaction which converts certain air pollutants in the exhaust gases into harmless substances.

CAV – Coasting Air Valve.

CBVV – Carburetor Bowl Vent Valve.

CC – Catalytic Converter.

CCC – Computer Command Control; Converter Clutch Control solenoid.

CCEGR – Coolant Controlled Exhaust Gas Recirculation

CCEVS – Coolant Controlled Engine Vacuum Switch.

CCIE – Coolant Controlled Idle Enrichment.

CCO – Converter Clutch Override.

CCT – Computer Controlled Timing.

CCV – Closed Crankcase Ventilation system.

CEC – Combined Emission Control system; Computerized Emission Control system.

Central Fuel Injection (CFI) – A computer-controlled fuel metering system which sprays atomized fuel into a throttle body mounted on the intake manifold.

CES – Clutch Engine Switch.

"Check Engine" light – A light on the instrument panel that lets the driver know of any detectable engine management system malfunctions. Also used as an emissions maintenance reminder light on some vehicles. Often, when this light is on, a trouble code is stored in the computer (see Chapter 2).

Check Valve (CV) – A one-way valve which allows a liquid or gas to flow in one direction only.

Choke Thermal Vacuum Switch (CTVS) – A switch used on some GM vehicles to deny vacuum to either the front or the auxiliary choke vacuum breaks. Its purpose is to slow the opening of the choke and to provide better driveability when the engine is cold.

CIS – Continuous Injection System.

CKV – Check Valve.

CLC – Converter Lock-up Clutch.

Closed Crankcase Ventilation (CCV) system – Vents crankcase pressure and vapors back into the engine where they are burned during combustion rather than being vented to the atmosphere (See "PCV system").

Closed loop mode – Once the engine has reached "warm-up" temperature, the engine management computer collects the precise data from all the sensors (coolant temperature sensor, throttle position sensor, oxygen sensor etc.) to determine the most efficient air/fuel mixture for combustion.

CO – Carbon Monoxide.

CO_2 – Carbon Dioxide.

COC – Conventional Oxidation Catalyst.

Cold Weather Modulator (CWM) – A vacuum modulator located in the air cleaner on some models. The modulator prevents the air cleaner duct door from opening to non-heated intake air when outside air is below 55-degrees F. Similar to a Temperature Vacuum Switch.

Computer – A device that takes information, processes it, makes decisions and outputs those decisions.

Computer Command Control (C-3) system – An earlier engine management system used on General Motors vehicles.

Computer Controlled Catalytic Converter (C-4) system – A later engine management system used on General Motors vehicles.

Computer Controlled Timing (CCT) – A system that feeds input from various engine sensors into a computer. The computer then matches spark timing exactly to engine requirements throughout its full range of operations.

Continuous Injection System (CIS) – A mechanical fuel injection system designed and manufactured by Bosch, used on many German vehicles. In a CIS system, the fuel injectors are always open (i.e. they emit a continuous spray of fuel into the intake ports). The amount of fuel sprayed is determined by the fuel pressure in the system, which in turn is determined by the position of the throttle.

Conventional Oxidation Catalyst (COC) – A catalyst which acts on the two major pollutants, HC and CO.

Coolant Controlled Exhaust Gas Recirculation (CCEGR) – A system that prevents exhaust gas recirculation until engine coolant temperature reaches a specific value.

Coolant Temperature Override (CTO) switch – A switch that prevents vacuum from reaching a component until coolant temperature reaches a certain value.

Coolant Temperature sensor – tells the computer the temperature of the engine coolant. This sensor helps the computer control fuel metering and spark timing.

CP – Crankshaft Position sensor.

CPRV – Canister Purge Regulator Valve. See "Canister purge valve."

CPS – Canister Purge Solenoid.

CPV – Canister Purge Valve.

Crankcase Breather – A port or tube that vents fumes from the crankcase. An inlet breather allows fresh air into the crankcase.

CRV – Coasting Richener Valve.

CSOV – Canister purge Shut Off Valve.

CSSA – Cold Start Spark Advance system.

CSSH – Cold Start Spark Hold system.

CTAV – Cold Temperature Activated Vacuum.

CTO – Coolant Temperature Override switch.

CTS – Coolant Temperature Sensor; Coolant Temperature Switch.

CTVS – Choke Thermal Vacuum Switch.

Curb idle – Normal idle rpm; computer controlled on many modern vehicles.

CV – Check Valve.

CWM – Cold Weather Modulator.

D

Dash-Pot (DP) – A diaphragm that controls the rate at which the throttle closes.

DBC – Dual-Bed Catalytic Converter.

DEFI – Digital Electronic Fuel Injection.

Delay Vacuum Bypass (DVB) system – An optional system used by Ford that bypasses the spark delay valve during cold operation to improve driveability.

Delay valve – Vacuum restriction used to retard or delay the application of a vacuum signal. Also referred to as a Vacuum Delay Valve (VDV).

Detonation – An uncontrolled explosion (after the spark occurs at the spark plug) which spontaneously combusts the remaining air/fuel mixture, resulting in a "pinging" noise.

DFS – Deceleration Fuel Shut-off.

Dieseling – The tendency of an engine to continue running after the ignition is turned off.

Digital Volt-Ohm Multimeter (DVOM) – A digital electronic meter that displays voltage and resistance.

DIS – Distributorless Ignition System.

Diverter Valve – Used in air injection systems to channel airflow to either the exhaust manifold or oxidation catalyst under different operating conditions.

DLV – Delay Valve.

DP – Dashpot.

DRB II – Diagnostic Readout Box II. See "Break-Out Box."

DS – Detonation Sensor. See "Knock sensor."

DSAV – Deceleration Spark Advance system.

Dual Catalytic Converter – A Three-Way Catalyst (TWC). Also referred to as a dual-bed converter.

Duty cycle – The electronic measurement expressed in the percent of total time (On Time) the fuel injector is activated. This duration, or pulse width, is related directly to the amount of fuel injected into the combustion chamber.

DV – Delay Valve.

DVOM – Digital Volt-Ohmmeter.

DVTRV – Diverter Valve.

E

EACV – Electronic Air Control Valve. A valve used in fuel-injection systems, usually computer controlled, that controls the amount of air bypassing the throttle during idle. The more air that bypasses the throttle, the higher the idle speed.

Early Fuel Evaporation (EFE) system – A device that heats the air/fuel mixture entering the intake manifold when the engine is cold.

ECA – Electronic Control Assembly.

ECC – Electronic Controlled Carburetor.

ECM – Electronic Control Module.

ECS – Evaporation Control System.

ECT – Engine Coolant Temperature sensor.

ECU – Electronic Control Unit.

EDIS – Electronic Distributorless Ignition System.

EDM – Electronic Distributor Modulator.

EEC – Evaporative Emission Controls.

EEC/EEC-I/EEC-II/EEC-III/EEC-IV – Ford Electronic Engine Control systems.

EEGR – Electronic EGR valve.

EFC – Electronic Feedback Carburetor; Electronic Fuel Control.

EFE – Early Fuel Evaporation.

EFE TVS – EFE Thermal Vacuum Switch.

EFI – Electronic Fuel Injection.

EGO – Exhaust Gas Oxygen sensor.

EGR – Exhaust Gas Recirculation system.

EGR control solenoid (EGRC) – Energizes to allow manifold vacuum to the EGR valve.

EGR cooler assembly – Heat exchanger using engine coolant to reduce exhaust gas temperature.

EGR-EPV – EGR External Pressure Valve.

EGR-EVR – EGR Electronic Vacuum Regulator.

EGR-FDLV – EGR Forward Delay Valve.

EGR-RSR – EGR Reservoir.

EGR-S/O – EGR Shut-Off

EGR-TVS – EGR Thermal Vacuum Switch.

EGRV – EGR Vent solenoid.

EGR vacuum – A vacuum source above the closed throttle plate; used for control of ported EGR valves. Vacuum is zero at closed throttle.

EGR valve – A valve used to introduce exhaust gases into the intake air stream. There are several types (see "Integral Backpressure Transducer EGR valve," "Ported EGR valve," Electronic EGR valve" and "Valve and Transducer Assembly").

EGR Valve Position (EVP) sensor – A potentiometric sensor used in electronically controlled EGR systems. Sensor wiper position is proportional to EGR valve pintle position, which allows electronic control assembly to determine actual EGR flow at any point in time.

EGR-VCV – EGR Vacuum Control Valve.

EGR Vent (EGRV) solenoid – Electrical solenoid that normally vents EGRC vacuum line. When EGRV is energized, EGRC can open the EGR valve.

EGR venturi vacuum amplifier – A device that uses a relatively weak venturi vacuum to control a manifold vacuum signal to operate the EGR valve. Contains a check valve and relief valve that open whenever the venturi vacuum signal is equal to or greater than manifold vacuum.

EGR-VSOL – EGR Vacuum Solenoid.

EGR-VVA – EGR Venturi Vacuum Amplifier.

EHCV – Exhaust Heat Control Valve.

EIS – Electronic Ignition System.

ELB – Electronic Lean Burn.

Electronic Control Assembly (ECA) – A Ford vehicle computer consisting of a calibration assembly containing the computer memory, its control program and processor assembly (the computer hardware).

Electronic Control Module (ECM) – A GM term and also a generic term referring to the computer. The ECM is the brain of the engine control systems receiving information from various sensors in the engine compartment. The ECM calculates what is required for proper engine operation and controls the different actuators to achieve it.

Electronic EGR valve – EGR valve used in engine management systems in which the EGR flow is controlled by the computer (usually by means of an EGR valve position sensor attached to the EGR valve). Operating vacuum is supplied by EGR solenoid valve(s).

Electronic Engine Control (EEC) system – Ford's computerized engine control systems. There are four versions: EEC-I controls engine timing. EEC-II controls engine timing and fuel (on engines with an FBC system). EEC-III-FBC is a refined version of EEC-II. EEC-III-CFI controls engine timing and fuel (on engines with an EFI system). EEC-IV is a refined version of the EEC-III system.

Electronic Fuel Injection (EFI) – A computer controlled fuel system that distributes fuel through an injector located in each intake port of the engine.

Electronic Spark Timing (EST) system – This replaces the vacuum or centrifugal mechanisms in the distributor and uses the computer to advance or retard the spark timing.

Enable – A microcomputer decision that results in an engine management system being activated and permitted to operate.

Energized – Having the electrical current or electrical source turned on.

Engine Coolant Temperature (ECT) sensor – The thermistor sensor that provides coolant temperature information to the computer. Used to alter spark advance and EGR flow during warm-up or an overheating condition.

EPC – Electronic Pressure Control.

EPR-SOL – Exhaust Pressure Regulator Solenoid.

EPR-VLV – Exhaust Pressure Regulator Valve.

ERS – Engine Rpm Sensor.

ESC – Electronic Spark Control system.

ESS – Engine Speed Sensor.

ESSM – Engine Speed Switch Module.

EST – Electronic Spark Timing system.

ETC – Electronic Throttle Control.

EVAP – Evaporative emission control system.

Evaporation Control System (ECS) – A system used to prevent the escape of gasoline vapors to the atmosphere from the fuel tank and carburetor.

Evaporative Emission Control (EVAP or EEC) system – Emission control system that prevents fuel vapors from entering the atmosphere, primarily from the fuel tank and the carburetor.

Evaporative emissions – Hydrocarbon emissions formed by the evaporation of gasoline or other automotive fuels.

Evaporative Emission Shed System (EESS) – A Ford evaporative emission control system introduced in 1978.

EVCR – Emission Vacuum Control Regulator.

EVP – EGR Valve Position sensor.

EVR – EGR Vacuum Regulator.

Exhaust Back Pressure Transducer Valve (BPV or BPS) – A device used to sense exhaust pressure changes and control vacuum to the EGR valve in response to these changes.

Exhaust emissions – The tail pipe products of incomplete combustion of gasoline or other fuels.

Exhaust Gas Check (EGC) valve – A device that allows air injection system air to enter the exhaust manifold, but prevents a reverse flow in the event of improper operation of other components.

Exhaust Gas Oxygen (EGO) sensor – A device that changes its output voltage as the exhaust gas oxygen content changes when compared to the oxygen content of the atmosphere. This constantly changing voltage signal is sent to the processor for analysis and adjustment to the air/fuel ratio.

Exhaust Gas Recirculation (EGR) system – A system used to control oxides of nitrogen. The exhaust gases are recirculated, lowering the engine combustion temperatures, thereby reducing engine pollutants.

Exhaust Heat Control Valve (HCV) – A valve which routes hot exhaust gases to the intake manifold heat riser during cold engine operation. Valve can be thermostatically controlled, vacuum operated or computer controlled.

F

FAP – Forced Air Pre-heat system.

FBC – Feedback Carburetor system.

FBCA – Feedback Carburetor Actuator.

FCS – Feedback Carburetor Solenoid, Feedback Control System, Fuel Control Solenoid or Fuel Control System.

FCV – Fuel Cut-off Valve.

Feedback Carburetor Actuator – A computer-controlled stepper motor that varies the carburetor air/fuel mixture.

Feedback Control System (FCS) – A computer-controlled fuel system employing a stepper motor or a dithering solenoid that controls air-fuel mixture by bleeding precise amounts of air (determined by the computer) into the main and idle systems of the carburetor.

FI – Fuel Injection.

Fuel tank vapor valve – A valve mounted in the top of the fuel tank. Vents excess vapor and pressure from the fuel tank into the evaporative emission control system.

Fuel Vapor Recovery (FVR) system – A valve responsible for venting excess fuel vapor and pressure from the fuel system to the EEC system.

Fuel-vacuum separator – Used to filter waxy hydrocarbons from the carburetor ported vacuum to protect the vacuum delay and distributor vacuum controls.

FVEC – Fuel Vapor Emission Control.

FVR – Fuel Vapor Recovery system.

G

GND – Ground.

GRND – Ground.

Ground (GND or GRND) – The negatively charged side of a circuit. A ground can be a wire, the negative side of the battery or even the vehicle chassis.

Ground circuit – The return side of an electric circuit.

H

HAC – High altitude compensator.

HAEC – High Altitude Emission Control system.

HAI – Heated Air Inlet system.

HC – Hydrocarbon.

HCV – Heat control valve.

Heat riser valve – See "Exhaust Heat Control Valve (HCV)".

Heated Air Inlet (HAI) system – A system that operates during cold weather and cold start. Brings warm, filtered air into the engine to control the volume of air entering the engine, vaporize the fuel better and reduce HC and CO emissions.

Heated Exhaust Gas Oxygen (HEGO) sensor – An EGO sensor with a heating element.

HEGO – Heated EGO sensor.

HEI – High Energy Ignition.

Hg – The chemical abbreviation for Mercury. Inches of Mercury (In-Hg) is the unit of measurement used for when measuring vacuum.

High Energy Ignition (HEI) – An electronic ignition system used by General Motors (GM).

HSC – High Swirl Combustion chamber.

Hydrocarbons (HC) – Any compound that is made up of Hydrogen (H) and Carbon (C) molecules. Gasoline, diesel fuel, and lubricating oils are made up of hydrocarbons.

Hydrogen (H) – Highly flammable elemental gas.

Hydrogen Sulfide (H_2S) – A flammable poisonous gas that has an odor suggestive of rotten eggs.

I

IAC – Idle Air Control.

IAS – Inlet Air Solenoid valve.

IAT – Intake Air Temperature.

IBP – Integral Back Pressure.

ICM – Ignition Control Module.

Idle Speed Control (ISC) – Maintains the idle speed of the engine at a minimum level. There are currently two types of computer controlled idle speed control: DC motor ISC and air bypass ISC.

Idle Tracking Switch (ITS) – Used on CFI vehicles to inform the EEC if the throttle is in contact with the DC motor.

Idle Vacuum Valve (IVV) – This device may be used in conjunction with other vacuum controls to dump air injection system air during extended periods of idle, to protect the catalyst.

IGN – Ignition.

IMVC – Intake Manifold Vacuum Control.

Infrared Analyzer – An instrument used to measure unburned hydrocarbons (HC) and carbon monoxide (CO) discharged from a vehicle exhaust pipe.

INJ – Injector.

Injector – An electronic fuel-injection solenoid. When energized, it allows fuel to flow into the throttle body (throttle body injection) or the intake port (port injection systems). On vehicles with Bosch CIS (continuous injection system), the injector is always open (i.e. it's not an electronically operated solenoid).

INJ GND – Injector Ground.

Integral Backpressure Transducer EGR valve – This type of EGR valve combines inputs of exhaust backpressure and EGR ported vacuum into one unit. It requires both inputs to operate on vacuum alone. There are two common designs – poppet and tapered pintle.

ISS – Idle Stop Solenoid.

ITC – Ignition Timing Control system.

ITS – Idle Tracking Switch.

IVV – Idle Vacuum Valve.

J

JAS – Jet Air Stream.

JCAV – Jet-Controlled Air Valve.

JVS – Jet Valve System.

K

KAM – Keep Alive Memory.

Keep Alive Memory (KAM) – A series of vehicle battery powered memory locations in the computer which allows it to store input failures identified during normal operation for use in later diagnostic routines. KAM even adopts some calibration parameters to compensate for changes in the vehicle system.

KNK – Knock Sensor.

Knock – A general term used to describe various noises occurring in an engine; can be used to describe noises made by loose or worn mechanical parts, preignition, detonation, etc.

Knock Sensor (KNK or KS) – A piezoelectric accelerometer designed to resonate at approximately the same frequency as the engine knock frequency and provide this information to the computer. The computer retards ignition timing when knock is detected.

KS – Knock Sensor.

L

Lead (Tetraethyl lead) – An extremely toxic additive used in leaded gasoline to increase octane and provide valve wear protection.

Lead oxides and lead halogenates – Compounds formed when leaded gasoline is burned in an engine. Present in the exhaust and as engine deposits.

M

MAF – Mass Air Flow sensor.

MAFTS – Manifold Air/Fuel Temperature Sensor.

MAJC – Main Air Jet Control.

Malfunction Indicator Light (MIL) – An electric circuit between the computer and the "CHECK ENGINE" or "SERVICE ENGINE SOON" light on the dash panel of computer-equipped vehicles.

Manifold Absolute Pressure (MAP) sensor – A pressure-sensitive disk capacitor used to measure air pressure inside the intake manifold. The MAP sensor sends a signal to the computer which uses this information to determine load conditions so it can adjust spark timing and fuel mixture.

Manifold Charge Temperature (MCT) sensor – Same as the Air Charge Temperature (ACT) sensor.

Manifold Control Valve (MCV) – A thermostatically operated valve in the exhaust manifold for varying heat to the intake manifold with respect to the engine temperature. Exhaust Heat Control Valve.

Manifold Vacuum – The vacuum that occurs below the throttle plate of a carburetor or throttle body; present throughout the intake manifold. Generated by the pumping action of the pistons. Manifold vacuum is high at idle and lowers as the throttle plates open.

MAP – Manifold Absolute Pressure sensor.

Mass Air Flow (MAF) sensor – A device that measures the mass (volume) of intake air entering the engine.

MAT – Manifold Absolute Temperature sensor.

MCS – Mixture Control System.

MCU – Microprocessor Control Unit.

MEC – Motronic Engine Control.

MCT – Manifold Charge Temperature sensor.

MCU – Microprocessor Control Unit.

MECS – Mazda Equipped Control System.

MFI – Mechanical Fuel Injection, or Multi-port Fuel Injection.

MHCV – Manifold Heat Control Valve. See "Exhaust Heat Control Valve (HCV)".

Microcomputer – A device that takes information, processes it, makes decisions and outputs those decisions.

Microprocessor – An integrated circuit within a microcomputer that controls information flow within the microcomputer. Also called the Central Processing Unit (CPU).

Microprocessor Control Unit (MCU) – An integral part of an electronically controlled feedback carburetor using a TWC catalyst. Various sensors monitor conditions. MCU is widely used on Ford vehicles for the control of air-fuel ratios.

MIL – Malfunction Indicator Light.

Mixture control solenoid – A solenoid used in feedback carburetor systems. The computer controls the solenoid to provide the optimum air/fuel ratio for the current operating conditions.

Modulator – A device that controls or regulates the intensity of a vacuum, electrical, or pressure signal.

Monolithic Substrate – The ceramic honeycomb structure as a base to be coated with a metallic catalyst material for use in the catalytic converter.

MTA – Managed Thermactor Air.

N

Neutral Drive Switch (NDS) – A sensor that provides information on transmission status to the computer.

Nitrogen (N) – An element gas which is inert. Seventy-eight percent of the air is nitrogen.

Nitrogen dioxide (NO_2) – The main ingredient in the brownish haze we have become accustomed to seeing hanging over our cities.

Nitrogen oxides (NOx) – A compound formed during the engine's combustion process when oxygen in the air combines with nitrogen in the air to form the nitrogen oxides which are agents in photochemical smog.

NO_2 – Nitrogen Dioxide.

NOx – Oxides of Nitrogen.

O

O – Oxygen.

O_3 – Ozone.

OC – Oxidation Catalyst.

Ohmmeter – An instrument that measures resistance of a conductor in units called ohms.

Oil Thermal Vacuum Switch (OTVS) – A switch used by some GM vehicles to shut off vacuum to the early evaporation (EFE) valve when oil temperature reaches 150 degrees Fahrenheit.

Open circuit – An electrical circuit whose path has been interrupted or broken – either accidentally (a broken wire) or intentionally (a switch turned off).

Open loop mode – Mode in which the computer operates without feedback from the oxygen sensor while the engine is in the cold running condition.

Open system – Descriptive term for a crankcase emissions control system which vents to the atmosphere.

Orifice Spark Advance Control (OSAC) – A device used by Chrysler to apply vacuum advance over a period of time. By limiting the timing advance rate, NOx is reduced.

OS – Oxygen sensor.

OSAC – Orifice Spark Advance Control.

OTVS – Oil Thermal Vacuum Switch.

Output driver – A transistor in the output control area of the computer that is used to turn various actuators on and off.

Oxides of Nitrogen (NOx) – Usually a combination of colorless and odorless nitrogen oxide (NO) and toxic, pungent red-brown nitrogen dioxide (NO_2). Nitrogen oxides are a major constituent of smog.

OXS – Oxygen Sensor system.

Oxygen (O_2) – A colorless, odorless gas that makes up about 20 percent of our atmosphere and is necessary for combustion or burning to occur.

Oxygen Sensor (OXS) system – Tells the computer if there is an excess or absence of oxygen in the engine exhaust. The computer then keeps the air/fuel mixture at levels that produce the least emissions.

Ozone (O_3) – A form of oxygen with a pungent odor. Formed naturally in the upper atmosphere, its color gives the sky its bluish hue. Ozone forms a protective layer that screens out many of the sun's harmful ultraviolet rays before they reach the earth. It is also a major constituent of photochemical smog.

P

PA – Pulse Air.

PACV – Pulse Air Check Valve.

PAF – Pulse Air Feeder.

PAI – Pulse Air Injection.

Paraffins, olefins, aromatic hydrocarbons – Unburned hydrocarbons (HC) that are formed when gasoline and other hydrocarbon fuels are only partially combusted.

Particulate matter – Solids such as carbon and some liquids that are found in exhaust gases.

Parts Per Million (PPM) – A unit of measurement in emission analysis.

PAS – Pulse Air System.

PCOV – Purge Control Valve.

PCV – Positive Crankcase Ventilation system.

PCVS – PCV Solenoid.

PCVV – Positive Crankcase Ventilation Valve.

PECV – Power Enrichment Control Valve.

PFE – Pressure Feedback EGR sensor.

PFI – Port Fuel Injection.

Piezoelectric – Electric polarity due to pressure in quartz.

PIP – Profile Ignition Pickup.

Photochemical – The chemical action of radiant energy, or sunlight (photo) on air pollutants (chemicals), which creates smog.

Ported EGR Valve – Operated by a vacuum signal from the carburetor EGR port. The port signal actuates the valve diaphragm. As vacuum increases, spring pressure is overcome, opening the valve and allowing EGR flow. The amount of the flow is dependent on the position of the tapered pintle or poppet whose position reflects the strength of the vacuum signal.

Ported Vacuum Advance (PVA) – A port for a vacuum connection that is located above the idle position of the throttle plates. When the throttle plates are in the idle position, there is no vacuum at the port. When the throttle is opened, a vacuum is available to the vacuum advance unit.

Ported Vacuum Switch (PVS) – A temperature actuated switch that changes vacuum connections when the coolant temperature changes. (Originally used to switch spark port vacuum; now used for any vacuum switching function that requires coolant temperature sensing).

Positive Crankcase Ventilation (PCV) system – An emission control system that routes engine crankcase fumes into the intake manifold or air cleaner, where they are drawn into the cylinders and burned along with the air-fuel mixture.

Positive Crankcase Ventilation Valve (PCVV) – A one-way valve which controls the flow of vapors from the crankcase into the engine.

POT – Potentiometer.

PPM or ppm – see "Parts Per Million.

Preignition – The earlier-than-intended ignition of the air-fuel mixture in the combustion chamber. For example, the explosive mixture being fired in a cylinder by a flake of incandescent carbon before the electric spark occurs.

Processor – The onboard computer that receives data from a number of sensors and other electronic components. Based on input data information programmed into the computer's memory, the processor outputs signals to control various engine functions.

Profile Ignition Pickup (PIP) – A "Hall Effect" vane switch that furnishes crankshaft position data to the Ford EEC-IV processor.

Program – A set of detailed instructions that a computer follows when controlling a system.

PROM – Programmable Read Only Memory.

PSOV – Canister Purge Shut-Off Valve.

PSPS – Power Steering Pressure Switch.

PSV – Pulse Air Shut-off Valve.

PTC – Positive Temperature Coefficient choke heater.

Pulse Air system – An exhaust emission control system that utilizes exhaust pulses to pull air into the exhaust system through a reed-type check valve.

Pulse width – The electronic measurement in milliseconds of the duration of the signal that activates the fuel injector. The duration or pulse width is related directly to the amount of fuel injected into the combustion chamber.

Purge Control Valve (PURGE CV) – Used to control the release of fuel vapors from the charcoal canister into the engine.

PURGE CV – Purge Control Valve.

Purge solenoid – A device used to control the operation of the purge valve in an evaporative control emission system.

PVA – Ported Vacuum Advance.

PVCS – Ported Vacuum Control System.

PVFFC – Pressure/Vacuum Fuel Filler Cap.

PVS – Ported Vacuum Switch.

PWR GND – Power Ground.

R

RAM – Random Access Memory.

Random Access Memory – A type of memory which is used to store information temporarily.

RC – Rear Catalytic Converter.

RDV – Retard Delay Valve or Reverse Delay Valve.

Read – A microcomputer operation wherein information is retrieved from memory.

Read Only Memory – A type of memory used to store information permanently. Information can't be written to ROM, it can only be read.

Reed valve – A check-valve used on some air injection systems. Prevents a reverse flow of air from the exhaust manifold to the intake air cleaner.

Reference Voltage – A voltage provided by a voltage regulator to operate potentiometers and other sensors at a constant level.

Relay – A switching device, operated by a low current circuit, that controls the opening and closing of another circuit of higher current capacity.

Relief valve – A pressure limiting valve located in the air injection system. It functions to relieve part of the airflow if the pressure exceeds a calibrated value.

Resistance – The opposition offered by a substance or body to the passage of electric current through it.

Retard Delay Valve (RDV) – Vacuum restriction used to retard or delay the application of a vacuum signal.

ROM – Read Only Memory.

RVSV – Rollover/Vapor Separator Valve.

S

Sampling – The act of periodically collecting information from a sensor. The computer samples input from various sensors in the process of controlling a system.

SAS – Secondary Air Supply system. See "Air Injection System".

SBC – Single Bed Catalytic converter.

SCC – Spark Control Computer.

SCS – Spark Control System, or Speed Controlled Spark.

SCVAC – Speed Control Vacuum Control.

SDV – Spark Delay Valve.

SEC ACT – Secondary Actuator.

SEFI – Sequential Electronic Fuel Injection.

Sensor – A device that measures an operating condition and provides an input signal to a microcomputer.

Sequential Electronic Fuel Injection (SEFI) – A computer controlled fuel system that distributes fuel through an injector located in each intake port of the engine. Each injector is fired separately and has individual circuits.

Separator Assembly – Fuel Vacuum (SA-FV) – Fuel Vacuum Separator FVS.

SFI – Sequential Fuel Injection. See "Sequential Electronic Fuel Injection."

SHED – Sealed Housing Evaporative Determination system.

Shift Indicator Light (SIL) – A system that provides a visual indication to the driver when to shift to the next higher gear to obtain optimum fuel economy.

Signal – A voltage condition that transmits specific information in an electronic system.

SIL – Shift Indicator Light.

SIS – Solenoid Idle Stop.

Smog – The air pollution created when unburned hydrocarbons (HC) and oxides of nitrogen (NOx) combine in the presence of sunlight. These then react to form ozone, nitrogen dioxide and nitrogen nitrate.

SO_2 – Sulfur Oxides.

SOL – Solenoid.

Solenoid – A wire coil with a moveable core that changes position by means of electro-magnetism when current flows through the coil.

Solenoid Vent Valve (SVV) – Energized by ignition switch to control fuel vapor flow to the canister. When the ignition is off, the valve is open.

Solid State Ignition (SSI) system – A system used by Ford.

SOL V – EGR Solenoid Vacuum Valve Assembly.

Spacer entry EGR system – Exhaust gases are routed directly from the exhaust manifold through a stainless steel tube to the carburetor base.

Spark Delay Valve (SDV) – A valve in the vacuum advance hose that delays the vacuum to the vacuum advance unit during rapid acceleration from idle or from speeds below 15 mph, and cuts off spark advance immediately on deceleration. Has an internal sintered orifice to slow air in one direction, a check valve for free air flow in the opposite direction and a filter.

Spark knock – Same as preignition.

SPFI – Single-Point Fuel Injection.

S-Port – A special carburetor port for ported vacuum.

SSI – Solid State Ignition system.

Stoichiometric – Chemically correct. An air/fuel mixture is considered stoichiometric when it's neither too rich nor too lean; stoichiometric ratio is 14.7 parts of air for every part of fuel.

Sulfur Oxides (SO_2) – A major air pollutant, formed when gasoline with sulfur impurities is combusted.

SVCBV – Solenoid Valve Carburetor Bowl Vent.

SVV – Solenoid Vent Valve.

T

TAB – Thermactor Air Bypass solenoid (Ford Systems).

TAC – Thermostatic Air Cleaner system (Ford Systems).

TAD – Thermactor Air Diverter solenoid.

TAV – Temperature Activated Vacuum system.

TC – Throttle Closer.

TCAC – Thermostatic Controlled Air Cleaner system.

TCC – Transaxle Converter Clutch system.

TCS – Throttle Control System, or Transmission Controlled Spark system or Transmission Converter Switch.

TDP – Throttle closing Dashpot.

TDS – Time Delay Solenoid.

TEC – Thermactor Exhaust Control system.

Temperature Vacuum Switch (TVS) – Controls vacuum to the EGR valve and/or canister purge valve based on coolant or intake air temperature. Canister purge and EGR do not typically operate when the engine is cold.

TES – Thermal Electric Switch.

TFI, TFI-IV – Thick Film Integrated Ignition (Ford systems).

Thermactor – An air injection type of exhaust emission control system used on Ford vehicles.

Thermactor II – Another name for Ford's Pulse Air System.

Thermactor Air Bypass solenoid (TAB) – An electrical solenoid that switches engine manifold vacuum to bypass the atmosphere.

Thermactor Air Control solenoid vacuum valve assembly – Used on Thermactor air control systems; consists of two normally open solenoid valves with vents.

Thermactor Air Control valve – Combines a bypass (dump) valve with a diverter (upstream/downstream) valve; controls the flow of the Thermactor air in response to vacuum signals to its diaphragms.

Thermactor Air Diverter (TAD) solenoid – An electrical solenoid that switches engine manifold vacuum; when energized, switches Thermactor air from downstream (past the oxygen sensor) to upstream (before the oxygen sensor).

Thermactor Exhaust Control (TEC) system – An air injection type of exhaust emission control system used by Ford.

Thermal Ignition Control (TIC) – A device used by Chrysler that shifts the vacuum advance vacuum source from ported vacuum to manifold vacuum when coolant temperature exceed 225-degrees Fahrenheit.

Thermal Reactor (TR) – A specially designed exhaust manifold that uses heat and air to burn the unburned hydrocarbons in the exhaust gases to reduce pollution.

Thermal Vacuum Switch (TVS) – A temperature sensitive switch that shifts the source of the advance from ported to manifold vacuum when coolant temperature reaches approximately 225-degrees Fahrenheit.

Thermal Vent Valve (TVV) – A temperature-sensitive valve assembly located in the canister vent line. The TVV closes when the engine is cold and opens when it's hot to prevent fuel tank vapors from being vented through the carburetor fuel bowl when the fuel tank heats up before the engine compartment.

Thermistor – A resistor that changes its resistance with temperature.

Thick Film Integrated (TFI) – A Ford electronic ignition system.

Three-Way Catalyst (TWC) – A catalyst designed to simultaneously convert hydrocarbons, carbon monoxide and oxides of nitrogen into relatively inert substances.

Throttle Kicker – A type of throttle position solenoid specifically designed to speed-up the engine rpm by pushing on the throttle plate on the carburetor.

Throttle Position Sensor (TPS) – A potentiometric sensor that tells the computer the position (angle) of the throttle plate. The sensor wiper position is proportional to throttle position. The computer uses this information to control fuel flow.

TIC – Thermal Ignition Control.

TIV – Thermactor Idle Vacuum valve.

TK – Throttle Kicker vacuum solenoid valve.

TKA – Throttle Kicker Actuator.

TKS – Throttle Kicker Solenoid.

TO – Thermal Override.

TOC – Throttle Opener Control system.

TP – Throttle Position, or Throttle Position Sensor.

TPI – Tuned Port Injection. A fuel injection system used by General Motors that combines port fuel injection with tuned intake runners for higher engine torque.

TPS – Throttle Position Sensor.

TR – Thermal Reactor.

Transducer – A device that receives a signal from one system (such as exhaust backpressure) and transfers that signal to another system (such as a vacuum source.

TSD – Throttle Solenoid.

TSP – Throttle Solenoid Positioner.

TVS – Temperature Vacuum Switch or Thermal Vacuum Switch.

TVSV – Thermostatic Vacuum Switch Valve.

TVV – Thermal Vacuum Valve, or Thermal Vent Valve.

TWC – Three-Way Catalyst.

U

UBC – Underbody Catalyst.

Unburned or partially burned hydrocarbons (HC) – Organic compounds such as paraffins, olefins, aromatics, aldehydes, keynotes and carboxylic acids that contain hydrogen and carbon in varying amounts. Many hydrocarbons are considered carcinogenic. Unburned hydrocarbons, always when an internal combustion engine does not operate at 100-percent efficiency, are another main component of photochemical smog.

V

VAC – Vacuum Advance Control.

Vacuum – A pressure that is LESS than atmospheric pressure. Vacuum is used extensively for control purposes and is generally measured in inches of mercury (Hg.). There are three types of vacuum: manifold, ported and venturi. The strength of these vacuums depends on the throttle opening, engine speed and load.

Vacuum Check Valve (VCK-V) – A one-way valve used to retain a vacuum signal in a line after the vacuum source is gone.

Vacuum Control Valve (VCV) – A ported vacuum switch; controls vacuum to other emission devices during engine warm-up.

Vacuum Delay Valve (VDV) – A valve used by GM to bleed ported vacuum to the vacuum advance unit through a small orifice and control vacuum advance rate.

Vacuum Differential Valve (VDV) – A device used in Thermactor systems with a catalyst that senses intake manifold vacuum and triggers the bypass valve to dump injected air to the atmosphere during deceleration.

Vacuum Operated Exhaust Heat Control valve (VHC) – A vacuum operated heat riser valve used by Ford to cause the exhaust to flow through the intake manifold crossover passage for preheating of the air-fuel mixture.

Vacuum Gauge – An instrument used to measure the amount of intake vacuum.

Vacuum pump/gauge – A hand-operated pump used to apply vacuum to a device or system in order to test it; a gauge on the pump indicates the amount of vacuum applied and can be used to measure how long a device or system can hold vacuum.

Vacuum Reducer Valve (VRV) – A valve used by GM on some models to reduce the vacuum signal to the vacuum advance unit by one and one-half to three inches Hg when coolant temperature is greater than 220-degrees to reduce detonation.

Vacuum Regulator – A device that provides constant vacuum output from the manifold when the vehicle is at idle.

Vacuum Regulator Valve (VRV) two-port – This vacuum regulator provides a constant output signal when the input level is greater than a preset level. At a lower input vacuum, the output equals the input.

Vacuum Regulator Valve (VRV) three and four-port – This type of vacuum regulator valve is used to control the vacuum advance to the distributor.

Vacuum Reservoir (VRESER) – Stores excess vacuum to prevent rapid fluctuations and sudden drops in a vacuum signal, such as during acceleration.

Vacuum Restrictor (VREST) – Controls the flow rate and/or timing in actions to the different emission control components.

Vacuum Retard Delay Valve (VRDV) – Delays a decrease in vacuum at the distributor vacuum advance unit when the source vacuum decreases. Used to delay release of vacuum from a diaphragm – a "momentary" vacuum trap.

Vacuum Switching Valve (VSV) – An electrically controlled vacuum switching valve used to control emission control devices.

Vacuum Transmitting Valve (VTV) – A valve used to limit the rate of vacuum advance.

Vacuum Vent Valve (VVV) – Controls the induction of fresh air into a vacuum system to prevent chemical decay of the vacuum diaphragm that can occur on contact with fuel.

VAF – Vane Air Flow meter.

Valve and Transducer Assembly – This type of EGR valve consists of a modified ported EGR valve and a remote transducer. Works the same way as an Integral Backpressure Transducer EGR valve.

Vane Air-Flow (VAF) meter – A sensor with a movable vane connected to a potentiometer calibrated to measure the amount of air flowing to the engine.

Vane Air Temperature (VAT) sensor – Located inside the vane airflow meter housing; senses the temperature of the air flowing into the engine.

Vapor-recovery system – Another name for the evaporative emission control system.

Variable Reluctance Sensor (VR or VRS) – A non-contact transducer that converts mechanical motion into electrical control signals.

VAT – Vane Air Temperature sensor.

VAV – Vacuum Advance Valve.

VB – Vacuum Break.

VBV – Vacuum Bias valve.

VCKV – Vacuum Check Valve.

VCS – Vapor Control System, or Vacuum Control Switch.

VCV – Vacuum Control Valve.

VDV – Vacuum Delay Valve or Vacuum Differential Valve.

VECI – Vehicle Emission Control Information decal.

Vehicle Emission Control Information (VECI) decal – Critical specifications for servicing the emission systems.

Venturi Vacuum – A weak vacuum signal that originates at the venturi of the carburetor. As engine speed increases, the venturi signal increases.

Venturi Vacuum Amplifier (VVA) – Used with some EGR systems so that carburetor venturi vacuum can control EGR valve operation; venturi vacuum is desirable because it's proportional to the airflow through the carburetor.

VHC – Vacuum Operated exhaust Heat Control valve.

VIN – Vehicle Identification Number.

VMV – Vacuum Modulator Valve.

Volatile Organic Compounds – Unburned hydrocarbon (HC) portions of gasoline.

Volt-Ohm Meter (VOM) – Used to measure voltage and resistance.

VOM – Volt-Ohm Meter.

VOTM – Vacuum-Operated Throttle Modulator.

VP – Vacuum Pump.

VRDV – Vacuum Retard Delay Valve.

VRESER – Vacuum Reservoir.

VREST – Vacuum Restrictor.

VR or VRS – Variable Reluctance Sensor.

VR/S – Vacuum Regulator Solenoid.

V-RSR – Vacuum reservoir.

V-RST – Vacuum restrictor.

VRDV – Vacuum Retard Delay Valve.

VRES – Vacuum Reservoir.

VREST – Vacuum Restrictor.

VRV – Vacuum Reducer Valve, or Vacuum Regulator Valve.

VS – Vacuum Switch.

VSA – Vacuum Switch Assembly.

VSC – Vehicle Speed Control sensor.

VSS – Vehicle Speed Sensor, or Vacuum Switch Solenoid.

VSV – Vacuum Solenoid Valve, or Vacuum Switching Valve.

VTM – Vacuum Throttle Modulator.

VTP – Vacuum Throttle Positioner.

VTV – Vacuum Transmitting Valve.

VVA – Venturi Vacuum Amplifier.

VVC – Variable Voltage Choke.

VVV – Vacuum Vent Valve.

W

WAC – Wide-open throttle Air conditioner Cutoff.

WOT – Wide-Open Throttle.

WOTS – Wide-Open Throttle Switch.

WOTV – Wide Open Throttle Valve.

Index

D

E

F

T

U

V

W

Haynes Automotive Manuals

NOTE: New manuals are added to this list on a periodic basis. If you do not see a listing for your vehicle, consult your local Haynes dealer for the latest product information.

ACURA

*1776 **Integra & Legend** all models '86 thru '90

AMC

Jeep CJ - *see JEEP (412)*
694 **Mid-size models,** Concord, Hornet, Gremlin & Spirit '70 thru '83
934 **(Renault) Alliance & Encore** all models '83 thru '87

AUDI

615 **4000** all models '80 thru '87
428 **5000** all models '77 thru '83
1117 **5000** all models '84 thru '88

AUSTIN

Healey Sprite - *see MG Midget Roadster (265)*

BMW

*2020 **3/5 Series** not including diesel or all-wheel drive models '82 thru '92
276 **320i** all 4 cyl models '75 thru '83
632 **528i** & 530i all models '75 thru '80
240 **1500 thru 2002** all models except Turbo '59 thru '77
348 **2500, 2800, 3.0 & Bavaria** all models '69 thru '76

BUICK

Century (front wheel drive) - *see GENERAL MOTORS (829)*
*1627 **Buick, Oldsmobile & Pontiac Full-size (Front wheel drive)** all models '85 thru '95 **Buick** Electra, LeSabre and Park Avenue; **Oldsmobile** Delta 88 Royale, Ninety Eight and Regency; **Pontiac** Bonneville
1551 **Buick Oldsmobile & Pontiac Full-size (Rear wheel drive)** **Buick** Estate '70 thru '90, Electra'70 thru '84, LeSabre '70 thru '85, Limited '74 thru '79 **Oldsmobile** Custom Cruiser '70 thru '90, Delta 88 '70 thru '85, Ninety-eight '70 thru '84 **Pontiac** Bonneville '70 thru '81, Catalina '70 thru '81, Grandville '70 thru '75, Parisienne '83 thru '86
627 **Mid-size Regal & Century** all rear-drive models with V6, V8 and Turbo '74 thru '87
Regal - *see GENERAL MOTORS (1671)*
Skyhawk - *see GENERAL MOTORS (766)*
Skylark '80 thru '85 - *see GENERAL MOTORS (38020)*
Skylark '86 on - *see GENERAL MOTORS (1420)*
Somerset - *see GENERAL MOTORS (1420)*

CADILLAC

*751 **Cadillac Rear Wheel Drive** all gasoline models '70 thru '93
Cimarron - *see GENERAL MOTORS (766)*

CHEVROLET

*1477 **Astro & GMC Safari Mini-vans** '85 thru '93
554 **Camaro V8** all models '70 thru '81
866 **Camaro** all models '82 thru '92
Cavalier - *see GENERAL MOTORS (766)*
Celebrity - *see GENERAL MOTORS (829)*
625 **Chevelle, Malibu & El Camino** all V6 & V8 models '69 thru '87
449 **Chevette & Pontiac T1000** '76 thru '87
550 **Citation** all models '80 thru '85
*1628 **Corsica/Beretta** all models '87 thru '95
274 **Corvette** all V8 models '68 thru '82
*1336 **Corvette** all models '84 thru '91
1762 **Chevrolet Engine Overhaul Manual**
704 **Full-size Sedans** Caprice, Impala, Biscayne, Bel Air & Wagons '69 thru '90
Lumina - *see GENERAL MOTORS (1671)*
Lumina APV - *see GENERAL MOTORS (2035)*
319 **Luv Pick-up** all 2WD & 4WD '72 thru '82
626 **Monte Carlo** all models '70 thru '88
241 **Nova** all V8 models '69 thru '79
*1642 **Nova and Geo Prizm** all front wheel drive models, '85 thru '92
420 **Pick-ups '67 thru '87** - Chevrolet & GMC, all V8 & in-line 6 cyl, 2WD & 4WD '67 thru '87; Suburbans, Blazers & Jimmys '67 thru '91
*1664 **Pick-ups '88 thru '95** - Chevrolet & GMC, all full-size pick-ups, '88 thru '95; Blazer & Jimmy '92 thru '94; Suburban '92 thru '95; Tahoe & Yukon '95
*831 **S-10 & GMC S-15 Pick-ups** all models '82 thru '93
*1727 **Sprint & Geo Metro** '85 thru '94
*345 **Vans - Chevrolet & GMC,** V8 & in-line 6 cylinder models '68 thru '95

CHRYSLER

2114 **Chrysler Engine Overhaul Manual**
*2058 **Full-size Front-Wheel Drive** '88 thru '93
K-Cars - *see DODGE Aries (723)*
Laser - *see DODGE Daytona (1140)*
*1337 **Chrysler & Plymouth Mid-size** front wheel drive '82 thru '93
Rear-wheel Drive - *see Dodge Rear-wheel Drive (2098)*

DATSUN

402 **200SX** all models '77 thru '79
647 **200SX** all models '80 thru '83
228 **B - 210** all models '73 thru '78
525 **210** all models '78 thru '82
206 **240Z, 260Z & 280Z** Coupe '70 thru '78
563 **280ZX** Coupe & 2+2 '79 thru '83
300ZX - *see NISSAN (1137)*
679 **310** all models '78 thru '82
123 **510 & PL521 Pick-up** '68 thru '73
430 **510** all models '78 thru '81
372 **610** all models '72 thru '76
277 **620 Series Pick-up** all models '73 thru '79
720 Series Pick-up - *see NISSAN (771)*
376 **810/Maxima** all gasoline models, '77 thru '84
Pulsar - *see NISSAN (876)*
Sentra - *see NISSAN (982)*
Stanza - *see NISSAN (981)*

DODGE

400 & 600 - *see CHRYSLER Mid-size (1337)*
*723 **Aries & Plymouth Reliant** '81 thru '89
1231 **Caravan & Plymouth Voyager Mini-Vans** all models '84 thru '95
699 **Challenger & Plymouth Saporro** all models '78 thru '83
Challenger '67-'76 - *see DODGE Dart (234)*
236 **Colt** all models '71 thru '77
610 **Colt & Plymouth Champ (front wheel drive)** all models '78 thru '87
*1668 **Dakota Pick-ups** all models '87 thru '93
234 **Dart, Challenger/Plymouth Barracuda & Valiant** 6 cyl models '67 thru '76
*1140 **Daytona & Chrysler Laser** '84 thru '89
*545 **Omni & Plymouth Horizon** '78 thru '90
*912 **Pick-ups** all full-size models '74 thru '91
*556 **Ram 50/D50 Pick-ups & Raider and Plymouth Arrow Pick-ups** '79 thru '93
2098 **Dodge/Plymouth/Chrysler** rear wheel drive '71 thru '89
*1726 **Shadow & Plymouth Sundance** '87 thru '93
*1779 **Spirit & Plymouth Acclaim** '89 thru '95
*349 **Vans - Dodge & Plymouth** V8 & 6 cyl models '71 thru '91

EAGLE

Talon - *see Mitsubishi Eclipse (2097)*

FIAT

094 **124 Sport Coupe & Spider** '68 thru '78
273 **X1/9** all models '74 thru '80

FORD

*1476 **Aerostar Mini-vans** all models '86 thru '94
788 **Bronco and Pick-ups** '73 thru '79
*880 **Bronco and Pick-ups** '80 thru '95
268 **Courier Pick-up** all models '72 thru '82
2105 **Crown Victoria & Mercury Grand Marquis** '88 thru '94
1763 **Ford Engine Overhaul Manual**
789 **Escort/Mercury Lynx** all models '81 thru '90
*2046 **Escort/Mercury Tracer** '91 thru '95
*2021 **Explorer & Mazda Navajo** '91 thru '95
560 **Fairmont & Mercury Zephyr** '78 thru '83
334 **Fiesta** all models '77 thru '80
754 **Ford & Mercury Full-size,** Ford LTD & Mercury Marquis ('75 thru '82); Ford Custom 500, Country Squire, Crown Victoria & Mercury Colony Park ('75 thru '87); Ford LTD Crown Victoria & Mercury Gran Marquis ('83 thru '87)
359 **Granada & Mercury Monarch** all in-line, 6 cyl & V8 models '75 thru '80
773 **Ford & Mercury Mid-size,** Ford Thunderbird & Mercury Cougar ('75 thru '82); Ford LTD & Mercury Marquis ('83 thru '86); Ford Torino, Gran Torino, Elite, Ranchero pick-up, LTD II, Mercury Montego, Comet, XR-7 & Lincoln Versailles ('75 thru '86)
*654 **Mustang & Mercury Capri** all models including Turbo. Mustang, '79 thru '93; Capri, '79 thru '86
357 **Mustang V8** all models '64-1/2 thru '73
231 **Mustang II** 4 cyl, V6 & V8 models '74 thru '78
649 **Pinto & Mercury Bobcat** '75 thru '80
1670 **Probe** all models '89 thru '92
*1026 **Ranger/Bronco II** gasoline models '83 thru '93
*1421 **Taurus & Mercury Sable** '86 thru '94
*1418 **Tempo & Mercury Topaz** all gasoline models '84 thru '94
1338 **Thunderbird/Mercury Cougar** '83 thru '88
*1725 **Thunderbird/Mercury Cougar** '89 and '93
344 **Vans** all V8 Econoline models '69 thru '91
*2119 **Vans** full size '92-'95

GENERAL MOTORS

*829 **Buick Century, Chevrolet Celebrity, Oldsmobile Cutlass Ciera & Pontiac 6000** all models '82 thru '93
*1671 **Buick Regal, Chevrolet Lumina, Oldsmobile Cutlass Supreme & Pontiac Grand Prix** all front wheel drive models '88 thru '95
*766 **Buick Skyhawk, Cadillac Cimarron, Chevrolet Cavalier, Oldsmobile Firenza & Pontiac J-2000 & Sunbird** all models '82 thru '94
38020 **Buick Skylark, Chevrolet Citation, Olds Omega, Pontiac Phoenix** '80 thru '85
1420 **Buick Skylark & Somerset, Oldsmobile Achieva & Calais and Pontiac Grand Am** all models '85 thru '95
*2035 **Chevrolet Lumina APV, Oldsmobile Silhouette & Pontiac Trans Sport** all models '90 thru '94
General Motors Full-size Rear-wheel Drive - *see BUICK (1551)*

GEO

Metro - *see CHEVROLET Sprint (1727)*
Prizm - *see CHEVROLET Nova (1642)*
*2039 **Storm** all models '90 thru '93
Tracker - *see SUZUKI Samurai (1626)*

GMC

Safari - *see CHEVROLET ASTRO (1477)*
Vans & Pick-ups - *see CHEVROLET (420, 831, 345, 1664)*

(Continued on other side)

** Listings shown with an asterisk (*) indicate model coverage as of this printing. These titles will be periodically updated to include later model years - consult your Haynes dealer for more information.*

Haynes North America, Inc., 861 Lawrence Drive, Newbury Park, CA 91320 • (805) 498-6703

Haynes Automotive Manuals (continued)

NOTE: New manuals are added to this list on a periodic basis. If you do not see a listing for your vehicle, consult your local Haynes dealer for the latest product information.

HONDA

351 **Accord CVCC** all models '76 thru '83
1221 **Accord** all models '84 thru '89
2067 **Accord** all models '90 thru '93
42013 **Accord** all models '94 thru '95
160 **Civic 1200** all models '73 thru '79
633 **Civic 1300 & 1500 CVCC** '80 thru '83
297 **Civic 1500 CVCC** all models '75 thru '79
1227 **Civic** all models '84 thru '91
*2118 **Civic & del Sol** '92 thru '95
*601 **Prelude CVCC** all models '79 thru '89

HYUNDAI

*1552 **Excel** all models '86 thru '94

ISUZU

*1641 **Trooper & Pick-up**, all gasoline models Pick-up, '81 thru '93; Trooper, '84 thru '91

JAGUAR

*242 **XJ6** all 6 cyl models '68 thru '86
*478 **XJ12 & XJS** all 12 cyl models '72 thru '85

JEEP

*1553 **Cherokee, Comanche & Wagoneer Limited** all models '84 thru '93
412 **CJ** all models '49 thru '86
50025 **Grand Cherokee** all models '93 thru '95
*1777 **Wrangler** all models '87 thru '94

LINCOLN

2117 **Rear Wheel Drive** all models '70 thru '95

MAZDA

648 **626** Sedan & Coupe (rear wheel drive) all models '79 thru '82
*1082 **626 & MX-6 (front wheel drive)** all models '83 thru '91
267 **B Series Pick-ups** '72 thru '93
370 **GLC Hatchback (rear wheel drive)** all models '77 thru '83
757 **GLC (front wheel drive)** '81 thru '85
*2047 **MPV** all models '89 thru '94
Navajo-see Ford Explorer (2021)
460 **RX-7** all models '79 thru '85
*1419 **RX-7** all models '86 thru '91

MERCEDES-BENZ

*1643 **190 Series** all four-cylinder gasoline models, '84 thru '88
346 **230, 250 & 280** Sedan, Coupe & Roadster all 6 cyl sohc models '68 thru '72
983 **280 123 Series** gasoline models '77 thru '81
698 **350 & 450** Sedan, Coupe & Roadster all models '71 thru '80
697 **Diesel 123 Series** 200D, 220D, 240D, 240TD, 300D, 300CD, 300TD, 4- & 5-cyl incl. Turbo '76 thru '85

MERCURY

See FORD Listing

MG

111 **MGB** Roadster & GT Coupe all models '62 thru '80
265 **MG Midget & Austin Healey Sprite Roadster** '58 thru '80

MITSUBISHI

*1669 **Cordia, Tredia, Galant, Precis & Mirage** '83 thru '93
*2097 **Eclipse, Eagle Talon & Plymouth Laser** '90 thru '94
*2022 **Pick-up & Montero** '83 thru '95

NISSAN

1137 **300ZX** all models including Turbo '84 thru '89
*1341 **Maxima** all models '85 thru '91
*771 **Pick-ups/Pathfinder** gas models '80 thru '95
876 **Pulsar** all models '83 thru '86
*982 **Sentra** all models '82 thru '94
*981 **Stanza** all models '82 thru '90

OLDSMOBILE

Bravada - *see CHEVROLET S-10 (831)*
Calais - *see GENERAL MOTORS (1420)*
Custom Cruiser - *see BUICK Full-size RWD (1551)*
*658 **Cutlass** all standard gasoline V6 & V8 models '74 thru '88
Cutlass Ciera - *see GENERAL MOTORS (829)*
Cutlass Supreme - *see GM (1671)*
Delta 88 - *see BUICK Full-size RWD (1551)*
Delta 88 Brougham - *see BUICK Full-size FWD (1551), RWD (1627)*
Delta 88 Royale - *see BUICK Full-size RWD (1551)*
Firenza - *see GENERAL MOTORS (766)*
Ninety-eight Regency - *see BUICK Full-size RWD (1551), FWD (1627)*
Ninety-eight Regency Brougham - *see BUICK Full-size RWD (1551)*
Omega - *see GENERAL MOTORS (38020)*
Silhouette - *see GENERAL MOTORS (2035)*

PEUGEOT

663 **504** all diesel models '74 thru '83

PLYMOUTH

Laser - *see MITSUBISHI Eclipse (2097)*
For other PLYMOUTH titles, see DODGE listing.

PONTIAC

T1000 - *see CHEVROLET Chevette (449)*
J-2000 - *see GENERAL MOTORS (766)*
6000 - *see GENERAL MOTORS (829)*
Bonneville - *see Buick Full-size FWD (1627), RWD (1551)*
Bonneville Brougham - *see Buick (1551)*
Catalina - *see Buick Full-size (1551)*
1232 **Fiero** all models '84 thru '88
555 **Firebird** V8 models except Turbo '70 thru '81
867 **Firebird** all models '82 thru '92
Full-size Front Wheel Drive - *see BUICK Oldsmobile, Pontiac Full-size FWD (1627)*
Full-size Rear Wheel Drive - *see BUICK Oldsmobile, Pontiac Full-size RWD (1551)*
Grand Am - *see GENERAL MOTORS (1420)*
Grand Prix - *see GENERAL MOTORS (1671)*
Grandville - *see BUICK Full-size (1551)*
Parisienne - *see BUICK Full-size (1551)*
Phoenix - *see GENERAL MOTORS (38020)*
Sunbird - *see GENERAL MOTORS (766)*
Trans Sport - *see GENERAL MOTORS (2035)*

PORSCHE

*264 **911** all Coupe & Targa models except Turbo & Carrera 4 '65 thru '89
239 **914** all 4 cyl models '69 thru '76
397 **924** all models including Turbo '76 thru '82
*1027 **944** all models including Turbo '83 thru '89

RENAULT

141 **5 Le Car** all models '76 thru '83
Alliance & Encore - *see AMC (934)*

SAAB

247 **99** all models including Turbo '69 thru '80
*980 **900** all models including Turbo '79 thru '88

SATURN

2083 **Saturn** all models '91 thru '94

SUBARU

237 **1100, 1300, 1400 & 1600** '71 thru '79
*681 **1600 & 1800** 2WD & 4WD '80 thru '89

SUZUKI

*1626 **Samurai/Sidekick and Geo Tracker** all models '86 thru '95

TOYOTA

1023 **Camry** all models '83 thru '91
92006 **Camry** all models '92 thru '95
935 **Celica Rear Wheel Drive** '71 thru '85
*2038 **Celica Front Wheel Drive** '86 thru '92
1139 **Celica Supra** all models '79 thru '92
361 **Corolla** all models '75 thru '79
961 **Corolla** all rear wheel drive models '80 thru '87
*1025 **Corolla** all front wheel drive models '84 thru '92
636 **Corolla Tercel** all models '80 thru '82
360 **Corona** all models '74 thru '82
532 **Cressida** all models '78 thru '82
313 **Land Cruiser** all models '68 thru '82
*1339 **MR2** all models '85 thru '87
304 **Pick-up** all models '69 thru '78
*656 **Pick-up** all models '79 thru '95
*2048 **Previa** all models '91 thru '93
2106 **Tercel** all models '87 thru '94

TRIUMPH

113 **Spitfire** all models '62 thru '81
322 **TR7** all models '75 thru '81

VW

159 **Beetle & Karmann Ghia** all models '54 thru '79
238 **Dasher** all gasoline models '74 thru '81
*884 **Rabbit, Jetta, Scirocco, & Pick-up** gas models '74 thru '91 & Convertible '80 thru '92
451 **Rabbit, Jetta & Pick-up** all diesel models '77 thru '84
082 **Transporter 1600** all models '68 thru '79
226 **Transporter 1700, 1800 & 2000** all models '72 thru '79
084 **Type 3 1500 & 1600** all models '63 thru '73
1029 **Vanagon** all air-cooled models '80 thru '83

VOLVO

203 **120, 130 Series & 1800 Sports** '61 thru '73
129 **140 Series** all models '66 thru '74
*270 **240 Series** all models '76 thru '93
400 **260 Series** all models '75 thru '82
*1550 **740 & 760 Series** all models '82 thru '88

TECHBOOK MANUALS

2108 **Automotive Computer Codes**
1667 **Automotive Emissions Control Manual**
482 **Fuel Injection Manual, 1978 thru 1985**
2111 **Fuel Injection Manual, 1986 thru 1994**
2069 **Holley Carburetor Manual**
2068 **Rochester Carburetor Manual**
10240 **Weber/Zenith/Stromberg/SU Carburetors**
1762 **Chevrolet Engine Overhaul Manual**
2114 **Chrysler Engine Overhaul Manual**
1763 **Ford Engine Overhaul Manual**
1736 **GM and Ford Diesel Engine Repair Manual**
1666 **Small Engine Repair Manual**
10355 **Ford Automatic Transmission Overhaul**
10360 **GM Automatic Transmission Overhaul**
1479 **Automotive Body Repair & Painting**
2112 **Automotive Brake Manual**
2113 **Automotive Detaiing Manual**
1654 **Automotive Eelectrical Manual**
1480 **Automotive Heating & Air Conditioning**
2109 **Automotive Reference Manual & Illustrated Dictionary**
2107 **Automotive Tools Manual**
10440 **Used Car Buying Guide**
2110 **Welding Manual**

SPANISH MANUALS

98905 **Códigos Automotrices de la Computadora**
98915 **Inyección de Combustible 1986 al 1994**
99040 **Chevrolet & GMC Camionetas** '67 al '87 Incluye Suburban, Blazer & Jimmy '67 al '91
99041 **Chevrolet & GMC Camionetas** '88 al '95 Incluye Suburban '92 al '95, Blazer & Jimmy '92 al '94, Tahoe y Yukon '95
99075 **Ford Camionetas y Bronco** '80 al '94
99125 **Toyota Camionetas y 4-Runner** '79 al '95

** Listings shown with an asterisk (*) indicate model coverage as of this printing. These titles will be periodically updated to include later model years - consult your Haynes dealer for more information.*

Over 100 Haynes motorcycle manuals also available

Haynes North America, Inc., 861 Lawrence Drive, Newbury Park, CA 91320 • (805) 498-6703